INTERDISCIPLINARY MATHEMATICS

VOLUME XX

CARTANIAN GEOMETRY, NONLINEAR WAVES,

AND CONTROL THEORY

PART A

ROBERT HERMANN

MATH SCI PRESS
53 JORDAN ROAD
BROOKLINE, MA 02146

Copyright © 1979 by Robert Hermann
All rights reserved

ISBN 0-915-69227-9

Library of Congress Cataloging in Publication Data

Hermann, Robert.
  Cartanian geometry, nonlinear waves, and control theory.

  (Interdisciplinary mathematics ; v. 20)
  Includes bibliographies.
  1. Geometry, Differential. 2. Wave-motion, Theory.
3. Nonlinear theories. 4. Control theory. I. Title.
QA649.H46   516'.36   79-764
ISBN 0-915-69227-9 (v. 1)

MATH SCI PRESS
53 JORDAN ROAD
BROOKLINE, MA 02146

Printed in the United States of America

PREFACE

I began in 1970 to write this series of books in order to develop a *unified* mathematical science and technology. After all, if subjects like category theory, logic, differential topology are accepted and integrated into the mathematical world, why not system theory, mathematical elementary particle theory, relativity, etc.? I had no master plan, but intended to write down what I could, as best I could, and see where it led.

Twenty volumes are now completed and I can say more definitively that the unifying theme is the role that *geometry* plays in physics and engineering. "Applied mathematics" is usually thought of as involving the more concrete parts of analysis and certain areas like numerical analysis and combinatorics, which interface computer science ; but my vision is quite different. To a large extent I am inspired by the historical example of the 19th century, where the basis of much of the fruitful interchange between mathematics and physics was precisely in the area we call "geometry" or "the geometric theory of differential equations".

"Geometry" is both the oldest and newest branch of mathematics, but also the most mysterious and challenging. What does it mean to "think geometrically" or to "have geometric intuition"? Does Nature (physical or biological) organize itself according to geometric patterns? If so, why? Is there any geometric coherence to the areas like economics, Operations Research, and Computer Science, which are *not* organized along the lines of natural laws? In today's mathematical world, geometry is fragmented, usually into what one might call *differential, algebraic* and *topological geometry*. It is a long way from Euclid, yet in some sense the classical is still very important. I cannot imagine someone doing research in, say, functional analysis or probability theory having to concern himself with the most obscure parts of 19th century mathematical literature--yet the most important subject for applications of geometry to nonlinear physical phenomena is now the *Bäcklund transformations*, which were first discovered in a geometric context in the 1880's and for which Darboux's *Theorie des Surfaces* is still the best reference. Why do entire areas of geometric mathematics sink into obscurity, to the extent that their key concepts and results are now lost to us? (This has happened to large parts of 19th century differential and algebraic geometry, and to 1920's - 1930's tensor analysis.)

Of course, my own field is differential geometry, and I am naturally most concerned with it. In the last thirty years there has been a turning away from the grand syntheses and generalizations of the 19th and first third of the 20th centuries in order to concentrate on "global" problems of Riemannian and complex analytic manifolds. The traditional contact with applications--even with General Relativity, which was *the* driving force for research in differential geometry in the earlier 1900's--was lost, and replaced with motivation arising from topology. I feel that much was given up in this process, and one of my main goals is to revive the older traditions. I believe also that research in the area of "general geometric structures" (which seemed very promising when I was a graduate student in the 1950's, but is today moribund) would be revitalized if it were to re-establish contact with areas of applications. The work on the global Riemannian and complex analysis problems which now occupies geometry is worthy, but somewhat dull by comparison to the classics. It essentially reduces differential geometry to the role of servant and follower of other branches of mathematics, whereas it should be at the head of the line, advancing into areas of *new* science. What has happened is that the requirements of the modern academic bureaucratic structure (tenure, publications, theses, etc.) have pushed geometers to sell their magnificent birthright for that proverbial ounce of pottage.

In certain applied fields there are extraordinary possibilities for infusion of differential geometric ideas, possibly even leading to what I like to call the "supernovae" of science. (The interaction between differential geometry and relativity is a historic example.) Certainly, we can now pick up "Phys. Rev. D" (the main journal for elementary particle physics) and see that the physicists are busy "inventing"—in their own inimitable way—differential-geometric ideas like fiber bundles, connections, infinite dimensional Lie groups, deformations, but giving them their own names. Mathematical system theory, which I believe is the mathematical and theoretical super-structure for much of technology, is now using geometric material in an interesting, but still modest way; I also see rather large possibilities here. And of course there is also biology, where the geometry is obviously *there*, but very diffuse, attracting such minds as d'Arcy Thompson, Thom, Gelfand, Piatetski-Shapiro and Zeeman like moths to a flame. Making "sense" of it (as Einstein did for the geometry in physics, for example) is very difficult and probably very long-range.

Curiously, these possibilities are often resisted strenuously by workers in the applied disciplines. Thus, elementary particle physicists are often quite knowledgable and sympathetic to functional analysis, certain computational and symbol-manipulation parts of Lie group theory, etc., but seem to be very skeptical that learning how to work with the "modern" geometric ideas might be useful and important. Of course, they may have been right in feeling that fancy mathematics should be fought tooth and claw; but my feeling is that, left to themselves, they have made rather a mess of things in the last twenty years. Certainly, the evolution seems to be toward Einstein's feeling that God must be a geometer, and I do not see why there should not be more systematic attempts to understand the underlying geometric structure of elementary particles.

The third area which I believe is ripe for a significant infusion of geometric ideas is that of nonlinear waves (solitons, inverse scattering, Korteweg-de Vries, Sine-Gordon, Toda lattices, Bäcklund ...). I have described these prospects in detail in Volumes 12, 14, 15, 18 and "Lie Groups" Volume 6; geometric ideas should come even more to the foreground as more complicated and sophisticated situations are investigated. This material represents a striking Renaissance of 19th century differential geometry (including much of Elie Cartan's work, which in a sense is the highest and best form of that material). It is a great plus that the possibility of significant application to scientific and technological phenomena is so formidable! From the point of view of physics, what I find especially interesting is that it represents an important revival of an old idea—look for *special* solutions of *special* differential equations to describe physical phenomena, rather than the "generaticity" approach which underlies the hope that "global analysis" might be useful in applications. It is very sobering for a mathematician to look at quantum mechanics and see that a vast edifice of applications and implications has been built up by the physicists on the foundation of the explicit solution of a handful of simple cases! I believe this is ultimately why Lie group theory is important in applications; it was invented by Lie precisely to guide the way toward the search for special relations as a sort of generalized Galois theory.

One of the main themes of this volume is that there are close relations between the theory of solitons-nonlinear waves and what the Systems Theorists call the theory of time-varying systems. One of the basic ideas in the former discipline is that of *isospectral deformation* of *differential operators* due to Kruskal and Lax. I now introduce a new concept, *isospectral deformation of systems*, which I expect to be significant in both disciplines.

There are obviously many formidable practical obstacles in the way of realization of my central vision of an "applied geometry" which will utilize the mathematical thought to its full potential but still keep in close touch with physics and engineering. It really requires a class of people who do not exist: *broadly trained* mathematicians who are capable of sitting down to read applied journals (this is the key test!) and scientists who are willing to invest substantial effort in learning globs of mathematics. A further difficult feature of differential geometry is that it involves all three branches

PREFACE   v

of mathematics. It was much more alive in the 19th century precisely because there were then no essential barriers between these branches and the specialization and fragmentation which is so characteristic of present-day mathematics had barely begun. It is sometimes possible to still see glimpses of a unified contemporary mathematics, but this overview is always at the expense of giving up the concreteness and amenity to calculation which is always so essential to "applied mathematics". It is characteristic that physicists are now often the only ones who carry on certain 19th century mathematical traditions, if only in a vestigial way.

The only hope I see under present conditions to develop any serious interchange in the classical style between mathematics and the frontiers of applications is to push ahead with the formation of a new scientific organization dedicated solely to this purpose. I proposed such a step several years ago, but was persuaded to try to integrate it within the AMS and SIAM. Such a possibility has now been definitively rebuffed by these groups, and I am now proceeding with what I propose to call the *Association for Physical and Systems Mathematics*. Our major activities will be the organization of a *Summer Workshop in Physical and Systems Mathematics*, which will attempt to bring together active workers in these areas (who badly need such an activity, since they are dispersed and isolated) and to spread the gospel through lectures, notes, seminars, etc. I hope that the first such Workshop will take place in the summer of 1980.

A main theme of this volume is the work I have been doing with Clyde Martin on the ramifications of *Lie group theory* and *algebraic geometry* in mathematical systems theory. Some beginnings of this are in "Interdisciplinary Mathematics", Volumes 3, 8 and 13. Our collaboration has mainly taken the form of a series of six papers, "Applications of Algebraic Geometry to Systems Theory, I-VI", written while we were both working at Harvard in 1975-76. Parts I and II have already been published in the *IEEE Control Systems Society Transactions* and the *IEEE Proceedings*. Parts 3 and 4, which were the most innovative, have unfortunately been delayed by SIAM journal bureaucracy. They should appear (consolidated into one paper) in the September 1978 issue of *SIAM Journal of Control*. Chris Byrnes, now of Harvard, has extended our methods (of course, combined with much of his own) to cover *delay systems*. Look for his work, which should be soon appearing. Parts 5 and 6 presented several tentative ideas for extending the methodology to cover *time-varying linear* and *infinite dimensional* systems. Some of this material was presented at the *Proceedings of the 1976 Ames Conference on Geometric Control*, which Math Sci Press published as "Lie Groups", Volume 7.

The heart of the method Martin and I used in our work was to convert a time-invariant, linear system into a seemingly more complicated mathematical gadget, a *holomorphic vector bundle over the Riemann sphere*. This seems to go counter to the observation that good applied mathematics should be more "concrete". However, it is even so an extremely useful transformation, since it sets a certain body of ideas which had been toyed with before in the system theory literature (and which are very important for the applications, since they involve "feedback" and "identification" most intimately) into shape for a definitive treatment using straightforward and powerful techniques of Lie group theory and algebraic geometry. What is probably most important--and which needs much more work--is that these techniques, or ones closely related to them, are probably *indispensable* for treating the two-dimensional systems problems which are now coming into the foreground, both for *theory* and *applications*. In fact, further work in this direction is presented in this volume, although so far still in fragmentary form. The material I present here in such "systems" is also in the closest contact with material from quantum field-theoretic *physics*.

Thus, I propose a dialog between *systems theorists* and *physicists*, two groups who, to paraphrase Winston Churchill, are profoundly separated by a common language (i.e., mathematics). "Physical mathematics" I *define* as the study and development of the *analytic tools* needed for the physical sciences (including a good deal of chemistry, biology, etc.). Similarly, "mathematical systems theory" is coming to serve the similar role of providing the analytical foundations for modern engineering, and the fields in which the engineering methodology plays a role (e.g., economics, computer sciences, etc.). Its main practitioners live in certain electrical engineering departments, and its main scientific societies are the IEEE Control Systems Society and the International Federation of Automatic Control. "System Theory" is a very confusing area, since it includes such a wide variety of people and subjects. Of course, I deal mainly with the mathematical side, but I must say that part of its intellectual attraction for me is precisely its complexity, richness and interface with so many applied areas. It can be thought of as the "new" *applied mathematics*, to be contrasted with such traditional fields as fluid mechanics and numerical analysis. However, it should be thought of *mainly* as a branch of *engineering* with certain overlap into *mathematics* (explicitly) and *physics* (implicitly), although there have been alarmingly and amazingly little contact and understanding between these groups.

There is also confusion in the scientific public's mind between *control* and *optimal control*. The latter is the narrower, mainly involving the classic tools of optimization and the calculus of variations, beefed up with certain "modern" touches. The former is concerned much more with the classical engineering problems of designing a system to meet certain performance criteria, possibly to adapt to certain types of changes in conditions, and to achieve a prescribed stability. This is a very complex business, often more art than science, but the hope is that there are certain mathematical principles and models which can be abstracted from engineering practice and studied intensively.

Much of my own work since 1975 has been motivated by my attempt to serve this mathematical function for control theory problems encountered in the work of people at Ames Research Center of the National Aeronautics and Space Administration who are concerned with the design of control systems of the more complicated sorts of aircraft (STOL, helicopters, etc.). I should explain to the reader who is not familiar with the state of aeronautical engineering (as I was not when I began in 1975) that Control Theory has long played a useful but modest role in airplane design based largely on the "classical" single input-single output linear control system. Now, the actual systems were in fact much more complicated than this, but drastic approximations have adequately served in the past. Of course, that marvelous "adaptive control system", the pilot's brain, takes up the slack! However, control of certain new aircraft is substantially more complex, and simply making modifications in the existing systems does not seem to be adequate. There are two aspects that suggest mathematical attention:

> The nonlinearity of the equations can no longer be ignored (as it is in classical aircraft control design). Does systems-theory-on-manifolds have any useful insights, and/or is it possible to extend the current theory to be useful?

> Is there an adequate theory of approximation-structural stability for multi-input and output systems?

These broad questions have motivated my work. I am very grateful to the Ames Research Center, in particular Brian Doolin, for the support they have given me.

Several chapters presented here were written as preliminary material for Technical Reports that I am writing concerning the aircraft problems, with the support of NASA-AMES grant No. NSG-2252. Much of this work was done in collaboration with Clyde Martin of Case Western Reserve University who also is associated with Ames. I would also like to thank Roger Brockett and Chris Byrnes, with whom I have discussed much of this material. Several other topics are also included. I have great hopes and plans for saying something substantial about numerical analysis using Lie theory and algebraic geometry. A start is made here on the Gauss elimination algorithm and the LR, QR algorithms. Finally, I would like to thank Karin Young and Michael Ackerman for their great help in this enterprise.

TABLE OF CONTENTS

	Page
PREFACE	iii

GROUP I
THE GEOMETRY AND LIE THEORY OF MATRIX RICCATI EQUATIONS ... 1

Chapter 1: LIE THEORY AND THE MATRIX RICCATI EQUATION ... 3

1. Introduction ... 3
2. Notation of the Theory of Differentiable Manifolds and Lie Group Theory ... 4
3. Stable Sets and Subgroups for Flows ... 7
4. The Grassmann Spaces and the Grassmann Coset Spaces ... 9
5. The Lie Algebra of the Symplectic Group ... 12
6. Limits of One-Parameter Group Orbits on Grassmann Manifolds ... 15
7. Grassmannian Orbits of the One-Parameter Symplectic Group Arising from the Linear Constant, Quadratic Performance Criterion, Regulator Problem ... 17

References ... 24

Chapter 2: THE LINEAR OPTIMAL CONTROL PROBLEM AND THE SYMPLECTIC GROUPS. HAMILTONIAN FEEDBACK ... 27

1. Introduction ... 27
2. Symmetry or Skew-Symmetry of the Hamiltonian Matrix ... 28
3. Systems which are Harmonic Oscillators with Constraints ... 29
4. The Feedback Group and the Linear Symplectic Groups ... 31

Chapter 3: THE LIE THEORY OF THE LAGRANGE-GRASSMANN MANIFOLD AND ITS GENERALIZATIONS ... 35

1. Introduction ... 35
2. The Cartan Subalgebra of $\mathscr{P}$ ... 39
3. A Method of Construction of Compact Grassmann Coset Spaces ... 47

Bibliography ... 52

Chapter 4: LOCAL ANALYSIS NEAR A ZERO POINT OF A VECTOR FIELD BELONGING TO A TRANSITIVE FINITE DIMENSIONAL LIE ALGEBRA OF VECTOR FIELDS ... 53

1. Introduction ... 53
2. The Tangent Bundle to a Homogeneous Space ... 55
3. The Linearization of a Vector Field Defined by an Element of the Lie Algebra ... 56
4. The Tangent Bundle to the Grassmann Manifold ... 56
5. The Tangent Bundle to Grassmannians in Terms of Partitioned Matrices. The Linearization of a Riccati Vector Field Near a Zero ... 58
6. Stability Properties of Fixed Points of Riccati Vector Fields ... 61

Bibliography ... 64

Page

Chapter 5: THE STATIONARY SOLUTIONS OF RICCATI EQUATIONS AS THE ORBITS OF CENTRALIZERS ... 65

1. Introduction ... 65
2. Notation of Manifold Theory ... 66
3. Stationary Solutions of Lie Systems ... 69
4. Some General Facts about Fixed Points of Vector Fields Belonging to Finite Dimensional Transitive Lie Algebras. The Centralizer Foliation ... 69
5. The Fixed Point Set of a Vector Field Arising from a Semisimple Element of a Finite Dimensional Lie Algebra. The Cartan Decomposition of a Semisimple Lie Algebra ... 72
6. Matrix Riccati Equations as Lie Systems on Grassmannians ... 73
7. The Fixed Point Set of Riccati Vector Fields Corresponding to Diagonalizable Linear Transformations ... 76
8. The Symplectic Matrix Riccati Equation ... 77

Bibliography ... 80

Chapter 6: MATRIX RICCATI EQUATIONS DEFINED BY LINEAR QUADRATIC DIFFERENTIAL GAMES ... 81

1. Introduction ... 81
2. Caratheodory's Method for Differential Games ... 81
3. Linear Quadratic, Two-Person, Zero-Sum Differential Games ... 83

GROUP II
HOLOMORPHIC VECTOR BUNDLES AND LINEAR SYSTEMS ... 87

Chapter 7: VECTOR BUNDLES DEFINED BY FINITE DIMENSIONAL, LINEAR, TIME-INDEPENDENT SYSTEMS ... 89

1. Introduction ... 89
2. Linear Input-Output Systems ... 90
3. The "Frequency Domain" Discription via a Vector Bundle ... 91
4. Grassmann Spaces and their Canonical Vector Bundles ... 93
5. Vector Bundles Defined by Linear Equations ... 95
6. The Vector Bundle on the Riemann Sphere Defined by an Input System. Controllability as a Bundle Property ... 96
7. Controllability and Observability as a Vector Bundle Property ... 100

Bibliography ... 102

Chapter 8: THE KRONECKER THEORY, VECTOR BUNDLES, AND INPUT-OUTPUT SYSTEMS ... 105

1. Introduction ... 105
2. The Kronecker Invariants of Input Systems ... 105
3. Input System of Kronecker Type ... 109
4. The Input Vector Bundle for Kronecker Input Systems ... 110

Chapter 9: THE KRONECKER THEORY ... 115

1. Introduction ... 115
2. Pairs of Linear Maps from the Kronecker Point of View ... 115
3. The Kronecker Vector Bundle Defined by a Pair of Linear Maps ... 116

		Page
4.	Polynomial Cross-Sections of the Kronecker Vector Bundle	117
5.	The Line Bundles Determined by the Kronecker Reduction Process	119
	Bibliography	125

Chapter 10: TRANSFER FUNCTIONS AND HOLOMORPHIC VECTOR BUNDLES FOR 1-D AND n-D SYSYTEMS — 127

1. Introduction — 127
2. The Usual One-Dimensional Transfer Function as a Meromorphic Differential in the Riemann Surface of Genus Zero — 127
3. The "Kronecker Pencil" form of the Input-Output Relations — 129
4. A Generalization to Partial Differential Equations — 130
5. An Example from Classical Mathematical Physics — 130
6. Vector Bundles on Orbit Spaces — 133
7. Vector Bundles Defined by Systems of Partial Differential Equations and Linear Groups of Symmetries — 134
8. $SO(3,\mathbb{C})$ -Bundles — 135
9. The Compactified Vector Bundles Determined by the Helmholtz Equation — 138
10. The Vector Bundles Defined by Maxwell's Equations — 142

Chapter 11: THE RIEMANN SURFACE AND TRANSFER FUNCTION VECTOR BUNDLE OF A LINEAR, TIME-VARYING SYSTEM DEFINED IN TERMS OF LINEAR DIFFERENTIAL OPERATOR SYMMETRIES — 147

1. Introduction — 147
2. Linear Differential Operator Symmetries of Linear, Time-Varying Systems — 147
3. Differential Operator Symmetries of Scalar Input Output Systems of Classical Type — 148
4. Old and New Facts about Commuting Scalar Linear Differential Operators. Relations with Algebraic Geometry — 149
5. The Vector Bundle Defined by a Differential Operator Symmetry of a Linear Time-Varying System — 152
6. The Input-Output Bundle for Time-Invariant Systems — 153
7. The Vector Bundles Defined by the Burchnall-Chaundy Theory for Certain Scalar Input-Output Systems — 154
8. The Burchnall-Chaundy Vector Bundle — 156

   Bibliography — 157

GROUP III
ISOSPECTRAL DEFORMATION OF INPUT-OUTPUT SYSTEMS — 159

Chapter 12: ISOSPECTRAL DEFORMATIONS OF LINEAR, TIME-VARYING SYSTEMS — 161

1. Introduction — 161
2. Isospectral Deformation of Finite Dimensional Linear Maps — 161
3. The Kruskal-Lax "Isospectral Deformation" Conditions for Linear Differential Operators — 164

		Page
4.	Deformations of Linear Systems	166
5.	Linear Differential Operators in One Independent Variable that are the Input-Output Relations of Linear Systems. Specialization to Sturm-Liouville	168
6.	Isospectral Deformation of Linear Systems and Isospectral Linear Differential Operators	170
7.	Isospectral Deformation and Input-Output Relations of the Systems-Theoretic Isospectral Deformation	172
	Bibliography	175

Chapter 13: RELATIONS BETWEEN ISOSPECTRAL DEFORMATIONS OF SYSTEMS AND DIFFERENTIAL OPERATORS — 177

1. Introduction — 177
2. Linear Systems Associated with Differential Operators — 177
3. Condition for Sturm-Liouville to be Associated with a 2×2 System — 178
4. Input-Output Systems and Covariant Differentiation — 180
5. Isospectral Deformation of the Systems and Input-Output Relations — 184
6. Eigenvectors of the Input-Output Relations — 185

   Bibliography — 187

Chapter 14: ISOSPECTRAL DEFORMATIONS OF BOUNDARY VALUE PROBLEMS OF LINEAR SYSTEMS — 189

1. Introduction — 189
2. Boundary Conditions — 190
3. "Local" Isospectral Deformation of Systems — 191
4. Isospectral Deformation with Boundary Conditions — 192

Chapter 15: THE KORTEWEG-DE VRIES-STURM-LIOUVILLE SYSTEM FROM THE POINT OF VIEW OF ISOSPECTRAL DEFORMATIONS OF SYSTEMS — 197

1. Introduction — 197
2. Isospectral Deformations of Sturm-Liouville — 197
3. The Sturm-Liouville, $n = 1$ Case — 200
4. Sturm-Liouville, $n = 2$ — 201
5. Sturm-Liouville, $n = 3$, and Korteweg-de Vries — 202

   Bibliography — 204

Chapter 16: PLANE WAVE SOLUTIONS OF THE ISOSPECTRAL DEFORMATION EQUATIONS — 205

1. Introduction — 205
2. The Plane Wave Solutions — 205
3. General Solution of the Ordinary Differential Equation for the Plane Wave Solutions — 206

TABLE OF CONTENTS xi

Page

GROUP IV
USEFUL CONCEPTS IN TIME-VARYING LINEAR SYSTEMS THEORY 209

Chapter 17: INFINITE DIMENSIONAL LIE GROUPS, HALPHEN-WILCZYNSKI
THEORY, AND LINEAR SYSTEMS 211

1. Introduction 211
2. Linear Time-Varying (Input-Output) Systems of Classical Type 212
3. Linear State-Time Transformations 213
4. Time-Preserving System Isomorphisms 215
5. Some Elementary Parts of the Wilczynski Theory 215
6. The Input-Output Form of Scalar Differential Operators 218
7. Linear Transformations of the Input-Output System Which are Gauge Transformations of the Input-Output Relations 220
8. Wilczynski Equivalent Time-Varying Linear Systems 224
9. The W-Group as a Transformation Group on the Space of Systems and its Differential Invariants 226
10. The Prolongations of a Linear, Time-Varying System 229
11. The (2×2) - Case 231

Chapter 18: TIME-VARYING INPUT-OUTPUT SYSTEMS GOVERNED BY LAPLACE
INTEGRAL OPERATORS WITH ALGEBRAIC "SYMBOLS" 237

1. Introduction 237
2. Input-Output Relations for Second Order, Linear Differential Equations 237
3. The Input-Output Relations in Laplace Integral Form 241
4. The Bessel Equations 243

Chapter 19: THE ASYMPTOTIC BEHAVIOR OF A LINEAR, TIME-VARYING SYSTEM
AND TRANSFER FUNCTION 247

1. Introduction 247
2. Asymptotic Behavior of Time-Independent Systems 247
3. Input-Output Behavior of Certain Equations whose Laplace Transforms are First Order 249
4. Input-Output Relations Governed by Laplace Equations 251
5. Linear Systems from the Point of View of Thomé 253

Bibliography 255

Chapter 20: CONSERVATIONS LAWS AND ASYMPTOTIC FORMAL POWER SERIES
SOLUTIONS OF DIFFERENTIAL EQUATIONS 257

1. Introduction 257
2. Order of Contact of Maps 257
3. The Jet Spaces and their Differential Geometry 260
4. The Contact Structure on $J^\infty(X,Y)$ 262
5. Prolongation of Vector Fields 264
6. Asymptotic Solutions of the Riccati Equation. (Liouville-Greene-Poincare-Langer-Jeffries-Wentzel-Kramers-Brilloin) 265
7. JWBK (and Liouville-Green-Birkhoff-Langer) Type Solutions to the Sturm-Liouville Equation 270
8. The Formal Power Series Algebra of an Algebra and its Differential Form Algebra 271
9. The Formal Power Series Algebra of an Algebra and its Differential Form Algebra 271

		Page
10.	The LGPBLJWKB-Solution to the Riccati Equation within the Context of Differential Form Algebras	274
11.	Asymptotic Maps Defined by Formal Power Series in the Manner of Poincaré and Birkhoff	276
12.	The LGPBLJWKB-Formulas and the Conservation Laws for the Korteweg-de Vries Equation	277

Chapter 21: THE POISSON-MOYAL BRACKET AND THE BURCHNALL-CHAUNDY THEORY OF LINEAR ORDINARY DIFFERENTIAL OPERATORS — 279

1. Introduction — 279
2. The Transvection Differential Operator — 279
3. The Transvection Applied to Polynomials — 282
4. The Lie Algebra of Linear Differential Operators and the Poisson-Moyal Lie Algebra — 284
5. Commutativity and Semi-Commutativity in the Poisson-Moyal Lie Algebra — 285
6. Semi-Commutativity for the Case of Second Order Polynomials — 287

Bibliography — 290

Chapter 22: THE BIRKHOFF-LANGER EXPANSION OF LINEAR SYSTEMS — 291

1. Introduction — 291
2. The Formal Asymptotic Series — 291
3. The Volterra-Birkhoff-Langer Series for Bilinear Systems — 292
4. Birkhoff-Langer Expansions for Linear Input Systems — 294

Chapter 23: THE BIRKHOFF-LANGER EXPANSION AND THE CONSERVATION LAWS OF NON-LINEAR WAVE EQUATIONS — 295

1. Introduction — 295
2. Completely Integrable (in the Sense of Mayer-Frobenius) Linear Systems of Two Independent Variables — 295
3. The Birkhoff-Langer Expansion — 297

GROUP V
SOME WORK IN GEOMETRIC CONTROL THEORY — 301

Chapter 24: BOLTZMANN-HAMEL EQUATIONS OF ANALYTICAL MECHANICS — 303

1. Introduction — 303
2. Differential Forms with Special Attention to those of First and Second Degree — 305
3. Lagrange's Equations in the Language of Differential Forms — 312
4. Quasicoordinates as Differential Forms — 316
5. The Rotating Rigid Body and the Rotation Group — 319
6. The Euler Angles — 324
7. The Boltzmann-Hamel Equations — 327
8. The Boltzmann-Hamel Equations on a Lie Group — 333

TABLE OF CONTENTS   xiii

                                                                                    Page

Chapter 25: LINEARIZATION OF NONLINEAR SYSTEMS FROM THE POINT OF VIEW
  OF CONTROL THEORY   337

  1. Introduction   337
  2. Linearization of a Nonlinear Input System   337
  3. An Input System as an Exterior Differential System   338
  4. The Linear Variational Equations in Terms of a "Non-Holonomic"
     or "Moving" Frame   339

Chapter 26: A GENERAL ALGEBRAIC AND GEOMETRIC FORMALISM FOR
  PERTURBATION THEORY OF NONLINEAR DIFFERENTIAL EQUATIONS AND
  INPUT-OUTPUT SYSTEMS   343

  1. Introduction   343
  2. van der Pol's Equation   345
  3. The Method of "Multiple Time-Scale" for the van der Pol
     Input-Output   348
  4. An Abstract Perturbation Theory for Curves in Lie Algebras   349

Chapter 27: LIE THEORY, THE METHOD OF "VARIATION OF PARAMETERS",
  VOLTERRA SERIES, PERTURBATION THEORY AND THE VOLTERRA PATH
  INTEGRAL   353

  1. Introduction   353
  2. Flows on Manifolds and their Infinitesimal Generators   353
  3. The Perturbation-Variation of Constants Formula   355
  4. The Picard Approximations   356
  5. The Differential Equations for the Curve $t \to C_t = \alpha_t^* B_t \alpha_t^{-1*}$ 359
  6. The Volterra Series and the Feynman-Dyson Expansion   360
  7. The "Variation of Constants" in Analytical Mechanics   363
  8. The First Order Approximation to the Analytical Mechanics
      Perturbation Problem   366
  9. Perturbation of the Harmonic Oscillator   367
  10. The Approach to Perturbation Theory via Lagrange's Original
      Method and the Hamilton-Jacobi Equation   368
  11. The Volterra Path Integral   372

Chapter 28: SOME POTENTIAL USES OF ALGEBRAIC GEOMETRY IN SYSTEMS
  THEORY AND NUMERICAL ANALYSIS   375

  1. Introduction   375
  2. Algebra in Linear Systems Theory: The Identification
     Problem   377
  3. Some Basic Ideas of Algebraic Geometry   380
  4. Calculus Properties of Algebraic Sets   385
  5. The Almost-Onto Property   391
  6. The Method of Gaussian Elimination as an Example of an
     Almost Onto Map   394
  7. A General Group-Theoretic and Geometric Framework for the
     Gauss Algorithm   398

Page

GROUP VI
SURVEYS AND DEVELOPMENT OF NEW GEOMETRIC METHODOLOGY FOR APPLICATIONS   401

Chapter 29: SCATTERING THEORY FOR THE RICCATI EQUATION   403

1. Introduction   403
2. Scattering Theory for the Riccati Equation   404
3. Scattering for the Schrödinger-Sturm-Liouville Equation and the Bäcklund Transformation for the Korteweg-de Vries Equation   407
4. Changes in the Scattering Function Induced by a Lie Algebra-Valued One-Form of Curvature Zero   410
5. Prolongations and Generalized Conservation Laws for Exterior Differential Systems   412
6. Estabrook-Wahlquist Prolongations and Generalized Conservation Laws   413
7. General Remarks   415

Bibliography   417

Chapter 30: EXTERIOR SYSTEMS, INPUT-OUTPUT SYSTEMS, AND VECTOR BUNDLES   419

1. Introduction   419
2. Exterior Differential Systems   420
3. Input-Output Systems as Exterior Differential Systems   425
4. The Kronecker Theory of Pencils of Linear Maps   426
5. The Vector Bundle Associated with Pairs of Linear Maps   431
6. The Kronecker-Grothendieck Invariants and the Lie-Cartan Equivalence Problem   434
7. The Holomorphic Vector Bundle Invariants of Linear, Time-Invariant Input Systems   435
8. Feedback Equivalence and the Kronecker Theory of Equivalence of Pairs of Linear Maps   438
9. Deformation of Exterior Differential Systems, Singular Perturbation Theory and Simplification of Complicated Systems   441
10. Final Remarks   444

Bibliography   445

Chapter 31: SOME GENERAL REMARKS ABOUT GROUPS, DIFFERENTIAL FORMS AND THEIR GENERALIZATIONS   447

1. Introduction   447
2. Some General Ideas   447
3. Graded Lie Algebras Defined by Exterior Derivative Relations in Differential Form Algebras   449
4. Derivations and Two-Derivations of Differential Algebras   451
5. Differential Form Algebras with Generating Subalgebras. "Geometric" Action of Graded Lie Algebras   453
6. Maurer-Cartan Forms   455
7. Taylor's Formula and the Hopf Algebra Structure for a Lie Group   458
8. Generalized Lie Groups Based on an Associative Algebra, Following Berezin and Kac   463
9. The Taylor Series for Transformation Group Actions   467
10. The Co-Algebraization of Transformation Semi-Group Actions, Following Berezin and Kac   468

Bibliography   469

TABLE OF CONTENTS

Page

Chapter 32: SOME FURTHER COMMENTS ABOUT DIFFERENTIAL FORMS OVER ASSOCIATIVE ALGEBRA — 471

1. Introduction — 471
2. Differential Forms over Arbitrary Associative Algebras — 471
3. "Twisted" Commutative Algebras — 473
4. Z_2-Graded Grassmann Algebras as Twisted Commutative Algebras — 475
5. The Heisenberg-Relation Associative Algebra. "Quantum" Hamiltonian Theory — 476

Chapter 33: GRADED LIE GROUPS FROM CARTAN'S POINT OF VIEW — 479

1. Introduction — 479
2. Exterior Derivative and Graded Lie Algebras — 479

Chapter 34: REMARKS ABOUT BOSONS AND FERMIONS — 483

1. Introduction — 483
2. Elementary Remarks about the Fermi Exclusion Principle — 483
3. Conditions on Dynamical Interaction — 486

Chapter 35: A RELATION BETWEEN THE LR AND QR ALGORITHMS OF MATRIX NUMERICAL ANALYSIS AND LIE THEORY — 491

1. Introduction — 491
2. The Lie Theory of the LR Algorithm — 491
3. The QR Algorithm — 496

Chapter 36: THE REDHEFFER *-OPERATION OF SCATTERING THEORY AND THE GEOMETRY OF GRASSMANIANS AND LIE SYSTEMS — 497

1. Introduction — 497
2. Working Backwards from the Linearization — 497
3. Introduction to Grassmanians — 499
4. Lie Systems and Linearizations — 500

Bibliography — 500

GROUP I

THE GEOMETRY AND LIE THEORY OF MATRIX RICCATI EQUATIONS

INTRODUCTION

The chapters in this group represent work that was done in collaboration with Clyde Martin as part of our "program" of refining the standard *linear* control theory on a Lie-group and algebro-geometric foundation. However, we were never able to complete the part of the program which dealt with the Riccati equations, and I am presenting this material, which was prepared as preparation for our more ambitious work, in this form.

A good deal of the success of linear control theory in engineering and other practical applications (e.g., aeronautics) is due to the relative ease of handling the matrix Riccati equations on computers and the relatively powerful theorems which are available to relate important properties of linear control systems to other properties of matrix Riccati equations. For example, one of these theorems (due to Yacubovich and Kalman) relates the search for *stable* feedback laws to certain types of solutions of the *algebraic* Riccati equation, i.e., to the *fixed points* of the vector field whose orbits are the solutions. The introduction of "geometry" is motivated by the fact that the matrix Riccati equations really *live on compact homogeneous spaces*, the so-called *Lagrange-Grassmann manifolds*, which are coset spaces of the symplectic group $Sp(n,R)$ by one of its *parabolic subgroups*. The material in the chapters in this group is mainly oriented toward describing background information about these spaces and the vector fields associated with the Riccati equations. We have not done all that we had hoped to do, especially the theory of *differential games*, which lies on the horizon and provides many challenging problems to Lie theory.

Chapter I

LIE THEORY AND THE MATRIX RICCATI EQUATION

1. INTRODUCTION

It has been pointed out in various places [1-4,6] that the *matrix Riccati equation* [7-12] of linear optimal control theory can usefully be regarded as what is classically called a *Lie system* of ordinary differential equations [13,14]. This means that solving the equations with given initial conditions is determined by a curve in a given Lie group G which acts on the space M of initial conditions. Now, in the matrix Riccati case, M is a coset space

$$G/H \quad ,$$

where G is a semi-simple non-compact Lie group, H is a closed subgroup and G/H is *compact*. In [15,16] methods have been developed for studying the asymptotic and stability properties of such Lie systems in case G/H is what we now call a *Grassmannian coset space*. Control theorists have already developed techniques for this purpose [7,17,18]. In this approach, the Liapounov functions are directly defined in the symmetric matrix space on which the control theorist's matrix Riccati equation lives. Following the methods of [15,16] other Liapounov functions can be defined using Lie-theoretic techniques. One can also use the methods of the modern theory of "dynamic systems" (in the topologist's, not the control-theorist's sense [23,24]) to study the properties of Riccati systems. We also study the "algebraic" matrix Riccati equation as a special one of the problems of finding the zero points of a vector field in G/H determined by the Lie algebra of $\mathcal{G}$. Again, we will show that using Lie theoretic ideas provides new and powerful methods for studying this question.

We believe that the introduction of these new Lie theoretic methods may ultimately be useful in extending the practical scope of the standard linear optimal control theory. Here are three areas where we can see further progress on the horizon and in which we plan further work in this series of publications:

a) Asymptotic properties and stability behavior of the matrix Riccati equations appearing in the theory of linear differential games [11,18,26,27],

b) Development of a "Lie" theory for *difference* equations which can deal in an analogous way with the matrix Riccati difference equations which appear in optimal control problems with *discrete* time,

c) Adaptation of the techniques of numerical analysis used to solve matrix Riccati equations to be compatible with their natural geometric and Lie theoretic structure.

Here is another "practical" reason for introducing Lie (and manifold) theoretic methods into the study of the matrix Riccati equation. It provides a curve

$$t \to P(t)$$

in the space of symmetric matrices satisfying an ordinary differential equation of the following form:

$$-\frac{dP}{dt} = A'P + PA' - PBR^{-1}B'P + Q$$

In this space solutions may go to infinity at finite time, or the limit as $t \to \infty$ of $P(t)$ may not exist. We shall exhibit the space of all P's as an open, dense subset of a compact coset space $M = Sp(n,R)/H$, where H is a certain subgroup of the Lie group $SL(n,R)$ of $2n \times 2n$ linear canonical (or "symplectic") matrices. Thus, the asymptotic behavior of $P(t)$ as $t \to \infty$ may involve going out of this open subset, but approaching a definite limit in M. This seems especially likely to happen in the "differential games" situation [18,26,27]. We can then provide a natural "compactification" for the situation, which may be useful as a geometric setting for the analytical results which are needed in control theory. It is also possible to use *topological* techniques (e.g., from the theory of dynamical systems [23,24]) to study these differential equations--again, a task facilitated by setting the differential equation on a *compact* coset space of a Lie group rather than the non-compact Euclidean space that is involved in the control theory literature.

2. NOTATION OF THE THEORY OF DIFFERENTIABLE MANIFOLDS AND LIE GROUP THEORY

In order to keep this paper within bounds and proceed as quickly as possible to the material that is of concrete interest to control theorists, we shall assume that the reader is familiar with the rudiments of manifold and Lie group theory [1,19,20,28,29]. This section will present the notation we shall use in the rest of the paper.

Differential manifolds will be denoted by the letters M, X, Y, Z,... If X is such a manifold, $\mathscr{F}(X)$ denotes the C^∞, real-valued functions on X. $\mathscr{F}(X)$ is (under point-wise addition and multiplication) a *commutative, associative algebra* (with the real numbers R as field of scalars). A *vector field* V is an R-linear map: $\mathscr{F}(X) \to \mathscr{F}(X)$ such that

LIE AND RICCATI

$$V(f_1 f_2) = V(f_1)f_2 + f_1 V(f_2)$$

for $f_1, f_2 \in \mathscr{F}(X)$.

$\mathscr{V}(X)$ denotes the set of such vector fields. $\mathscr{V}(X)$ admits both multiplication by elements of $\mathscr{F}(X)$ --which makes it an $\mathscr{F}(X)$-module--and a Lie algebra operation

$$(V_1, V_2) \to [V_1, V_2] = V_1 V_2 - V_2 V_1 ,$$

called the *Jacobi bracket*.

Remark. In the control theory literature, this is often called the *Lie bracket*. This is not historically correct, in fact, Lie himself [13] calls it the Jacobi bracket!

Let t be an additional real parameter, and let $t \to V(t)$ be a curve in $\mathscr{V}(X)$. One can then define the *orbit curves of this curve* as the curves $t \to x(t)$ in X such that

$$\frac{d}{dt} f(x(t)) = V(t)(f)(x(t)) \tag{2.1}$$

for all $f \in \mathscr{F}(X)$.

One readily sees that in local coordinates, (2.1) is equivalent to a time-dependent system of ordinary differential equations of the form

$$\frac{dx}{dt} = v(x,t) . \tag{2.2}$$

One proves in this way that, given $p_0 \in X$, $t_0 \in R$, there is a unique solution of (2.1) such that

$$x(t_0) = p_0 .$$

(This statement must be understood modulo "finite escape time" phenomena, which we ignore, i.e., we assume that solutions to (2.1) exist over $-\infty < t < \infty$.) For fixed t, if we follow the solution of (2.1) from 0 to t, and, using the initial conditions p_0, we obtain a diffeomorphism

$$\phi_t : X \to X$$

This defines a one-parameter family $t \to \phi_t$ of diffeomorphisms of X such that

$$\phi_0 = \text{identity map}$$

Such an object is called a *flow* on X, and the curve $t \to V(t)$ in $\mathcal{V}(X)$ is called its *infinitesimal generator*. In particular, $V(t)$ is *independent of* t if and only if

$$\phi_{t_1+t_2} = \phi_{t_1} \phi_{t_2}$$

for $t_1, t_2 \in R$.

Giving a *Lie group structure* on a set G amounts to giving the following sort of data:

a) A group structure, i.e., a map $G \times G \to G$, multiplication denoted as $(g_1, g_2) \to g_1 g_2$ and satisfying the usual group axioms,

b) A manifold structure such that the group operations $G \times G \to G$ are smooth (i.e., C^∞) maps.

A Lie group G acts as a *transformation group* on a manifold X if one is given a smooth map $G \times X \to X$, denoted as $(g,x) \to gx$, such that

$$g_1(g_2 x) = (g_1 g_2) x$$

for $g_1, g_2 \in G$.

Let G be a Lie group. A *one parameter subgroup of* G is a map $t \to g(t)$ of $R \to G$ such that

$$g(t_1, t_2) = g(t_1) g(t_2)$$

for $t_1, t_2 \in R$.

If G acts as a transformation group on the manifold X and if $t \to g(t)$ is a one-parameter subgroup, the action

$$\phi_t(x) = g(t)(x)$$

on X has a vector field V as its infinitesimal generator. Let $\mathcal{G}$ denote the set of one-parameter subgroups of G. The transformation group action of G on X then defines a mapping

$$\mathcal{G} \to \mathcal{V}(X)$$

that may be thought of as the "infinitesimal" action of G. Its study was the main object of Lie's work [13,30]. Here is a main theorem.

Theorem 2.1. $\mathcal{G}$ has a Lie structure such that the maps $\mathcal{G} \to V(X)$ defined by transformation group actions are Lie algebra homomorphisms.

If the Lie group G acts on X, a flow $t \to \phi_t$ generated by (2.1) or (2.2) is said to be a *Lie system* of G if, for each t, $\phi_t \in G$. If $t \to V(t)$ is the infinitesimal generator of the flow, it follows from the standard results of Lie theory that

$$V(t) \in (\text{image of } \mathcal{G} \text{ in } \mathcal{V}(X) \text{ defined via the transformation group action})\ .$$

Thus, the qualitative study of Lie systems is closely linked to the study of properties of Lie transformation groups. In particular, the matrix Riccati equations are Lie systems for a coset space

$$X = G/H\ ,$$

$G = Sp(n,R)$, X = a *compact* manifold [1-3,5].

3. STABLE SETS AND SUBGROUPS FOR FLOWS

Let X be a manifold, t a real variable $0 \leq t < \infty$, and $t \to \phi_t$ a flow on X, i.e., a one-parameter family of diffeomorphisms of X such that:

$$\phi_0 = \text{identity map}\ .$$

The map $(x,t) \to \phi_t(x)$ is a C^∞ map $[0,\infty): X \to X$.

Definition. Let x_0 be a point of X. The set of all $x \in X$ such that

$$\lim_{t \to \infty} \phi_t(x) = x_0 \qquad (3.1)$$

is called the *stable set* of x_0, denoted by $S(x_0)$.

Of course, at this level of generality, there is not too much we can say of any significance about these sets. Let us make further assumptions. Suppose that

$$g: X \to X$$

is a diffeomorphism, and

$$x \in S(x_0)\ ,$$

i.e., (3.1) is satisfied. Then,

$$\phi_t(gx) = (\phi_t g \phi_t^{-1})(\phi_t x) \qquad (3.2)$$

In order to analyze the meaning of this formula for the asymptotic behavior, let us make further assumptions. Namely:

There is a Lie group G which acts as a Lie transformation group on X such that $g \in G$ and, for each t, $\phi_t \in G$.

Let G^{x_0} be the *isotopy subgroup* of G at x_0, i.e.,

$$G^{x_0} = \{g \in G: gx_0 = x_0\} . \qquad (3.3)$$

Suppose further that:

$$\lim_{t \to \infty} \phi_t g \phi_t^{-1} \in G^{x_0} \qquad (3.4)$$

It follows from (3.1), (3.2), and (3.3) that:

$$\lim_{t \to \infty} \phi_t(gx) = x_0 \qquad (3.5)$$

We can then sum up as follows:

<u>Theorem 3.1</u>. Let G be a Lie group which acts as a Lie transformation group on the manifold X. (In particular, the transformation group action itself defines a map $G \times X \to X$ which is *continuous* with respect to the manifold topology. It is this property that is used most heavily.) Let

$$SS(x_0) = \left\{ g \in G: \lim_{t \to \infty} \phi_t g \phi_t^{-1} \in G^{x_0} \right\} \qquad (3.6)$$

Then $SS(x_0)$ is a subgroup of G. It is called the *stable subgroup* of G at x_0 (relative to the given flow $t \to \phi_t$). Let $S(x_0)$ be the stable set of the flow at x_1. Then, the *action of* $SS(x_0)$ *preserves the set* $S(x_0)$.

<u>Proof</u>. That $SS(x_0)$ as defined by (3.6) is a subgroup of G should be obvious, since G^{x_0} is a subgroup. (For example, if $g_1, g_2 \in SS(x_0)$, then

$$\lim_{t \to \infty} (\phi_t g_1 g_2 \phi_t^{-1}) = \lim_{t \to \infty} \phi_t g_1 \phi_t^{-1} \phi_t g_2 \phi_t^{-1}$$

$$= \lim_{t \to \infty} (\phi_t g_1 \phi_t^{-1}) \lim_{t \to \infty} (\phi_t g_2 \phi_t^{-1})$$

$\in G^{x_0}, G^{x_0} \subset G^{x_0}$.) That $SS(x_0)$ satisfies

$$SS(x_0) \ S(x_0) \subset S(x_0)$$

follows from formula (3.5).

Remark. Notice how necessary it is to use the "topological" structure of G and X, and the action of G on X. It does not seem (at least a priori) to follow from this abstract set-up that $SS(x_0)$ is a *closed* (or even a *Lie*) subgroup of G. In all cases we have computed this is so, but we do not know any definitive answers about this.

The material in this section is abstract. In order to provide interesting examples where it may be applied, we now turn to the study of a class of spaces studied (but not named) in [15,16].

4. THE GRASSMANN SPACES AND THE GRASSMANN COSET SPACES

Let W be a finite dimensional vector space with a field F as scalars. (F will be either R or $\mathbb{C}$, the real or complex numbers.) L(W) denotes the set of linear maps: W → W. GL(W) denotes the subset of L(W) consisting of the invertible maps. GR(W) denotes the space of linear subspaces of W. GR(W) is called the *Grassmann space* associated with W. GL(W) acts as a transformation group on GR(W).

For $g \in GL(W)$, $\gamma \in GR(W)$,

$g(\gamma)$ is the linear subspace $\{gw: w \in \gamma'\}$.

The orbits of GL(W) on GR(W) are the *Grassmann manifolds*, the set of linear subspaces of W of a given dimension.

Let G be a Lie group. A homomorphism

$\rho: G \to GL(W)$

is called a *linear representation* of G. It defines a transformation group action of G on GR(W). For $\gamma \in GR(W)$ the orbit is

$G\gamma'$ = set of linear subspaces of W which are transforms under $\rho(g)$ of γ, as g runs through G.

Let

H = isotropy subgroup of G at γ, i.e., the set of $g \in G$ such that $g\gamma = \gamma$.

The orbit $G\gamma$ is then identified with the coset space G/H.

Definition. A coset space G/H of a Lie group is said to be a *Grassmann coset space* if it can be exhibited in this way as an orbit in GR(W) for *some* linear representation of G.

Much information about this class of coset spaces can be obtained by abstraction from the results proved in [16,17]. Here is one.

Theorem 4.1. Suppose that G is a real, semisimple Lie group with a finite center. Let K be a maximal compact subgroup of G and let G/H be a *compact* Grassmann coset space. Then K acts transitively on G/H.

Here is the main example of a Grassmann coset space which is of interest for the study of matrix Riccati equations of control theory. Let W be a real vector space of dimension $2n$ and let

$$\omega: W \times W \to R$$

be a skew-symmetric, non-degenerate bilinear form on W. The pair (W, ω) is said to be a *linear symplectic manifold*. Let $Sp(\omega)$ be the group of linear maps $g: W \to W$ such that

$$\omega(gw_1, gw_2) = \omega(w_1, w_2)$$

$$\text{for } w_1, w_2 \in W \quad .$$

A linear subspace γ is said to be a *Lagrange subspace* of the symplectic structure if:

$$\dim W = 2 \dim \gamma$$

$$\omega(\gamma, \gamma) = 0$$

The set of all Lagrange subspaces then forms a compact manifold. It is easy to see using standard linear algebra that $Sp(\omega)$ acts transitively on the space of Lagrange subspaces, which then is a coset space $Sp(\omega)/H$, with the property of forming a *compact* Grassmann coset space in terms of the above definition. K, the maximal compact subgroup of $Sp(\omega)$, is found by fixing a positive definite quadratic form on W, and letting K be the subgroup of elements of $Sp(\omega)$ which also preserve this quadratic form. (In terms of matrices, $Sp(\omega)$ is $Sp(n, R)$, K is $U(n)$, the group of complex unitary matrices, considered as $2n \times 2n$ real matrices.)

Here is the more traditional way of looking at this. Realize W as $R^n \times R^n$, the set of all pairs

$$\begin{pmatrix} x \\ y \end{pmatrix}$$

with x, y n-vectors. The inner product ω is defined as

$$\omega\left(\begin{pmatrix} x \\ y \end{pmatrix}, \begin{pmatrix} x_1 \\ y_1 \end{pmatrix} \right) = x'y_1 - y'x_1$$

If P is an $n \times n$ symmetric matrix, set:

$$\gamma_P = \text{set of all vectors of the form } \begin{pmatrix} x \\ Px \end{pmatrix} .$$

This defines a map $P \to \gamma_P$ of $R^{n(n+1)/2}$ ($\equiv$ set of $n \times n$ symmetric matrices) *onto* $Sp(\omega)/H$. It is readily seen that the image under this map is an open, dense subset. Thus, the manifold of all Lagrange subspaces of $R^n \times R^n$ form natural *compactifications* of the space of all $n \times n$ symmetric matrices.

Recall how the matrix Riccati equation is generated from an optimal control problem. Suppose our "performance index" is an integral of the form

$$\int (x'Qx + u'Ru) \, dt ,$$

where Q, R are symmetric $n \times n$ and $m \times m$ matrices. The *Hamiltonian* is then:

$$H = \frac{1}{2}(x'Qx + u'Ru) + y(Ax + Bu) ,$$

with

$$u = R^{-1}B'y' .$$

The extremal equations are

$$\frac{d}{dt}\begin{pmatrix} x \\ y \end{pmatrix} = \begin{pmatrix} A, & -BR^{-1}B' \\ -Q, & -A' \end{pmatrix} \begin{pmatrix} x \\ y \end{pmatrix}$$

The matrix Riccati equation is

$$\frac{dP}{dt} = -PBR^{-1}B'P + PA + A'P + Q = 0 .$$

Let $g(t)$ be the $2n \times 2n$ matrix function of t (i.e., a curve on $Sp(n,R)$) such that:

$$\frac{dg}{dt} = \begin{pmatrix} A, & -BR^{-1}B' \\ -Q, & -A' \end{pmatrix} g ,$$

$$g(0) = I .$$

Then,

$$\gamma_{P(t)} = g(t)(\gamma_{P(0)}) .$$

This formula exhibits the solution of the matrix Riccati equation as the orbits of flows on $Sp(n,R)$, i.e., a "Lie system", with $Sp(n,R)$ acting on the *compact* coset space $Sp(n,R)/H$. If a solution $P(t)$ exhibits "finite escape time", i.e., the solution does not exist for all t, this may mean geometrically that the curve $t \to \gamma_{P(t)}$ is leaving the open subset defined by the image of the set of symmetric matrices. In other words, G/H is a compact manifold within an open dense subset of the form γ_P parameterized by one "chart" as t varies. $P(t)$ may leave this "chart" and enter another one.

5. THE LIE ALGEBRA OF THE SYMPLECTIC GROUP

In [15,16] certain geometric methods are presented for studying Lie systems on Grassmann homogeneous spaces, based on certain information of a Lie algebraic nature. Typically, one wants to know certain information about the Lie algebra $\mathcal{G}$ and (in case it is non-compact and semisimple) its Cartan decomposition. In this section we will develop the material that is most relevant for control theory purposes, where $\mathcal{G}$ is the Lie algebra of the Lie group $Sp(n,R)$.

We deal with $2n \times 2n$ real matrices, partitioned into $n \times n$ blocks

$$\alpha = \begin{pmatrix} \alpha_{11} & \alpha_{12} \\ \alpha_{21} & \alpha_{22} \end{pmatrix} \tag{5.1}$$

Let

$$J = \begin{pmatrix} 0, & I_n \\ -I_n, & 0 \end{pmatrix} \tag{5.2}$$

α is *infinitesimal symplectic* if

$$\alpha J + J\alpha' = 0 \tag{5.3}$$

(' means the transpose of a matrix). Now,

$$\begin{pmatrix} \alpha_{11} & \alpha_{12} \\ \alpha_{21} & \alpha_{22} \end{pmatrix} \begin{pmatrix} 0 & I \\ -I & 0 \end{pmatrix} = \begin{pmatrix} -\alpha_{12}, & \alpha_{11} \\ -\alpha_{22}, & \alpha_{21} \end{pmatrix}$$

$$J\alpha' = \begin{pmatrix} 0 & I \\ -I & 0 \end{pmatrix} \begin{pmatrix} \alpha'_{11} & \alpha'_{21} \\ \alpha'_{12} & \alpha'_{22} \end{pmatrix} = \begin{pmatrix} \alpha'_{12}, & \alpha'_{22} \\ -\alpha'_{11}, & \alpha'_{21} \end{pmatrix}$$

Hence, condition (5.3) means that:

$$\alpha_{12} = \alpha'_{12}$$
$$\alpha_{21} = \alpha'_{21} \tag{5.4}$$
$$\alpha_{11} = \alpha'_{22}$$

Let $\mathcal{G}$ be the Lie algebra of all $2n \times 2n$ matrices α satisfying (5.4). Let $\mathcal{K}$ be the set of all $\alpha \in \mathcal{G}$ such that

$$\alpha J = J\alpha \quad . \tag{5.5}$$

(5.4) and (5.5) *together* combine to give

$$\alpha + \alpha' = 0 \tag{5.6}$$

or

$$0 = \begin{pmatrix} \alpha_{11} & \alpha_{12} \\ \alpha_{21'} & -\alpha_{11} \end{pmatrix} + \begin{pmatrix} \alpha_{11} & \alpha_{12} \\ \alpha_{21'} & -\alpha_{11} \end{pmatrix}'$$

$$= \begin{pmatrix} \alpha_{11} & \alpha_{12} \\ \alpha_{21'} & -\alpha_{11} \end{pmatrix} + \begin{pmatrix} \alpha'_{11} & \alpha_{21} \\ \alpha_{12'} & \alpha'_{11} \end{pmatrix}$$

or

$$\alpha'_{11} = -\alpha_{11}$$
$$\alpha_{12} = -\alpha_{21} \tag{5.7}$$

i.e., α belongs to the Lie algebra of the orthogonal group $SO(2n,\mathbb{R})$. $\mathcal{K}$ is therefore a *compact* Lie algebra, and in fact is the *maximal* compact Lie subalgebra of $\mathcal{G}$.

For $\alpha \in \mathcal{G}$, set:

$$\sigma(\alpha) = -\alpha' \quad . \tag{5.8}$$

Notice that $\sigma(\alpha)$ again belongs to $\mathcal{G}$, i.e., σ defines an *automorphism* of period 2 of $\mathcal{G}$. Also,

$$\mathcal{K} = \{\alpha \in \mathcal{G} : \sigma(\alpha) = \alpha\} \tag{5.9}$$

This means that $\mathcal{K}$ is a *symmetric Lie subalgebra of* $\mathcal{G}$ in the sense of E. Cartan. (G/K is a *Hermitian symmetric space* in the terminology of [19]. This symmetric space plays a basic role in many parts of mathematics; it is often called the *Siegel upper half plane*.)

Set:

$$\mathcal{P} = \{\alpha \in \mathcal{G} : \sigma(\alpha) = -\alpha\} \qquad (5.10)$$

$$= \{\alpha \in \mathcal{G} : \alpha' = \alpha\}$$

$$= \begin{pmatrix} \alpha_{11'} & \alpha_{12} \\ \alpha_{12'} & -\alpha_{11} \end{pmatrix} : \alpha'_{11} = \alpha_{11}; \ \alpha'_{12} = \alpha_{12}$$

Then, $\mathcal{P}$ is a linear subspace of $\mathcal{G}$, such that:

$$\mathcal{G} = \mathcal{K} \oplus \mathcal{P} \qquad \text{(of vector spaces)}$$

$$[\mathcal{K}, \mathcal{P}] \subset \mathcal{P}$$
$$\qquad \qquad \qquad \qquad (5.11)$$
$$[\mathcal{P}, \mathcal{P}] \subset \mathcal{K}$$

This is the *Cartan decomposition* of the non-compact, real Lie algebra $\mathcal{G}$. If G/H is the space of Lagrangian subspaces of $R^n \times R^n$, let us determine $\mathcal{H}$, the Lie algebra of H. Suppose

$$\alpha = \begin{pmatrix} \alpha_{11} & \alpha_{12} \\ \alpha_{21'} & -\alpha_{11} \end{pmatrix} \in \mathcal{G}$$

Let it act on a vector $R^3 \times R^3$ of the form $\begin{pmatrix} x \\ 0 \end{pmatrix}$.

$$\alpha \begin{pmatrix} x \\ 0 \end{pmatrix} = \begin{pmatrix} \alpha_{11} x \\ \alpha_{21} x \end{pmatrix}$$

Thus

$$\mathcal{H} = \text{set of } \alpha \in \mathcal{G} \text{ such that}$$

$$\alpha_{21} = 0 \qquad (5.12)$$

$$= \text{set of } 2n \times 2n \text{ matrices of the form } \begin{pmatrix} \alpha_{11} & \alpha_{12} \\ 0 & -\alpha_{11} \end{pmatrix}$$

$$(51.3)$$

<u>Remark</u>. Notice that this implies that the eigenvalues of α come in pairs, each the negative of the other. One can, in fact, prove (by a generalization that every matrix is similar to a matrix in triangular form) that every matrix in $\mathcal{G}$ is similar to a *complex* one of form (5.13). This then gives

an easy proof of the well known fact that the eigenvalues of infinitesimal symplectic matrices come in pairs, each the negative of the other. In turn, this fact prevents infinitesimal symplectic matrices from being "stable" in the usual sense.

6. LIMITS OF ONE-PARAMETER GROUP ORBITS ON GRASSMANN MANIFOLDS

Let W be a vector space with either the real or complex numbers as field of scalars. Let

$$GR^m(W)$$

denote the Grassmann manifold of m-dimensional linear subspaces of W. Let $GL(W)$ denote the group of linear automorphisms of W. Let $L(W)$ denote the Lie algebra (under commutation) of linear maps: $W \to W$. Thus, for $\alpha \in L(W)$, the mapping

$$t \to \exp(t\alpha) = \alpha + t\alpha + \frac{t^2 \alpha^2}{2!} + \cdots$$

defines a one-parameter group of linear automorphisms of W.

$GL(W)$ acts on $GR^m(W)$ as a transformation group--if $g \in GL(W)$, $\gamma \in GR^m(W)$, i.e., if γ is an m-dimensional linear subspace of W, then

$$g\gamma$$

is the transform of the subspace γ by g.

Now, $GR^m(W)$ is a manifold, and thus has a natural topology. Further, it is *compact* in this topology. Our goal in this section is to investigate the limits as $t \to \infty$ of the orbits

$$t \to \exp(t\alpha)(\gamma)$$

of one-parameter subgroups of $GL(w)$.

Now, when $\exp(t\alpha)$ is made explicit as an $n \times n$ matrix, the functions of t occurring as the matrix elements are not "arbitrary". In fact, one sees--via the Cayley-Hamilton theorem--that they are solutions of constant coefficient, linear ordinary differential equations of order n. Thus, they are linear combinations of functions of the form:

$$t^j e^{at} \cos bt$$

$$t^j e^{at} \sin bt \quad ,$$

(6.1)

with $1 \leq j \leq n$, and $a \pm b\sqrt{-1}$, the eigenvalues of α. Thus, we can write $\exp(t\alpha)$ in the following form:

$$\exp(t\alpha) = f_k(t)\alpha_1 + \cdots + f_r(t)\alpha_r \qquad (6.2)$$

where $\alpha_1,\ldots,\alpha_r$ are linear maps: $W \to W$ and $f_1(t),\ldots,f_r(t)$ are functions of the form (6.1).

Let $\gamma \in GR^m(w)$. Let $w_1,\ldots,w_m$ be a basis for γ. Then,

$$w_i(t) = \exp(t\alpha)(w_i) , \qquad 1 \leq i \leq m ,$$

are a basis for $\exp(t\alpha)(\gamma)$. In order to assure the convergence of $\exp(t\alpha)(\gamma)$ we must find an $m \times n$ matrix function of t, $(\beta_i^j(t))$ with nonzero determinant such that the linearly independent elements

$$w_i'(t) = \sum_{j=1}^m \beta_i^j(t) w_j(t)$$

converges as $t \to \infty$ and *remain linearly independent at* $t = \infty$. The $\lim_{t \to \infty} \exp(t\alpha)(\gamma)$ will then be the linear subspace of W spanned by those limiting elements.

Now, the simplest way to choose this matrix β is to impose a positive definite inner product on w (i.e., to make it a real Hilbert space) and then to choose $w_i'(t)$ as the vectors obtained by the known Gram-Schmidt orthogonalization process. The elements of $\beta_i^j(t)$ are then essentially just rational functions of the functions $f_1(t),\ldots,f_r(t)$ which appears in (6.2). There is no possibility of deducing that these will have limits as $t \to \infty$ (without further assumptions) if the sine and cosine actually appear in (6.1), i.e., if α has one eigenvalue *with a nonzero, pure imaginary part*. However, if all the eigenvalues are real, then it is easy to see that the Gram-Schmidt formula will provide vectors which converge as $t \to \infty$. We can then sum up as follows.

Theorem 6.1. If α is a linear map: $W \to W$ which only has real eigenvalues, and if γ is an m-dimensional linear subspace of W, then

$$\lim_{t \to \infty} \exp(t\alpha)(\gamma)$$

converges to an element of $G^m(W)$. If α does not have real eigenvalues, the method described above gives an indeterminate answer about the limit.

Of course, we may still be able to say something about the existence of the limits by other methods, particularly if we are given more information about α and γ. We shall now consider a situation of this type that arises from control theory. Later on, we shall discuss purely differential geometric methods.

7. GRASSMANNIAN ORBITS OF THE ONE-PARAMETER SYMPLECTIC GROUP ARISING FROM THE LINEAR CONSTANT, QUADRATIC PERFORMANCE CRITERION, REGULATOR PROBLEM

One of the key sequences of results in modern optimal control theory is that which relates the properties of linear quadratic regulators, solutions of the "algebraic" Riccati equation, and stabilization by state feedback. (These results are primarily due to R. Kalman [17]. An excellent review is given in the article by Jan Williams [..) In this section we shall show how these results can be interpreted in a Lie theoretic context-- and indeed, how they provide important information about the action of the real symplectic group $G = Sp(n,R)$ and the space G/H of "Lagrange" linear subspaces of R^{2n}.

For this section we shall use notation which is closer to that used in the optimal control literature. x denotes an element of R^n considered as a column vector. x' is its transpose, a row vector. Points of R^{2n} will be denoted as

$$\begin{pmatrix} x \\ y \end{pmatrix} \quad x,y \in R^n \quad .$$

A linear map $\alpha: R^{2n} \to R^{2n}$ is a 2×2 partitioned matrix

$$\alpha = \begin{pmatrix} \alpha_{11} & \alpha_{12} \\ \alpha_{21} & \alpha_{22} \end{pmatrix}$$

whose entries are $n \times n$ matrices. Thus,

$$\alpha \begin{pmatrix} x \\ y \end{pmatrix} = \begin{pmatrix} \alpha_{11}x + \alpha_{12}y \\ \alpha_{21}x + \alpha_{22}y \end{pmatrix}$$

$$J = \begin{pmatrix} 0 & I \\ -I & 0 \end{pmatrix}$$

The *symplectic form* on R^{2n} is the bilinear map $R^{2n} \times R^{2n} \to R$ associated with J.

$$(x_1 y_1) J \begin{pmatrix} x \\ y \end{pmatrix} \qquad (7.1)$$

α is *infinitesimally symplectic* of *Hamiltonian* if

$$\alpha' J + J\alpha = 0 \qquad (7.2)$$

$\mathcal{G}$ denotes the Lie algebra (under commutation) of $2n \times 2n$ matrices satisfying this condition. G denotes the Lie group of $2n \times 2n$ matrices generated by this algebra. (In Lie theory, it is denoted as $Sp(n,R)$; it may also be defined as the $2n \times 2n$ invertible matrices which leave invariant the symplectic form (7.1).)

Let M be the (compact) submanifold of $GR^n(R^{2n})$ consisting of those n-dimensional linear subspaces on which the symplectic form is identically zero. M is an orbit space of G, i.e., M is a *compact Grassmann coset space* of G. Let γ_0 be the element of M defined as follows

$$\gamma_0 = \left\{ \begin{pmatrix} x \\ 0 \end{pmatrix} : x \in R^{2n} \right\} \tag{7.3}$$

Let H be the isotropy subgroup of G at γ_0, H is the set of all $g \in G$ of the form

$$g = \begin{pmatrix} g_{11} & g_{12} \\ 0 & g_{22} \end{pmatrix} \tag{7.4}$$

$\mathcal{H}$, its Lie algebra, is the set of all matrices of the form

$$\alpha = \begin{pmatrix} \alpha_{11} & \alpha_{12} \\ 0 & -\alpha_{11} \end{pmatrix} \tag{7.5}$$

with

$$\alpha'_{12} = \alpha_{12} \quad . \tag{7.6}$$

If P is a *symmetric*, $n \times n$ real matrix, set:

$$\gamma_P = \left\{ \begin{pmatrix} x \\ Px \end{pmatrix} : x \in R^n \right\} \tag{7.7}$$

Notice that $\gamma_P \in M$, i.e., the symplectic form, is zero on γ_P (because P is symmetric). Let M_0 be the set of all elements of M which can be written in this way.

<u>Theorem 7.1.</u> M_0 is an open, dense subset of M. The map

$$P \to \gamma_P$$

is a diffeomorphism between M_0 and the set of symmetric matrices, i.e., $R^{n(n+1)/2}$.

Consider a $2n \times 2n$ real matrix of the form

$$g_p = \begin{pmatrix} I & 0 \\ P & I \end{pmatrix} \qquad (7.8)$$

Then,

$$Jg_p = \begin{pmatrix} 0 & I \\ -I & 0 \end{pmatrix}\begin{pmatrix} I & 0 \\ P & I \end{pmatrix} = \begin{pmatrix} P & I \\ -I & 0 \end{pmatrix}$$

$$g_p'J = \begin{pmatrix} I & P' \\ 0 & I \end{pmatrix}\begin{pmatrix} 0 & I \\ -I & 0 \end{pmatrix} = \begin{pmatrix} -P' & I \\ -I & 0 \end{pmatrix}$$

Thus, the condition that $g_p \in G$, i.e., that

$$Jg + g'J = 0 \quad,$$

is

$$P' = P \quad. \qquad (7.9)$$

Let N be the subgroup of elements of G of the form (7.8)-(7.9). (N stands for "nilpotent". In fact, N is abelian, but for many general Lie-theoretic situations N would be non-abelian, but nilpotent.)

Theorem 7.2. M_0 is the orbit

$$N\gamma_0$$

of the subgroup N of G on the point $\gamma_0 \in M$.

Proof.

$$g_p\gamma_0 = \text{set of } \begin{pmatrix} I & 0 \\ P & I \end{pmatrix}\begin{pmatrix} x \\ 0 \end{pmatrix} : x \in R^n$$

$$= \text{set of } \begin{pmatrix} x \\ Px \end{pmatrix}$$

Q.E.D.

Theorem 7.2. Let $\alpha \in \mathcal{G}$. The one-parameter subgroup $t \to \exp(t\alpha)$ of G, its generator, has a fixed point in M_0 if and only if there is an $n \times n$ real matrix P such that

$$g_p \alpha g_{-p} \in \mathcal{H} \quad . \qquad (7.10)$$

Proof. H = set of $g \in G$ such that

$$g\gamma_0 = \gamma_0 \quad .$$

Suppose

$$\exp(t\alpha)(g_p \gamma_0) = g_p \gamma_0$$

or

$$g_p^{-1} \exp(t\alpha) g_p \gamma_0 \quad ,$$

or

$$g_p^{-1} \alpha \exp(t\alpha) g_p \in H \quad ,$$

but

$$g_p^{-1} \exp(t\alpha) g_p = \exp(g_p^{-1} \alpha g_p) \quad .$$

is the Lie algebra of H, i.e., the set of matrices tangent to the one-parameter subgroups of H, thus we have

$$g_p^{-1} \alpha g_p \in \mathcal{H}$$

From (7.8) we see that

$$g_p^{-1} = g_{-p} \quad . \tag{7.11}$$

Formula (7.10) now follows. All of these steps are reversible, to finish the proof of Theorem 7.2.

Let us now make (7.10) explicit. Write α as

$$\alpha = \begin{pmatrix} \alpha_{11} & \alpha_{12} \\ \alpha_{21} & -\alpha_{11}' \end{pmatrix}$$

with $\alpha_{12}' = \alpha_{12}$; $\alpha_{21}' = \alpha_{21}$.

$$g_p \alpha = \begin{pmatrix} I & 0 \\ P & I \end{pmatrix} \begin{pmatrix} \alpha_{11} & \alpha_{12} \\ \alpha_{21} & -\alpha_{11}' \end{pmatrix}$$

$$= \begin{pmatrix} \alpha_{11} & \alpha_{12} \\ P\alpha_{11} + \alpha_{21}, & P\alpha_{12} - \alpha_{11}' \end{pmatrix}$$

$$g_p \alpha g_{-p} = \begin{pmatrix} \alpha_{11} & \alpha_{12} \\ P\alpha_{11} + \alpha_{21}, & P\alpha_{12} - \alpha'_{11} \end{pmatrix} \begin{pmatrix} I & 0 \\ -P & I \end{pmatrix}$$

$$= \begin{pmatrix} \alpha_{11} - \alpha_{12}P, & \alpha_{12} \\ P\alpha_{11} + \alpha_{21} - P\alpha_{12}P + \alpha'_{11}P, & P\alpha_{12} - \alpha'_{11} \end{pmatrix}$$

Hence, condition (7.10) is equivalent to the following condition:

$$P\alpha_{11} + \alpha_{21} - P\alpha_{12}P + \alpha'_{11}P = 0 \tag{7.12}$$

In the control theory literature (7.12) is called the *algebraic Riccati equation*. Its solution (numerically) is often a key step in the practical applications. In the later sections we shall indicate certain general results one can deduce about such equations using Lie theory. In order to put Equation (7.12) in more familiar form, consider a linear quadratic optimal control problem of the usual form:

$$\frac{dx}{dt} = Ax + Bu \ . \tag{7.13}$$

Performance criterion:

$$\frac{1}{2} \int (x'Qx + u'Ru) \, dt \tag{7.14}$$

Then,

$$\alpha = \begin{pmatrix} A, & -BR^{-1}B' \\ -Q, & -A \end{pmatrix} \tag{7.15}$$

Thus, (7.12) takes the following familiar form:

$$PA + A'P - Q + PBR^{-1}B'P = 0 \tag{7.16}$$

If Q and R are positive definite, and the input system (7.13) is controllable, it is known [7,17] that (7.16) admits a *unique* positive definite solution P. However, for the moment we shall not make this assumption.

Let us return to the general situation. Suppose that P_0 is a solution of (7.12). Then

$$g_{P_0} \alpha g_{P_0}^{-1} = \begin{pmatrix} \alpha_{11} - \alpha_{12}P_0, & \alpha_{12} \\ 0 & P_0\alpha_{12} - \alpha'_{11} \end{pmatrix} \tag{7.17}$$

Our goal is to say as much as possible about

$$\lim_{t \to \infty} \exp(t\alpha)(\gamma) \quad ,$$

for an arbitrary $\gamma \in M$. Now, using (7.17), we have:

$$\exp(t\alpha) = g_{-P_0} \exp\left(t \begin{pmatrix} \alpha_{11} - \alpha_{12}P_0', & \alpha_{12} \\ 0 & P_0\alpha_{12} - \alpha_{11}' \end{pmatrix}\right) g_{P_0}$$

Set:

$$\beta = \begin{pmatrix} \alpha_{11} - \alpha_{12}P_0', & \alpha_{12} \\ 0 & -\alpha_{11}' + P_0\alpha_{12} \end{pmatrix}$$

$$= \begin{pmatrix} \beta_{11}', & \beta_{12} \\ 0, & \beta_{22} \end{pmatrix} \quad (7.18)$$

Now,

$$\beta \begin{pmatrix} x \\ y \end{pmatrix} = \begin{pmatrix} \beta_{11}x + \beta_{12}y \\ \beta_{22}y \end{pmatrix}$$

Suppose that

$$\frac{d}{dt}\begin{pmatrix} x \\ y \end{pmatrix} = \beta \begin{pmatrix} x \\ y \end{pmatrix} = \begin{pmatrix} \beta_{11}x + \beta_{12}y \\ \beta_{22}y \end{pmatrix}$$

or

$$\frac{dy}{dt} = \beta_{22}y$$

$$\frac{dx}{dt} = \beta_{11}x + \beta_{12}y$$

or

$$y(t) = \exp(\beta_{22}t)y(0)$$

$$x(t) = \exp(\beta_{22}y)z(t)$$

or

$$\beta_{11}x + \beta_{12}y = \beta_{11}x + \exp(\beta_{11}t)\frac{dz}{dt}$$

or

$$\frac{dz}{dt} = \exp(-\beta_{11}t)\beta_{12}\exp(\beta_{22}t)y(0) = \exp(-\beta_{11}t)\beta_{12}\exp(-\beta_{11}t)'y(0)$$

In order to solve this, set

$$\phi(\beta_{12}) = \beta_{11}\beta_{12} - \beta_{12}\beta_{11}' \tag{7.19}$$

Notice that ϕ is a linear map of the space of symmetric matrices into itself. Also,

$$\exp(-\beta_{11}t)\beta_{12}\exp(-\beta_{11}t)' = \exp(t\phi)(\beta_{12}) \quad . \tag{7.20}$$

Hence

$$\begin{aligned}
z(t) &= \int_0^t \frac{dz}{dt}\,dt + z(0) \\
&= \int_0^t \exp(t\phi)(\beta_{12})y(0)\,dt + z(0) \\
&= (\phi^{-1}\exp(t\phi))(\beta_{12})y(0)\Big|_{t=0}^{t} + z(0) \\
&= \phi^{-1}\exp(t\phi)(\beta_{12})y(0) - \phi^{-1}(\beta_{12})(y(0)) + z(0) \\
&= (\phi^{-1}(\exp(t\phi)-1))(\beta_{12})(y(0)) + z(0) \\
&= \phi^{-1}(\exp(t\phi)-1)(\beta_{12})(y(0)) + x(0)
\end{aligned}$$

Hence,

$$x(t) = \exp(\beta_{11}t)(\phi^{-1}(\exp(t\phi)-1)(\beta_{12})(y(0)) + \exp(\beta_{11}t)x(0)$$

We can then read from these formulas the following Lie-theoretic formula:

$$\exp(t\beta) = \begin{pmatrix} \exp(\beta_{11}t), & \exp(\beta_{11}t)\phi^{-1}(\exp(t\phi)-1))\beta_{12} \\ 0, & \exp(-\beta_{11}'t) \end{pmatrix} \tag{7.21}$$

$$= \begin{pmatrix} \exp(\beta_{11}t), & 0 \\ 0, & \exp(-\beta_{11}'t) \end{pmatrix}\begin{pmatrix} I, & \phi^{-1}(\exp(t\phi)-1)(\beta_{12}) \\ 0, & I \end{pmatrix}$$

$$\tag{7.22}$$

References

1. R. Hermann, Differential Geometry and the Calculus of Variations, Academic Press, New York, 1968 (Second Edition in preparation, to be published by Math Sci Press, Brookline, Mass.).

2. R. Hermann, Algebraic Topics Useful in Systems Theory (Interdisciplinary Mathematics, Vol. 3), Math Sci Press, Brookline, Mass., 1973.

3. C.R. Schneider, "Global Aspects of the Matrix Riccati Equation", Math. Systems Theory, 7 (1973), 281-286.

4. H. Sussman, "Orbits of Families of Vector Fields and Integrability of Systems with Singularities", Bull. Am. Math. Soc. 79 (1973), 197-199.

5. C. Martin, "Grassmann Manifolds and Global Properties of the Riccati Equation", Proceedings of the International Symposium on Operator Theory of Networks and Systems, Vol. 2, August 17-19, 1977, Lubbock, Texas, pp. 82-85.

6. R. Hermann and C. Martin, "Lie Theoretic Aspects of the Riccati Equation", Proceedings of the 1977 CDC Conference, New Orleans, La.

7. J.C. Williams, "Least Squares Stationary Optimal Control and the Algebraic Riccati Equation", IEEE Trans. Auto. Control, AC-16 (1971), 621-634.

8. J. Radon, "Zum Probleme va Lagrange", Hamburg Math. Einzebschriften, No. 2, B.G. Teubner, Leipzig (1928).

9. W.A. Coppel, "Matrix Quadratic Equations", Bull. Aust. Math. Soc. 10 (1974), 377-401.

10. B.D.O. Anderson and J.B. Moore, Linear Optimal Control, Prentice-Hall, New York.

11. M. Intriligator, Mathematical Optimization and Economic Theory, Prentice-Hall, Englewood Cliffs, N.J., 1971.

12. E.B. Lee and L. Markuz, Foundations of Optimal Control Theory, Wiley, New York, 1967.

13. S. Lie, Transformationsgruppen, Chelsea Publishing Co., New York.

14. E. Vessiot, Encyclopedie des Sciences Mathematiques (1910), Tome II, Vol. 3, 58-170.

15. R. Hermann, "Geometric Aspects of Potential Theory in Symmetric Spaces, III", Math. Analen, 153 (1964), 384-394.

16. R. Hermann, "Compactification of Homogeneous Spaces, I", J. Math. Mech. 14 (1965), 655-678.

17. R. Kalman, "Contributions to he Theory of Optimal Control", Bol. Soc. Mat. Mexicana, 1960, 102-119.

18. A.C.M. Van Switen, "Qualitative Behavior of Dynamical Games with Feedback Strategies", Ph.D. thesis, University of Groningen, 1977.

19. S. Helgason, Differential Geometry and Symmetric Spaces, Academic Press, New York, 1962.

20. W. Boothby, An Introduction to Differentiable Manifolds and Riemannian Geometry, Academic Press, New York, 1975.

21. R. Hermann, Geometric Structure of Systems-Control Theory and Physics, Part A (Vol. 9 of Interdisciplinary Mathematics), Math Sci Press, Brookline, Mass.

22. R. Hermann, Geometric Structure of Systems-Control Theory and Physics, Part B (Vol. 11 of Interdisciplinary Mathematics), Math Sci Press, Brookline, Mass.

23. P. Hartman, Ordinary Differential Equations, Wiley, New York, 1964.

24. S. Smale, "Differential Dynamical Systems", Bull. Am. Math. Soc. 73 (1967), 747-817.

25. H. Kwakemaak and R. Siuan, Linear Optimal Control Systems, Wiley, New York, 1972.

26. V.A. Jakubovic, "Solution of an Algebraic Problem Encountered in Control Theory", Soviet Math. Dakl. 11 (1970), 882-886.

27. V.A. Jacubovic, "On the Synthesis of Optimal Controls in a Linear Differential Game with Quadratic Pay-Off", Soviet Math. Dakl. 11 (1970), 1478-1481.

28. R. Gilmore, Lie Groups, Lie Algebras, and Some of Their Applications, Wiley, New York, 1974.

29. A. Sagle and R. Walde, Introduction to Lie Groups and Lie Algebras, Academic Press, New York, 1973.

30. R. Hermann, Sophus Lie's 1880 Transformation Group Paper, Lie Groups: History, Frontiers and Applications, Vol. 1, Math Sci Press, Brookline, Mass., 1975.

31. R. Hermann, Lie Groups for Physicists, W.A. Benjamin, Reading, Mass., 1966.

Chapter 2

THE LINEAR OPTIMAL CONTROL PROBLEM AND THE SYMPLECTIC GROUPS. HAMILTONIAN FEEDBACK

1. INTRODUCTION

Consider the linear input system

$$\frac{dx}{dt} = Ax + Bu \tag{1.1}$$

together with the "performance criterion"

$$\int (x'Qx + u'Ru)\, dt \tag{1.2}$$

$$x, y \in R^n, \quad u \in R^m \quad .$$

Following the well-known rules of the calculus of variations, one knows that the extremals are the curves on R^{2n}.

$$t \to (x(t), y(t)) \quad ,$$

which are orbits of the group

$$t \to \exp(t\alpha) \quad ,$$

with

$$\alpha = \begin{pmatrix} A & -BR^{-1}B' \\ -Q & -A \end{pmatrix} \quad . \tag{1.3}$$

Let $\mathscr{G}$ be the Lie algebra of $Sp(n, R)$, i.e., the set of real 2×2 partitioned $n \times n$ matrices

$$\beta = \begin{pmatrix} \beta_{11} & \beta_{12} \\ \beta_{21} & -\beta'_{11} \end{pmatrix} \tag{1.4}$$

with

$$\beta'_{12} = \beta_{12} ; \quad \beta'_{21} = \beta_{21} \quad .$$

Then $\alpha \in \mathscr{G}$.

Our main problem is to see how α sits in $\mathscr{G}$, and how it acts on $G/H \equiv$ Lagrange-Grassmann manifold M. A complete solution to this question does not yet seem to be available. For example, one might want to know under what conditions on A, B, Q and R, the α defined by (1.3) has no imaginary eigenvalues, when it is diagonalizable, etc. Approached from this direct algebraic viewpoint, there seem to be two obstacles:

a) The orbits of G acting in $\mathscr{G}$ via the adjoint representation are only imperfectly known and parameterized although there has long been a solution "in principle"

b) It seems to be very difficult to relate the matrices A, B, Q and R to the algebraic invariants of the elements of $\mathscr{G}$.

Our aim in this chapter is to develop some elementary facts about the relation between the optimal control problem (1.1)-(1.2) and the element $\alpha \in \mathscr{G}$ given by (1.3).

2. SYMMETRY OR SKEW-SYMMETRY OF THE HAMILTONIAN MATRIX

The most obvious question one can ask about an element of $\mathscr{G}$ is whether it belongs to $\mathscr{K}$ or $\mathscr{P}$, when

$$\mathscr{G} = \mathscr{K} \oplus \mathscr{P}$$

is the Cartan decomposition of $\mathscr{G}$. In previous chapters we have shown that $\mathscr{K}$ consists of the β of form (1.4) which are skew-symmetric, while $\mathscr{P}$ consists of the symmetric elements. Thus, for α of the form (1.3), we have:

$$\alpha' = \begin{pmatrix} A' & -Q \\ -BR^{-1}B & -A' \end{pmatrix} \qquad (2.1)$$

Thus,

$$\alpha' = -\alpha$$

if and only if

$$\boxed{\begin{aligned} A' &= -A \\ Q &= -BR^{-1}B' \end{aligned}} \qquad (2.2)$$

$\alpha' = \alpha$ if and only if

LINEAR OPTIMAL CONTROL   29

$$\boxed{\begin{aligned} A' &= A \\ Q &= BR^{-1}B' \end{aligned}} \qquad (2.3)$$

(2.2) then determines $\mathcal{K}$, while (2.3) determines $\mathcal{P}$. The elements $\alpha \in \mathcal{G}$ satisfying (2.2) have pure imaginary eigenvalues, while those satisfying (2.3) have real eigenvalues. Both conditions imply that α is diagonalizable (hence that $\text{Ad } \alpha$ is completely reducible), since α is *normal*, i.e., commutes with its transpose.

We can realize such Hamiltonians by appropriate choice of the variational system (1.1)-(1.2). We now turn to this question.

One simple-minded way to satisfy these relations is to set:

$$A = 0 \; ; \qquad B = I \quad .$$

Then,

$$\alpha = \begin{pmatrix} 0 & \pm R \\ -R & 0 \end{pmatrix} \qquad (2.1)$$

The system equations (1.1)-(1.2) take the following form

$$\frac{dx}{dt} = u$$

$$\int (\pm x'Rx + \dot{x}'R\dot{x}) \, dt \qquad (2.2)$$

Physically, these are harmonic oscillators, "stable" for the minus sign, "unstable" for the plus sign. α is essentially the "energy" operator, e.g., in quantum mechanics. (2.1) indicates that it sits in $\mathcal{G}$ in a "nilpotent" way; this indicates a close relation between the quantum mechanics of harmonic oscillators and the theory of *nilpotent* Lie groups. Now, a common method of quantum field theory is to regard a "quantum field" as a "perturbation" of an "infinite dimensional harmonic oscillator". At this point, we see a Lie group-theoretic way of studying this, but it will be pursued at another place.

3. SYSTEMS WHICH ARE HARMONIC OSCILLATORS WITH CONSTRAINTS

Here is a "physical" method for generating a more general class of systems. Start off with a harmonic oscillator, whose configuration vector is q, and whose Lagrangian is

$$\int (q'Mq + \dot{q}'N\dot{q}) \, dt \qquad (3.1)$$

Introduce "constraints" as sets of linear relations between q and $\dot{q}$. For example, suppose q is partitioned as follows:

$$q = \begin{pmatrix} q_1 \\ q_2 \end{pmatrix},$$

and the constraints are of the form

$$(\gamma_1, \gamma_2) \begin{pmatrix} q_1 \\ q_2 \end{pmatrix} + (\delta_1, \delta_2) \begin{pmatrix} \dot{q}_1 \\ \dot{q}_2 \end{pmatrix} = 0$$

i.e.,

$$q_1 + \gamma q_2 + \delta_1 \dot{q}_1 + \delta_2 \dot{q}_2 = 0 \quad .$$

Set:

$$x = q_2$$

$$u = \begin{pmatrix} \dot{q}_1 \\ \dot{q}_2 \end{pmatrix}$$

Thus,

$$\frac{dx}{dt} = (0, I)u \equiv Bu \quad . \qquad (3.2)$$

When these substitutions are made in (3.1), we end up with Lagrangians of the form:

$$\int (x'Qx + u'Ru + x'Su) \, dt \quad . \qquad (3.3)$$

The Hamiltonians can readily be calculated for the system (3.2)-(3.3) using the Pontryagin principle:

$$h = x'Qx + u'Ru + x'Su + y'Bu$$

$$h_u = 2u'Ru + x'S + y'B = 0$$

or

LINEAR OPTIMAL CONTROL 31

$$2u' = -\frac{1}{2}(x'S + y'B)R^{-1}$$

$$u = -\frac{1}{2}R^{-1}(S'x + B'y) \quad .$$

The extremal equations are:

$$\frac{dx}{dt} = Bu = -\frac{1}{2}BR^{-1}(S'x + B'y)$$

$$\frac{dy}{dt} = -h_{x'} = -2Qx - Su = -2Qx + \frac{1}{2}SR^{-1}(S'x + B'y) \quad ,$$

$$\frac{d}{dt}\begin{pmatrix} x \\ y \end{pmatrix} = \alpha \begin{pmatrix} x \\ y \end{pmatrix}$$

with

$$\alpha = \begin{pmatrix} -\frac{1}{2}BR^{-1}S', & -\frac{1}{2}BB' \\ \frac{1}{2}SR^{-1}S', & \frac{1}{2}SR^{-1}B' \end{pmatrix} \quad (3.4)$$

This material gives a brief indication of how the linear symplectic Lie algebra can be parameterized by an appropriate choice of variational problem.

4. THE FEEDBACK GROUP AND THE LINEAR SYMPLECTIC GROUPS

The basic ideas of the calculus of variations--and Hamilton-Jacobi theory--provide a correspondence between "variational problems"

$$\frac{dx}{dt} = f(x,u)$$

$$\int L(x,u) \, dt \quad (4.1)$$

and Hamiltonian systems:

$$\frac{dx}{dt} = H_y$$

$$\frac{dy}{dt} = -H_x \quad . \quad (4.2)$$

Now, each of these systems admits quite different sorts of "symmetries". The variational system (4.1) is adapted to the "feedback" transformation:

$$x \to F(x)$$

$$u \to G(x,u)$$

while the Hamiltonian system (4.2) is adapted to transformations

$$(x,y) \to (F(x,y), G(x,y))$$

which are "canonical", i.e., which preserve the two-form

$$dy \wedge dx \quad .$$

Now, both of these classes of transformations form *groups*. The obvious question arises of the relation between them. In this section we shall consider this question for the case of *linear* systems.

First we must formalize the nature of the relation between Hamiltonian systems and input systems. We restrict attention to *linear* systems.

<u>Definition</u>. A linear *Hamiltonian system*

$$\frac{dx}{dt} = \alpha_{11} x + \alpha_{12} y$$

$$\frac{dy}{dt} = \alpha_{21} x - \alpha'_{11} x \quad , \tag{4.3}$$

$\alpha'_{12} = \alpha_{12}$, $\alpha'_{21} = \alpha_{21}$, and a linear *input system*

$$\frac{dx}{dt} = Ax + Bu \tag{4.4}$$

are said to be *associated* if the following conditions are satisfied:

$$\alpha_{11} = A \quad . \tag{4.5}$$

For every solution $t \to (x(t), y(t))$ of (4.3), there is a curve $t \to u(t)$ such that

$$Bu(t) = \alpha_{12} y(t) \tag{4.6}$$

for all t .

Consider the "feedback" transformation for the system (4.4)

$$u = v - Fx \quad . \tag{4.7}$$

Also consider a transformation for the system (4.3):

LINEAR OPTIMAL CONTROL   33

$$z = y + Gx \qquad (4.8)$$

with $G' = G$.

(It is readily verified that the transformation

$$\begin{pmatrix} x \\ y \end{pmatrix} \rightarrow \begin{pmatrix} x \\ y \end{pmatrix}$$

is *canonical*.) We look for the conditions G must satisfy in order that it lead to the input system results from the feedback transform (4.7). Then

$$\frac{dx}{dt} = Ax + \alpha_{12} y$$

$$= (A - \alpha_{12} G) x + \alpha_{12} z$$

We must then have:

$$\alpha_{12} G = BF \qquad . \qquad (4.9)$$

Suppose that:

$$Bu = \alpha_{12} y$$

We must then also have:

$$\alpha_{12} z = Bv ,$$

which are again associated.

The simplest situation is that where α is the Hamiltonian matrix corresponding to the performance index

$$\int (x'Qx + u'Ru) \, dt$$

We know that

$$\alpha = \begin{pmatrix} A & -BR^{-1}B' \\ -Q & -A' \end{pmatrix}$$

i.e.,

$$\alpha_{12} = -BR^{-1}B' \qquad . \qquad (4.10)$$

Relation (4.9) then takes the form:

$$-BR^{-1}B'G = BF \qquad (4.11)$$

It can be satisfied by setting:

$$F = -R^{-1}B'G \quad . \qquad (4.12)$$

We can then sum up as following:

<u>Theorem 4.2.</u> Suppose

$$\alpha = \begin{pmatrix} A, & -BR^{-1}B' \\ -Q, & -A' \end{pmatrix}$$

is the Hamilton matrix of a quadratic variational problem. Transform α via the following canonical transformation:

$$\alpha \to \begin{pmatrix} I & 0 \\ -G & I \end{pmatrix} \alpha \begin{pmatrix} I & 0 \\ G & I \end{pmatrix} \qquad (4.13)$$

with $G' = G$. Let F be defined by (4.12). Then, the same effect can be made in the associated input system via the "state feedback" transformation

$$u \to u - Fx \quad .$$

We will call the transformation (4.13) *Hamiltonian feedback*.

Chapter 3

THE LIE THEORY OF THE LAGRANGE-GRASSMANN
MANIFOLD AND ITS GENERALIZATIONS

1. THE SYMPLECTIC GROUP AS A REAL, NON-COMPACT, SEMISIMPLE LIE GROUP

One of the most studied objects in mathematics is a *semisimple Lie group* [1-5]. (It is defined as a Lie group whose Lie algebra has no abelian ideals.) There are two types--the *compact* and *non-compact*. As we have seen, the *symplectic group* $Sp(n,R)$ --and its transitive action on the Lagrange-Grassmann manifold--plays the key geometric role in the study of the control-theoretic matrix Riccati equation. Our goal in this chapter is to describe these objects as far as possible in terms of the general theory of non-compact semisimple Lie groups.

Let G be the Lie group $Sp(n,R)$, the $2n \times 2n$ real symplectic matrices. Write such matrices in 2×2 partitioned form with entries which are $n \times n$:

$$g = \begin{pmatrix} g_{11} & g_{12} \\ g_{21} & g_{22} \end{pmatrix} .$$

The condition that $g \in G \equiv Sp(n,R)$ is that

$$g'Jg = J , \qquad (1.1)$$

with

$$J = \begin{pmatrix} 0 & I \\ -I & 0 \end{pmatrix} \qquad (1.2)$$

In accordance with general Lie group principles, $\mathscr{G}$, the Lie algebra of G, is the set of $2n \times 2n$ matrices of the form

$$\alpha = \begin{pmatrix} \alpha_{11} & \alpha_{12} \\ \alpha_{21} & \alpha_{22} \end{pmatrix}$$

such that

$$\alpha'J + J\alpha' \equiv 0 \qquad (1.3)$$

This is equivalent to the following conditions:

$$\alpha_{22} = -\alpha'_{11} \tag{1.4}$$

$$\alpha'_{12} = \alpha_{12} \ ; \quad \alpha'_{21} = \alpha_{21}$$

The *Killing form* (a symmetric bilinear form on $\mathcal{G}$)

$$(\alpha,\beta) \to B(\alpha,\beta)$$

is defined as follows

$$B(\alpha,\beta) = \text{trace}(\alpha\beta) \ . \tag{1.5}$$

It is known (because $\mathcal{G}$ is semisimple) that B is non-degenerate, i.e.,

$$B(\alpha,\beta) = 0$$

for all $\beta \in \mathcal{G} \Rightarrow \alpha = 0$

Now, $B(\alpha,\alpha)$ has no particular sign. A fundamental algebraic invariant of such a bilinear form is its *signature*, which is the number of *positive* minus the number of *negative* eigenvalues. In this case, the signature also has a group-theoretic significance.

Set:

$$\mathcal{K} = \{\alpha \in \mathcal{G} : \alpha' = -\alpha\} \tag{1.6}$$

$$\mathcal{P} = \{\alpha \in \mathcal{G} : \alpha' = \alpha\} \tag{1.7}$$

Then, we have the following commutation relations:

$$\mathcal{G} = \mathcal{K} \oplus \mathcal{P} \quad ;$$

$$[\mathcal{K},\mathcal{K}] \subset \mathcal{K}$$

$$[\mathcal{P},\mathcal{P}] \subset \mathcal{K} \tag{1.8}$$

$$[\mathcal{K},\mathcal{P}] \subset \mathcal{K}$$

Then say that $\mathcal{K}$ is a *symmetric Lie subalgebra of* $\mathcal{G}$. Further,

$$B(\alpha,\alpha) < 0 \qquad \text{for } \alpha \in \mathcal{K}-(0)$$

$$B(\alpha,\alpha) > 0 \qquad \text{for } \alpha \in \mathcal{K}-(0)$$

$$B(\mathcal{K},\mathcal{P}) = 0 \tag{1.9}$$

These relations describe the *Cartan decomposition for the non-compact semi-simple Lie algebra* $\mathcal{G}$.

This decomposition can also be described at the group level. Define a map $\sigma: G \to G$ as follows:

$$\sigma(g) = (g')^{-1} . \tag{1.10}$$

σ maps G into G.

Proof. If $g \in G$, it satisfies

$$g'Jg = J . \tag{1.11}$$

Here,

$$\sigma(g')J\sigma(g) = g^{-1}J(g')^{-1} . \tag{1.12}$$

Now, $J^{-1} = -J = J'$. Hence, taking inverses of both sides of (1.10) produces the relation

$$\sigma(g')'J\sigma(g) = J ,$$

which shows that $\sigma(g) \in G$.

σ maps G into G. It is a *group automorphism*, i.e.,

$$\sigma(g_1 g_2) \equiv \sigma(g_1)\sigma(g_2) .$$

Set:

$$K = \{g \in G: \sigma(g) = g\}$$
$$P = \{g \in G: \sigma(g) = g^{-1}\}$$

K is a subgroup of G. It is a *symmetric subgroup*. G/K is a *symmetric space*; K is a *compact group*; G/K is a *non-compact symmetric space*;

$$G = KP , \tag{1.13}$$

in the sense that each $g \in G$ can be written uniquely and continuously as

$$g = kp ,$$

with $k \in K$, $p \in P$.

Exp: $\mathcal{F} \to P$

is a *diffeomorphism*. G/K is diffeomorphic to P, i.e., G/K is diffeomorphic to Euclidean space. The elements of P are called *transvections*.

Let us determine K more precisely. For $k \in K$, $\sigma(k) = k'^{-1} = k$, i.e., k *is an orthogonal matrix*. K is a subgroup of $O(2n,R)$, the orthogonal group is $2n$ real variables. However, K is not all of $O(2n,R)$ since it satisfies another condition by virtue of its inclusion in G: the conditions

$$k'Jk = J,$$

$$k' = k^{-1}$$

together imply that

$$Jk = kJ \quad . \tag{1.14}$$

Put R^{2n} into correspondence with $\mathbb{C}^n$:

$$\begin{pmatrix} x \\ y \end{pmatrix} \to x + iy \tag{1.15}$$

for $x, y \in R^{2n}$.

Under this correspondence, k, as a linear map $R^{2n} \to R^{2n}$, goes over to a linear map $\mathbb{C}^n \to \mathbb{C}^n$ which is *complex linear*. Further, this linear map preserves the Hermitian norm on $\mathbb{C}^n$, i.e.,

$$K \text{ is isomorphic to } U(n), \text{ the group of } n \times n \tag{1.16}$$
$$\text{unitary matrices.}$$

It is instructive to compute more explicitly the Lie algebra of $\mathcal{K}$ and $\mathcal{P}$. If $\alpha \in \mathcal{G}$, it is of the form

$$\alpha = \begin{pmatrix} \alpha_{11} & \alpha_{12} \\ \alpha_{21} & -\alpha_{11} \end{pmatrix} \tag{1.17}$$

with $\alpha'_{12} = \alpha_{12}$, $\alpha'_{21} = \alpha_{21}$. Now,

$$\alpha' = \begin{pmatrix} \alpha'_{11} & \alpha'_{21} \\ \alpha'_{12} & -\alpha'_{11} \end{pmatrix} \tag{1.18}$$

hence: $\alpha \in \mathcal{K}$, i.e., $\alpha = -\alpha'$ if and only if

$$\alpha'_{11} = -\alpha_{11}$$
$$\tag{1.19}$$
$$\alpha_{12} = -\alpha'_{21}$$

$\alpha \in \mathscr{P}$, i.e., $\alpha' = \alpha$ if and only if

$$\alpha'_{11} = \alpha_{11}$$
$$\alpha_{12} = \alpha'_{21}$$
(1.20)

2. THE CARTAN SUBALGEBRA OF $\mathscr{P}$

Let $\mathscr{G}$ be a semisimple, non-compact Lie algebra,

$$\mathscr{G} = \mathscr{K} \oplus \mathscr{P}$$

the Cartan decomposition with $\mathscr{K}$ a compact Lie subalgebra. Because of the commutation relations $[\mathscr{P},\mathscr{P}] \subset \mathscr{K}$, any Lie subalgebra $\mathscr{A} \subset \mathscr{P}$ must be *abelian*, i.e.,

$$[\mathscr{A},\mathscr{A}] = 0 \qquad (2.1)$$

A *Cartan subalgebra* (for the pair $(\mathscr{G},\mathscr{K})$) is a *maximal* abelian subalgebra of $\mathscr{P}$, i.e., contained in no larger abelian algebra. E. Cartan proved that two such are conjugate under Ad K, i.e., if $\mathscr{A},\mathscr{A}' \subset \mathscr{P}$, both satisfy this condition: there is a $k \in K$ such that

$$\text{Ad } k(\mathscr{A}) = \mathscr{A}' \quad .$$

We shall now compute the Cartan subalgebra $\mathscr{A}$ in case $\mathscr{G}$ = Lie algebra of $Sp(n,R)$, $\mathscr{K}$ = Lie algebra of $U(n)$.

As we have seen (formula (1.19)), $\mathscr{P}$ consists of the $2n \times 2n$ partitioned matrices of the form:

$$\alpha = \begin{pmatrix} \alpha_{11} & \alpha_{12} \\ \alpha'_{12} & -\alpha_{11} \end{pmatrix} \qquad (2.2)$$

with $\alpha'_{11} = \alpha_{11}$. Let

$\mathscr{A}$ = set of $\alpha \in \mathscr{P}$ such that α is diagonal,

i.e.,

$$\alpha = \begin{pmatrix} \alpha_{11} & 0 \\ 0 & -\alpha_{11} \end{pmatrix}$$

$$\alpha_{11} = \begin{pmatrix} \lambda_1 & 0 \\ & \ddots & \\ 0 & & \lambda_n \end{pmatrix} \equiv \text{diag}(\lambda_1,\ldots,\lambda_n) \quad ,$$
(2.3)

with $\lambda_1,\ldots,\lambda_n \in R$. It is clear that $\mathscr{A}$ is abelian, i.e., $[\mathscr{A},\mathscr{A}] = 0$.

Theorem 2.1. $\mathscr{A}$ is a *maximal* abelian subalgebra of $\mathscr{P}$, i.e., $\mathscr{A}$ is a Cartan subalgebra for the pair $(\mathscr{G}, \mathscr{K})$.

Proof. Suppose

$$\mathscr{P} \ni \beta = \begin{pmatrix} \beta_{11} & \beta_{12} \\ \beta'_{12} & -\beta'_{11} \end{pmatrix} \; ; \quad \beta'_{11} = \beta_{11}$$

commutes with all $\alpha \in \mathscr{A}$ of the form (2.3)

$$\alpha\beta = \begin{pmatrix} \alpha_{11} & 0 \\ 0 & -\alpha'_{11} \end{pmatrix} \begin{pmatrix} \beta_{11} & \beta_{12} \\ \beta'_{12} & -\beta'_{11} \end{pmatrix} = \begin{pmatrix} \alpha_{11}\beta_{11}, & \alpha_{11}\beta_{12} \\ -\alpha'_{11}\beta'_{12}, & \alpha'_{11}\beta'_{11} \end{pmatrix}$$

$$\beta\alpha = \begin{pmatrix} \beta_{11} & \beta_{12} \\ \beta'_{12} & -\beta'_{11} \end{pmatrix} \begin{pmatrix} \alpha_{11} & 0 \\ 0 & -\alpha'_{11} \end{pmatrix} = \begin{pmatrix} \beta_{11}\alpha_{11}, & -\beta_{12}\alpha'_{11} \\ \beta'_{12}\alpha_{11} & \beta'_{11}\alpha'_{12} \end{pmatrix}$$

$\alpha\beta = \beta\alpha \Rightarrow \alpha_{11}\beta_{11} = \beta_{11}\alpha_{11}$. Since α_{11} is an arbitrary diagonal matrix, this implies that β_{11} is also diagonal. Also,

$$\alpha_{11}\beta_{12} = -\beta_{12}\alpha'_{11} \tag{2.4}$$

Thus,

$$\alpha_{11}^2 \beta_{12} = -\alpha_{11}\beta_{12}\alpha'_{11} = \beta_{12}\alpha'^2_{11} \; .$$

Again, this forces β_{12} to be diagonal, and (2.4) then forces

$$\beta_{12} = 0 \; .$$

This proves that $\beta \in \mathscr{A}$, and finishes the proof of Theorem 2.1.

Let H be the subgroup of $g \in G$ of the form

$$g = \begin{pmatrix} g_{11} & g_{12} \\ 0 & g_{22} \end{pmatrix} \tag{2.5}$$

(Recall that H is the subgroup such that G/H - the Lagrange-Grassmann manifold = the *compact* space in which the matrix Riccati equations of control theory "live".) The Lie algebra $\mathscr{H}$ of H consists of the matrices $\alpha \in \mathscr{G}$ of the form

$$\alpha = \begin{pmatrix} \alpha_{11} & \alpha_{12} \\ 0 & -\alpha'_{11} \end{pmatrix} \tag{2.6}$$

with $\alpha'_{12} = \alpha_{12}$. Set:

$$\alpha_0 = \begin{pmatrix} I & 0 \\ 0 & -I \end{pmatrix} \tag{2.7}$$

We shall compute

$$[\alpha_0, \alpha]$$

with $\alpha \in \mathcal{H}$

$$\alpha_0 \alpha = \begin{pmatrix} I & 0 \\ 0 & -I \end{pmatrix} \begin{pmatrix} \alpha_{11} & \alpha_{12} \\ 0 & -\alpha'_{11} \end{pmatrix}$$

$$= \begin{pmatrix} \alpha_{11} & \alpha_{12} \\ 0 & \alpha'_{11} \end{pmatrix}$$

$$\alpha \alpha_0 = \begin{pmatrix} \alpha_{11} & \alpha_{12} \\ 0 & -\alpha'_{11} \end{pmatrix} \begin{pmatrix} I & 0 \\ 0 & -I \end{pmatrix}$$

$$= \begin{pmatrix} \alpha_{11} & -\alpha_{12} \\ 0 & \alpha'_{11} \end{pmatrix}$$

Hence:

$$[\alpha_0, \alpha] = 2 \begin{pmatrix} 0 & \alpha_{12} \\ 0 & 0 \end{pmatrix} \tag{2.8}$$

Here are some consequences of this formula: set:

$$\mathcal{H}_0 = \left\{ \begin{pmatrix} \alpha_{11} & 0 \\ 0 & -\alpha'_{11} \end{pmatrix} \right\} \tag{2.9}$$

$$\mathcal{H}_+ = \left\{ \begin{pmatrix} 0 & \alpha_{12} \\ 0 & 0 \end{pmatrix} \right\} \tag{2.10}$$

Then,

$$[\alpha_0, \mathcal{H}_0] = 0 \tag{2.11}$$

$$[\alpha_0, \alpha] = 2\alpha$$

$$\text{for } \alpha \in \mathcal{H}_+ \tag{2.12}$$

Remark. This explains the notation we have chosen. $\mathcal{H}_0$ consists of the eigenvectors of Ad α_0 for eigenvalue zero, i.e., the *centralizer of* α_0 *in* $\mathcal{G}$. $\mathcal{H}_+$ consists of the eigenvectors of Ad α_0 with positive eigenvalues. (In fact, the eigenvalues of Ad α_0 are $0, \pm 2$.) The following result is a consequence. It is displayed in this way because it is a key result in our "geometric" method of studying asymptotic behavior.

Theorem 2.2. The element $\alpha_0 \in \mathcal{A}$ defined by formula (2.7) has the property that $\mathcal{H}$ is the subspace of $\mathcal{G}$ consisting of the eigenvectors of Ad α_0 with nonnegative eigenvalues.

Let K be the maximal compact subgroup of G, i.e., the set of $k \in G$ such that

$$k^{-1} = k'$$

$$Jk = kJ$$

Let us now look for the elements $k \in K$ such that

$$\text{ad } K(\alpha_0) \equiv k\alpha_0 k^{-1} \in \mathcal{H} . \tag{2.13}$$

(2.13) requires that there be an

$$\alpha = \begin{pmatrix} \alpha_{11} & \alpha_{12} \\ 0 & -\alpha'_{11} \end{pmatrix}$$

such that

$$k\alpha_0 k' = \begin{pmatrix} \alpha_{11} & \alpha_{12} \\ 0 & -\alpha'_{11} \end{pmatrix} \tag{2.14}$$

Let us examine the consequences of unitarity of k. Suppose

$$k = \begin{pmatrix} k_{11} & k_{12} \\ k_{21} & k_{22} \end{pmatrix}$$

Then,

THE LAGRANGE-GRASSMANN MANIFOLD

$$\begin{pmatrix} 0 & I \\ -I & 0 \end{pmatrix} \begin{pmatrix} k_{11} & k_{12} \\ k_{21} & k_{22} \end{pmatrix} = \begin{pmatrix} k_{21} & k_{22} \\ -k_{11} & -k_{12} \end{pmatrix}$$

$$= \begin{pmatrix} k_{11} & k_{12} \\ k_{21} & k_{22} \end{pmatrix} \begin{pmatrix} 0 & I \\ -I & 0 \end{pmatrix}$$

$$= \begin{pmatrix} -k_{12} & k_{11} \\ -k_{22} & k_{21} \end{pmatrix}$$

or

$$k_{21} = -k_{12} \, ; \qquad k_{11} = k_{22} \tag{2.15}$$

or

$$k = \begin{pmatrix} k_{11} & k_{12} \\ -k_{12} & k_{11} \end{pmatrix} .$$

Now,

$$k\alpha_0 = \begin{pmatrix} k_{11} & k_{12} \\ -k_{12} & k_{11} \end{pmatrix} \begin{pmatrix} I & 0 \\ 0 & -I \end{pmatrix}$$

$$= \begin{pmatrix} k_{11} & -k_{12} \\ -k_{12} & -k_{11} \end{pmatrix}$$

$$k\alpha_0 k' = \begin{pmatrix} k_{11} & -k_{12} \\ -k_{12} & -k_{11} \end{pmatrix} \begin{pmatrix} k'_{11} & -k'_{12} \\ k'_{12} & k'_{11} \end{pmatrix}$$

$$= \begin{pmatrix} k_{11}k'_{11} - k_{12}k'_{12}, & -k_{11}k'_{12} - k_{12}k'_{11} \\ -k_{12}k'_{11} - k_{11}k'_{12}, & k_{12}k'_{12} - k_{11}k'_{11} \end{pmatrix}$$

$$= \begin{pmatrix} \alpha_{11} & \alpha_{12} \\ 0 & -\alpha'_{11} \end{pmatrix} \tag{2.16}$$

This requires that

$$k_{12}k'_{11} = -k_{11}k'_{12} \qquad (2.17)$$

Use orthogonality of k:

$$I = kk' = \begin{pmatrix} k_{11} & k_{12} \\ -k_{12} & k_{11} \end{pmatrix} \begin{pmatrix} k'_{11} & -k'_{12} \\ k'_{12} & k'_{11} \end{pmatrix}$$

$$= \begin{pmatrix} k_{11}k'_{11} + k_{12}k'_{12}, & -k_{11}k'_{12} + k_{12}k'_{11} \\ -k_{12}k'_{11} - k_{11}k'_{12}, & k_{12}k'_{12} + k_{11}k'_{11} \end{pmatrix}$$

In particular,

$$k_{12}k'_{11} = k_{11}k'_{12} \qquad (2.18)$$

Putting together (2.17) and (2.18) gives

$$k_{11}k'_{12} = 0 \quad . \qquad (2.19)$$

Also,

$$k_{11}k'_{11} + k_{12}k'_{12} = I \quad . \qquad (2.20)$$

Thus, (2.16) takes the following form:

$$k\alpha_0 k' = \begin{pmatrix} 2k_{11}k'_{11} - I, & 0 \\ 0 & I - 2k_{11}k'_{11} \end{pmatrix} \qquad (2.21)$$

We can now apply this to compute the fixed points of the one parameter group $t \to \exp(tA_0)$ acting on $G/H = K/K \cap H$. Realize G/H as the Lagrange-Grassmann manifold, the set of all n-dimensional linear subspaces of R^{2n} on which the symplectic form is zero. Let

$$\gamma_0 = \left\{ \begin{pmatrix} x \\ 0 \end{pmatrix} : x \in R^n \subset R^{2n} \right\}$$

Thus,

$$k\gamma_0 = \left\{ \begin{pmatrix} k_{11} & k_{12} \\ -k_{12} & k_{11} \end{pmatrix} \begin{pmatrix} x \\ 0 \end{pmatrix} \right\}$$

$$= \left\{ \begin{pmatrix} k_{11}x \\ -k_{12}x \end{pmatrix} : x \in R^n \right\} \qquad (2.22)$$

Condition (2.19) and (2.20) mean that:

For $\begin{pmatrix} x_1 \\ y_1 \end{pmatrix}, \begin{pmatrix} x_2 \\ y_2 \end{pmatrix} \in k\gamma_0$, y_1 is perpendicular (2.23)

to x_2 in the usual inner product for R^n.

We can then sum up as follows:

<u>Theorem 2.3</u>. For each orthogonal projection map $\pi: R^n \to R^n$, consider the following n-dimensional linear subspace of R^{2n}:

$$\gamma_\pi = \left\{ \begin{pmatrix} x \\ y \end{pmatrix} : \pi(x) = 0 = (I-\pi)(y) \right\} \qquad (2.24)$$

Then each γ_π is a fixed point for the action of the group $t \to \exp(t\alpha_0)$ and every such fixed point is of this form.

<u>Proof</u>. To complete the proof, we shall prove that formula (2.24) does actually provide an n-dimensional Lagrangian subspace of R^{2n} which is left invariant by α_0.

First, check Lagrangianness: For

$$\begin{pmatrix} x \\ y \end{pmatrix}, \begin{pmatrix} x_1 \\ y_1 \end{pmatrix} \in \gamma_\pi$$

$$(x',y')J \begin{pmatrix} x_1 \\ y_1 \end{pmatrix} = (x',y') \begin{pmatrix} 0 & I \\ -I & 0 \end{pmatrix} \begin{pmatrix} x_1 \\ y_1 \end{pmatrix}$$

$$= (x'y') \begin{pmatrix} y_1 \\ -x_1 \end{pmatrix}$$

$$= x'y_1 - y'x_1$$

$$= 0 \quad ,$$

since $\pi(R^n)$ and $(I-\pi)(R^n)$ are orthogonal.

Second, check that γ_π is invariant under α_0:

$$\alpha_0 \gamma_\pi = \left\{ \begin{pmatrix} I & 0 \\ 0 & -I \end{pmatrix} \begin{pmatrix} x \\ y \end{pmatrix} : \pi(x) \equiv 0 \equiv (I-\pi)y \right\}$$

$$= \left\{ \begin{pmatrix} x \\ -y \end{pmatrix} \right\} = \gamma_\pi$$

q.e.d.

This result illustrates an interesting phenomenon that we will investigate in greater generality later on. Notice that

$$\gamma_0 = \left\{ \begin{pmatrix} x \\ 0 \end{pmatrix} : x \in R^n \right\}$$

is an *isolated* fixed point of the group $t \to \exp(t\alpha_0)$; others are not isolated, but form the union of connected manifolds, parameterized by an integer r, $0 < r < n$, with $n = \text{rank } \pi$. In fact, each of these manifolds is itself diffeomorphic to a Grassmann manifold, but in an n-dimensional vector space. These submanifolds can also be described as *orbits* in a way that will also be described in more detail later on. Set:

$$K_0 = \{k \in K : \alpha_0 k = k\alpha_0\}$$

We can describe K_0 more explicitly in terms of partitioned matrices. As we have seen, $k \in K$ can be written as

$$k = \begin{pmatrix} k_{11} & k_{12} \\ -k_{12} & k_{11} \end{pmatrix}$$

$$\alpha_0 k = \begin{pmatrix} I & 0 \\ 0 & -I \end{pmatrix} \begin{pmatrix} k_{11} & k_{12} \\ -k_{12} & k_{11} \end{pmatrix} = \begin{pmatrix} k_{11} & k_{12} \\ k_{12} & -k_{11} \end{pmatrix}$$

$$k\alpha_0 = \begin{pmatrix} k_{11} & k_{12} \\ -k_{12} & k_{11} \end{pmatrix} \begin{pmatrix} I & 0 \\ 0 & -I \end{pmatrix} = \begin{pmatrix} k_{11} & -k_{12} \\ -k_{12} & -k_{11} \end{pmatrix}$$

Hence, $k \in K_0$ means that

$$k_{12} = 0 \tag{2.25}$$

We see that the *connected components of the fixed point set of* α_0 *in* G/H *are the orbits of* K_0. We shall now see the general reason for this phenomenon.

3. A METHOD OF CONSTRUCTION OF COMPACT GRASSMANN COSET SPACES

The material presented in the previous section has a very general setting, which we can describe with the help of the differential-geometric technqiue developed in [6,7].

Let G be a connected Lie group with a finite center. Suppose that G is semisimple and non-compact. Let K be a maximal compact subgroup. Let $\mathcal{G}$ and $\mathcal{K}$ be the Lie algebras of G and K. Let $\mathcal{P}$ be the orthogonal complement of $\mathcal{K}$ on $\mathcal{G}$ with respect to the Killing form of $\mathcal{G}$. Then

$$\mathcal{G} = \mathcal{K} \oplus \mathcal{P} \tag{3.1}$$

(The Killing form is negative definite on $\mathcal{K}$, so that $\mathcal{G}$ is a direct sum as a *vector space* of $\mathcal{K}$ and $\mathcal{P}$.) Then the following relations are satisfied:

$$[\mathcal{K},\mathcal{K}] \subset \mathcal{K}$$
$$[\mathcal{K},\mathcal{P}] \subset \mathcal{P} \tag{3.2}$$
$$[\mathcal{P},\mathcal{P}] \subset \mathcal{K}$$

$$\text{For } \alpha \in \mathcal{P}, \text{ Ad } \alpha \text{ acting on } \mathcal{G} \text{ is diagonalizable and has } real \text{ eigenvalues.} \tag{3.3}$$

A *Cartan subalgebra* for $(\mathcal{G},\mathcal{K})$ is a maximal abelian subalgebra $\mathcal{A}$ of $\mathcal{P}$. Suppose one such is fixed. Pick an element α_0 of $\mathcal{A}$. Set:

$\mathcal{H}$ = linear subspace of $\mathcal{G}$ spanned by the eigenvectors Ad α_0 with nonnegative eigenvalues

$\mathcal{H}_+$ = linear subspace of $\mathcal{G}$ spanned by the eigenvectors of Ad α_0 with positive eigenvalues

$\mathcal{H}_-$ = linear subspace of $\mathcal{G}$ spanned by the eigenvectors of Ad α_0 with negative eigenvalues

$\mathcal{H}_0$ = set of $\alpha \in \mathcal{G}$ such that $[\alpha_0,\alpha] = 0$

Then, the following relations are satisfied:

$$\mathcal{G} = \mathcal{H}_+ \oplus \mathcal{H}_- \oplus \mathcal{H}_0 \tag{3.4}$$

(direct sum as vector spaces)

$$\mathcal{H}_+, \mathcal{H}_-, \mathcal{H}_0 \text{ are Lie subalgebras of } \mathcal{H} \tag{3.5}$$

$$[\mathcal{H}_-, \mathcal{H}_+] \subset \mathcal{H}_+ \tag{3.6}$$

$$\mathcal{H} = \mathcal{H}_+ \oplus \mathcal{H}_0 \tag{3.7}$$

(semi-direct sum of Lie algebras).

$$\mathcal{H}_+, \mathcal{H}_- \text{ are } \textit{nilpotent} \text{ Lie algebras} \tag{3.8}$$

Let $\mathcal{H}$ be the connected subgroup of G corresponding to the Lie subalgebra $\mathcal{H}$.

Theorem 3.1. $\mathcal{H}$ is its normalizer in $\mathcal{G}$, i.e., if $\alpha \in \mathcal{G}$ satisfies

$$[\alpha, \mathcal{H}] \subset \mathcal{H} \tag{3.9}$$

then $\alpha \in \mathcal{H}$.

Proof. Let $\mathcal{N}$ be the set of $\alpha \in \mathcal{G}$ which satisfies (3.9). ($\mathcal{N}$ is called the *normalizer of* $\mathcal{H}$ *in* $\mathcal{G}$.) One sees--using the Jacobi identity-- that $\mathcal{N}$ itself is a Lie algebra. Hence, $\operatorname{Ad} \alpha_0(\mathcal{N}) \equiv [\alpha_0, \mathcal{N}] \subset \mathcal{N}$. Suppose $\mathcal{H} \neq \mathcal{N}$. $\operatorname{Ad} \alpha_0$ is (by hypothesis) completely reducible, hence there is a linear subspace $\mathcal{M} \subset \mathcal{G}$ such that $\mathcal{N} = \mathcal{M} + \mathcal{H}$; $[\alpha_0, \mathcal{M}] \subset \mathcal{M}$. However, since $\mathcal{M}$ normalizes $\mathcal{H}$, we have

$$[\alpha_0, \mathcal{M}] \subset \mathcal{H} \quad .$$

These two relations together force

$$[\alpha_0, \mathcal{M}] = 0 \quad .$$

But $\mathcal{H}$ was constructed so as to include all elements of $\mathcal{G}$ which commute with $\mathcal{G}$; in particular,

$$\mathcal{M} \subset \mathcal{H} \quad .$$

This is a contradiction to our supposition that $\mathcal{H} \neq \mathcal{M}$, and we have

$$\mathcal{N} = \mathcal{H} \quad .$$

q.e.d.

THE LAGRANGE-GRASSMANN MANIFOLD   49

<u>Remark</u>. Theorem 3.1 does not require the semisimplicity of $\mathcal{G}$ nor that $\mathcal{H}$ be defined in precisely this way. It seems to work if

  a) $\mathcal{G}$ is an arbitrary finite dimensional Lie algebra
  b) $\mathcal{H}$ is a Lie subalgebra which contains an element α_0 such that Ad α_0 is completely reducible and $\mathcal{H}$ contains all $\alpha \in \mathcal{G}$ such that $[\alpha, \alpha_0] = 0$

Now, let H be the subgroup of G consisting of the $g \in G$ such that

$$\text{Ad } g(\mathcal{H}) = \mathcal{H} \qquad (3.10)$$

H is a closed subgroup of G, hence

$$G/H$$

is a manifold. The Lie subalgebra of $\mathcal{G}$ corresponding to H, i.e., the collection of $\alpha \in \mathcal{G}$ such that

$$t \to \exp(\alpha t) \in H$$

is obtained as follows:

$$\text{Ad } \exp(t\alpha)/\mathcal{H} = \mathcal{H}$$

  for all t

hence

$$\left(\frac{d}{dt} \text{ Ad } \exp(t\alpha)\right)(\mathcal{H}) \subset \mathcal{H}$$

i.e.,

$$[\alpha, \mathcal{H}] \subset \mathcal{H} .$$

By Theorem 3.1, $\alpha \in \mathcal{H}$. Thus, we have proved:

<u>Theorem 3.2</u>. The Lie subalgebra of $\mathcal{G}$ corresponding to the subgroup H is $\mathcal{H}$.

<u>Theorem 3.3</u>. G/H is a Grassmann coset space, i.e., there is a linear representation ρ of G by linear transformations in a real vector space W such that G/H is an orbit for the action of G in the Grassmann space GR(w).

  <u>Proof</u>. The choice ρ is the simplest one; ρ = adjoint representation of $\mathcal{G}$, i.e., $w = \mathcal{G}$, $\rho(g)(\alpha) = \text{Ad } g(\alpha)$. By our construction, G/H is the orbit of G on the element of GR($\mathcal{G}$) corresponding to the linear subspace $\mathcal{H}$.

Theorem 3.4. G/H is compact. K, the maximal compact subgroup of G, acts transitively on G/H.

Proof. Let p_0 be the element of G/H corresponding to the identity element of G. The isotropy subgroup of G at p_0 is then H. Regard $\mathcal{G}$ as a Lie algebra of vector fields on G/H. The subalgebra of $\mathcal{G}$ consisting of the vector fields which are zero at p_0 is then $\mathcal{H}$ itself.

Here is where we need to use the detailed structure of $\mathcal{G}$, $\mathcal{K}$, $\mathcal{P}$, $\alpha_0 \in \mathcal{P}$. The eigenvalues of Ad α_0 are real, and Ad α_0 is completely reducible. Let $\alpha \in \mathcal{G}$ be such an eigenvector, i.e.,

$$[\alpha_0, \alpha] = \lambda \alpha \tag{3.11}$$

with $\lambda \in R$. Set:

$$\alpha = \alpha_{\mathcal{K}} + \alpha_{\mathcal{P}}$$

with $\alpha_{\mathcal{K}} \in \mathcal{K}$; $\alpha_{\mathcal{P}} \in \mathcal{P}$. It follows from (3.11) (and the commutation relations (3.2)) that

$$[\alpha_0, \alpha_{\mathcal{K}}] = \lambda \alpha_{\mathcal{P}}$$
$$[\alpha_0, \alpha_{\mathcal{P}}] = \lambda \alpha_{\mathcal{K}} \tag{3.12}$$

To prove: The projection of $\mathcal{H}_+$ and $\mathcal{H}_-$ on $\mathcal{K}$ are one-one. Its image in $\mathcal{K}$ is the orthogonal complement (with respect to the Killing form) of $\mathcal{K} \cap \mathcal{H}_0$. \hfill (3.13)

Proof. Let $\alpha_1, \ldots, \alpha_n$ be elements of $\mathcal{H}_+$, which are eigenvectors of Ad α_0, with distinct eigenvalues $\lambda_1, \ldots, \lambda_n$. Let $\alpha = \alpha_1 + \cdots + \alpha_n$. Suppose that the projection on $\mathcal{K}$ is zero, i.e.,

$$\alpha_1 + \cdots + \alpha_n \in \mathcal{P} \tag{3.14}$$

Write

$$\alpha_1 = \alpha_{i,\mathcal{K}} + \alpha_{i,\mathcal{P}}$$

with $\alpha_{i,\mathcal{K}} \in \mathcal{K}$; $\alpha_{i,\mathcal{P}} \in \mathcal{P}$,

$$i = 1, \ldots, n \quad ,$$

such that

$$[\alpha_0, \alpha_{i,\mathcal{K}}] = \lambda_i \alpha_{i,\mathcal{P}}$$

$$[\alpha_0, \alpha_{i,\mathcal{P}}] = \lambda_i \alpha_{i,\mathcal{K}}$$

(3.14) implies that

$$\alpha_{1,\mathcal{K}} + \cdots + \alpha_{n,\mathcal{K}} = 0 \qquad (3.15)$$

Apply Ad α_0 to both sides of (3.15):

$$\lambda_1 \alpha_{1,\mathcal{P}} + \cdots + \lambda_n \alpha_{n,\mathcal{P}} = 0 \qquad (3.16)$$

Apply Ad α_0 to (3.16):

$$\lambda_1^2 \alpha_{1,\mathcal{K}} + \cdots + \lambda_n^2 \alpha_{n,\mathcal{K}} = 0 \qquad (3.17)$$

We can then eliminate $\alpha_{n,\mathcal{K}}$ between (3.17) and (3.15), obtaining a linear dependence relation between $\alpha_{1,\mathcal{K}}, \ldots, \alpha_{n-1,\mathcal{K}}$. Continuing by induction on n, we can eventually see that

$$\alpha_{i,\mathcal{K}} = 0 \quad ,$$

which is the contradiction, since then also $\alpha_{i,\mathcal{P}} = 0$, hence $\alpha_i = 0$.

Consider

$$(\text{Ad } \alpha_0)^2$$

acting in $\mathcal{K}$. It is a symmetric linear map with respect to the Killing form (which is negative definite, since $\mathcal{K}$ is compact). Hence, $\mathcal{K}$ can be written as the direct sum (orthogonal with respect to the Killing form) of the zero and nonzero eigenspaces. The direct sum of the nonzero eigenvectors is the projection of $\mathcal{H}_+$, while the zero eigenspace is clearly $\mathcal{H}_0 \cap \mathcal{K}$. Equation (3.13) then gives the following dimension identities

$$\dim \mathcal{H}_+ = \dim \mathcal{H}_-$$

$$\dim \mathcal{G} = \dim \mathcal{H}_+ \; \dim \mathcal{H}_- + \dim \mathcal{H}_0$$

$$= 2 \dim \mathcal{H}_+ + \dim \mathcal{H}_0$$

$$= \dim \mathcal{K} + \dim \mathcal{P} \; .$$

Let us compute the dimension of the orbit Kp_0, which is equal to

$$\dim \mathcal{K} - \dim(\mathcal{K} \cap \mathcal{H}_0) \quad .$$

We want to prove that the orbit Kp_0 is *open*; since it is compact (and G/H is connected) this will prove that K acts transitively. This then requires proving that:

$$\dim \mathcal{K} - \dim(\mathcal{K} \cap \mathcal{H}_0) = \dim \mathcal{G} - \dim \mathcal{H}$$

$$= \dim \mathcal{H}_+ \qquad (3.18)$$

(3.18) now follows from (3.13).

q.e.d.

Bibliography

1. R. Hermann, <u>Lie Groups for Physicists</u>, W.A. Benjamin, Reading, Mass., 1966.

2. S. Helgason, <u>Differential Geometry and Symmetric Spaces</u>, Academic Press, New York, 1961.

3. A. Sagle and R. Walde, <u>Introduction to Lie Groups and Lie Algebras</u>, Academic Press, New York, 1973.

4. H. Samelson, <u>Notes on Lie Algebra</u>, Van Nostrand Reinhald, New Jersey, 1969.

5. R. Gilmore, <u>Lie Groups, Lie Algebras and Some of Their Applications</u>, Wiley, New York, 1924.

6. R. Hermann, "Geometric Aspects of Potential Theory in Symmetric Spaces, III", <u>Math. Ann.</u> <u>153</u> (1964), 384-394.

7. R. Hermann, "Compactification of Homogeneous Spaces, I", <u>J. Math. Mech.</u> <u>14</u> (1965), 655-678.

Chapter 4

LOCAL ANALYSIS NEAR A ZERO POINT OF A VECTOR FIELD
BELONGING TO A TRANSITIVE FINITE DIMENSIONAL
LIE ALGEBRA OF VECTOR FIELDS

1. INTRODUCTION

Let A be a vector field on a manifold X. It generates (let us assume) a one-parameter group $t \to \phi_t$ of diffeomorphisms of X, i.e., a *stationary flow* in X. The theory of *dynamical systems* [1,2] deals with this situation *in general*. Of particular importance are the *zero* or *stationary points*, $x_0 \in X$ such that $A(x_0) = 0$, and the properties of the limit

$$\lim_{t \to \infty} \phi_t(x) \quad .$$

One knows that the linearization of A in the neighborhood of the fixed points x_0 plays an important role. This can be described as follows.

Let (x^i) be a local coordinate system for X whose values at the zero point of A are zero. Then, if

$$A = A^i \frac{\partial}{\partial A^i} \quad ,$$

$$A^i(0) = 0 \quad .$$

The linear vector field

$$A_0 = \frac{\partial A^i}{\partial x^j}(0) \, x^j \frac{\partial}{\partial x^i} \tag{1.1}$$

is the *linearization* of A at x_0. Its stability properties often determine the behavior of the orbits of A in the neighborhood of x_0. (This is a topic that is extensively developed in the theory of ordinary differential equations.)

Here is an invariant way of describing this. If $x_0 \in X$ is a zero point of A, then the

$$B \to [A,B](x_0)$$

map depends (as is readily verified) only on the value that the vector field B takes at x_0. This defines a linear transformation

$$\ell(A, x_0): X_{x_0} \to X_{x_0}$$

of the tangent spaces. (This is the linear transformation determined by the matrix $(\partial A^i/\partial x^j(0))$ in formula (1.1).) We call it the *linearization* of A near the zero point.

Here are some typical results:

a) If all the eigenvalues of $\ell(A,x_0)$ have negative real parts, then x_0 is an attractor for the group ϕ_t in the sense that there is an open subset U of x_0 such that

$$\lim_{t \to \infty} \phi_t(x) = x_0$$

for all $x \in U$.

b) If all the eigenvalues of $\ell(A,x_0)$ have nonzero real parts, with p of them having positive real parts, q of them negative real parts, then there are submanifolds Y_+, Y_- of X defined in a neighborhood such that:

$$\lim_{t \to \infty} \phi_t(x^+) = x_0$$

for $x^+ \in Y_+$;

$\phi_t(x^-)$ goes out of the neighborhood as $t \to \infty$;

$x_0 \in Y_+ \cap Y_-$;

$\dim Y_+ + \dim Y_- = \dim X$;

Y_+ and Y_- intersect clearly at x_0 ;

Y_+, Y_- are called the *stable* and *unstable* manifolds at x_0 .

These properties (and much more, of course) are basically general properties of vector fields on manifolds. Often one can, using the tools of differential topology, make global statements as well. However, in applications (e.g., the matrix Riccati equation of Control Theory) one often knows a good deal more, particularly that A sits inside of a transitive, finite dimensional Lie algebra of vector fields which act on X. In this chapter we shall describe certain general features of the Lie theory which are related to the problem of analyzing the behavior of A in the neighborhood of its zeros, then analyze the linearization for the case of a Riccati vector field acting on a Grassmann manifold.

LOCAL ANALYSIS NEAR A ZERO POINT

2. THE TANGENT BUNDLE TO A HOMOGENEOUS SPACE

If X is a manifold, a basic object of differential geometry and topology is its tangent vector bundle

$$T(X) \ .$$

Usually, it is defined in terms of the algebra $\mathscr{F}(X)$ of C^∞ real-valued functions on X. For our problem--the study of properties of orbits of vector fields which lie in transitive finite dimensional Lie algebras of vector fields-- it is more convenient to have T(X) described in terms of the Lie structure. Such a definition will now be presented.

Let G be a Lie group, K a closed subgroup. It is known (theorem of E. Cartan) that the coset space

$$X = G/K$$

can be given a manifold structure, with the natural transformation group structure

$$G \times X \to X$$

given by smooth (e.g., C^∞, or real analytic) maps. Therefore, we can construct its tangent bundle T(X) using manifold theory. G also acts on T(X), i.e., T(X) → X forms a *homogeneous bundle of* G.

We shall now give a construction of T(X) that only involves the Lie group structure G. (It is, in a sense, "purely algebraic", hence might be appropriate for generalizations to various algebraic situations.) Let $\mathscr{G}$ be the Lie algebra of G. For g ∈ G, let Ad g: $\mathscr{G} \to \mathscr{G}$ be the adjoint representation. (See Sagle-Walde [4], Helgason [5] or *Differential Geometry and the Calculus of Variations* for the standard Lie group-theoretic notation and ideas which I will be using here.) Let $\mathscr{K}$ be the Lie subalgebra of $\mathscr{G}$ corresponding to the subgroup K (i.e., $\mathscr{K}$ consists of the one-parameter subgroups of G which lie in K.)

For each g ∈ G, consider the vector space

$$V(G) = \mathscr{G}/\text{Ad } g(\mathscr{K}) \ .$$

Note that

$$V(gk) = V(g) \ ,$$

i.e., the assignment g → V(g) is invariant under right translation by K. Hence, it passes to the quotient to define a mapping

$$x \to V(X)$$

of

$$X \to \text{(space of finite dimensional real vector spaces)} \ .$$

Namely, if

$$x = gx_0 ,$$

where x_0 is the identity coset of G/K, $g \in G$, then

$$V(x) = V(g) .$$

The action of G on $T(g/K)$ can now be readily defined:

$$g_0(V_x \equiv \mathscr{G}/\mathrm{Ad}\, g(\mathscr{K})) = \mathscr{G}/\mathrm{Ad}(g_0 f)(\mathscr{K}) .$$

3. THE LINEARIZATION OF A VECTOR FIELD DEFINED BY AN ELEMENT OF THE LIE ALGEBRA

Continue with the notation of Section 2. Suppose $A \in \mathscr{G}$. We can assign a cross-section map $x \to A(x)$ of $X \equiv G/K \to T(G)$ as follows

$$A(gx_0) = \text{image of } A \text{ in } \mathscr{G}/\mathrm{Ad}\, g(\mathscr{K}) .$$

This defines A as a *vector field* on G/K.

Thus, a point $x = gx_0$ is a *zero point* for the vector field A if and only if

$$A \in \mathrm{Ad}\, g(\mathscr{K}) . \tag{3.1}$$

We can define a linear map

$$\ell(A,x): \mathscr{G}/\mathrm{Ad}\, g(\mathscr{K}) \to \mathscr{G}/\mathrm{Ad}\, g(\mathscr{K})$$

as follows

$$\ell(A,x) = \mathrm{Ad}\, A \text{ passed to the quotient to act in } \mathscr{G}/\mathrm{Ad}\, g(\mathscr{K}) .$$

It should be clear that

$$\ell(A,x)$$

is essentially the *linearization* of the vector field on G/K determined by A. Its eigenvalues determine the stability properties of the vector field A in the neighborhood of x, as explained in Section 1.

4. THE TANGENT BUNDLE TO THE GRASSMAN MANIFOLD

We can now apply this general formalism to the case where:

$G =$ group of linear automorphisms of a finite dimensional vector space V;

X = Grassman manifold of m-dimensional linear subspaces of V.

As usual, $\mathscr{G}$ ($\equiv$ Lie algebra of G) is to be identified with $L(V)$, the space of linear maps $V \to V$.

Let γ_0 be a fixed m-dimensional linear subspace of V.

K = subgroup of $g \in GL(V)$ such that $g(\gamma_0) \subset \gamma_0$.

$G^m(V)$, the Grassman manifold, is then the coset space G/K.

Let $\gamma \in G^m(V)$. Set

$$K(\gamma) = \{g \in GL(V) : g(\gamma) \subset \gamma\} \quad .$$

Obviously, $K(\gamma)$ is the isotropy subgroup of the action of G on $G^m(V)$. Let

$$L(V,\gamma) = \{A \in L(V) : A(\gamma) \subset \gamma\} \quad .$$

$L(V,\gamma)$ is the Lie algebra of $K(\gamma)$.

By the general theory,

$$L(V)/L(V,\gamma) \equiv G^m(V)_\gamma$$

is the tangent space to $G^m(V)$ at γ.

<u>Theorem 4.1.</u> $L(V)/L(V,\gamma)$ can be naturally identified with the vector space:

$$L(\gamma, V/\gamma) \quad ,$$

the space of linear maps $\gamma \to V/\gamma$.

<u>Proof.</u> For each $A \in L(V)$, construct

$$\bar{A}: \gamma \to V/\gamma$$

as follows:

$$\bar{A}(v) = \text{image of } A(v) \text{ in } V/\gamma \quad ,$$

for $v \in \gamma$.

<u>Exercise.</u> Show that this assignment

$$A \to \bar{A}$$

is the appropriate one to define the isomorphism of $L(V)/L(V,\gamma)$ with $L(\gamma, V/\gamma)$.

This result gives us a new way to define the tangent bundle to $G^m(V)$.

> For each $\gamma \in G^m(V)$, $G^m(V)_\gamma$ is the vector space $L(\gamma, V/\gamma)$.

In other words, a *tangent vector* to $G^m(V)$ is a pair

$$(\gamma, \alpha) \quad ,$$

where γ is an m-dimensional linear subspace of V and α is a linear map $\gamma \to V/\gamma$. If $A \in L(V)$, *the value at γ of the vector field* $A(\gamma)$ defined by A on $G^m(V)$ is the linear map

$$v \to \text{(image of } A(v) \text{ in } V/\gamma) \quad .$$

In particular, $A(\gamma) = 0$ if and only if

$$A(\gamma) \subset \gamma \quad .$$

Exercise. Start afresh, taking this as definition of $T(G^m(V))$. Show that it can be given a manifold structure *directly* by exhibiting an appropriate set of coordinate charts.

Remark. This formulation might also be appropriate for generalization to *infinite dimensional situations*.

5. THE TANGENT BUNDLE TO GRASSMANNIANS IN TERMS OF PARTITIONED MATRICES. THE LINEARIZATION OF A RICCATI VECTOR FIELD NEAR A ZERO

Section 4 is done from the "pure" linear algebraic viewpoint. For calculations and applications, it is often important to be able to work in the more traditional way using *partitioned matrices*. I shall now describe this veiwpoint.

Let V be a vector space. Suppose it is a direct sum of two linear subspaces:

$$V = V_1 \oplus V_2 \quad .$$

Write a $v \in V$ as a column vector

$$v = \begin{pmatrix} v_1 \\ v_2 \end{pmatrix} \quad , \quad v_1 \in V_1 \; ; \quad v_2 \in V_2 \quad .$$

LOCAL ANALYSIS NEAR A ZERO POINT   59

In other words,

$$V_2 = \left\{ \begin{pmatrix} v_1 \\ 0 \end{pmatrix} \right\}$$

$$V_1 = \left\{ \begin{pmatrix} 0 \\ v_2 \end{pmatrix} \right\}$$

A linear map $A: V \to V$ can then be written as:

$$A \begin{pmatrix} v_1 \\ v_2 \end{pmatrix} = \begin{pmatrix} A_{11} & A_{12} \\ A_{21} & A_{22} \end{pmatrix} \begin{pmatrix} v_1 \\ v_2 \end{pmatrix} \qquad (5.1)$$

with $A_{11} \in L(V_1, V_1)$; $A_{12} \in L(V_2, V_1)$; $A_{21} \in L(V_1, V_2)$; $A_{22} \in L(V_2, V_2)$. Thus we associate with a linear map A a 2×2 matrix

$$\begin{pmatrix} A_{11} & A_{12} \\ A_{21} & A_{22} \end{pmatrix}$$

of linear maps. All of the usual rules of matrix multiplication carry over, if one properly takes into account the fact that we are dealing with "matrices" whose elements do not commute. For example, we can write (5.1) as:

$$\begin{aligned}(Av)_1 &= A_{11}v_1 + A_{12}v_2 \\ (Av)_2 &= A_{21}v_1 + A_{22}v_2 \end{aligned} \qquad (5.2)$$

Now, suppose that

$$V_1 \in G^m(V) \quad .$$

We see that $\mathcal{K}$, the Lie subalgebra of Riccati vector fields which vanish at v_1, consists of linear maps A such that

$$A_{21} = 0 \quad ,$$

i.e., those whose partitioned matrices are in "triangular" form:

$$\begin{pmatrix} A_{11} & A_{12} \\ 0 & A_{22} \end{pmatrix} \qquad (5.3)$$

The tangent space to $G^m(V)$ at V_1 is then identified with the quotient vector space

$$L(V)/\mathcal{K} \quad .$$

We see that this vector space is identified with the matrices of the form

$$B = \begin{pmatrix} 0 & 0 \\ B_{21} & 0 \end{pmatrix}$$

with B_{21} a linear map

$$V_1 \to V_2 \equiv V/V_1 \quad .$$

This gives the result found earlier--the tangent space to $G^m(V)$ at V_1 is

$$L(V_1, V/V_1 \equiv V_2) \quad .$$

To find the linearization of a vector field A in $G^m(V)$, i.e., of form (5.3), compute the commutator:

$$[A,B] = \begin{pmatrix} A_{11} & A_{12} \\ 0 & A_{22} \end{pmatrix} \begin{pmatrix} 0 & 0 \\ B_{21} & 0 \end{pmatrix} - \begin{pmatrix} 0 & 0 \\ B_{21} & 0 \end{pmatrix} \begin{pmatrix} A_{11} & A_{12} \\ 0 & A_{22} \end{pmatrix}$$

$$= \begin{pmatrix} A_{12}B_{21} & 0 \\ A_{22}B_{21} & 0 \end{pmatrix} - \begin{pmatrix} 0 & 0 \\ B_{21}A_{11} & B_{21}A_{12} \end{pmatrix}$$

$$= \begin{pmatrix} A_{12}B_{21} & 0 \\ A_{22}B_{21} - B_{21}A_{11}, & -B_{21}A_{12} \end{pmatrix}$$

$$\begin{pmatrix} 0 & 0 \\ A_{22}B_{21} - B_{21}A_{11}, & 0 \end{pmatrix}$$

when projected mod $\mathcal{K}$. This shows that the linearization of $A \in \mathcal{K}$ is essentially the linear map

$$B_{21} \to A_{22}B_{21} - B_{21}A_{11} \quad .$$

We can now sum up as follows:

LOCAL ANALYSIS NEAR A ZERO POINT   61

<u>Theorem 5.1.</u> Let V_1 be an m-dimensional linear subspace V and let
$A \in L(V)$ be such that the Riccati vector field it determines on $G^m(V)$
vanishes at V, i.e., $A(V_1) \subset V_1$. Identify the tangent space to $G^m(V)$ at
V_1 with

$$L(V, V/V_1) .$$

Then

$$\ell(A): L(V,V/V_1) \to L(V,V/V_1)$$

is the following linear map:

$$\ell(A)(B)(v_1) = \bar{A}B(v_1) = BAv_1$$
$$\text{for } v_1 \in V_1 ,$$

where $\bar{A}: V/V_1 \to V/V_1$ is the quotient map of A, i.e., the map leading to a
commutative mapping diagram

$$\begin{array}{ccc} V & \xrightarrow{A} & V \\ \downarrow & & \downarrow \\ V/V_1 & \xrightarrow{\bar{A}} & V/V_1 \end{array}$$

where the vertical arrows are natural quotient maps.

6. STABILITY PROPERTIES OF FIXED POINTS OF RICCATI VECTOR FIELDS

Let V be an n-dimensional complex vector space. Let $G^m(V)$ be the
Grassmann manifold of m-dimensional linear subspaces of V. $G^m(V)$ is acted
on transitively by $GL(V)$, the group of automorphisms of V. The Lie algebra
of $GL(V)$ is $L(V)$. Suppose given an element $A \in L(V)$, i.e., a linear map
$V \to V$. We can consider A as a vector field on $G^m(V)$; it is called a *Riccati
vector field*. In this section we shall briefly describe the qualitative
stability properties of such A's which satisfy two sorts of conditions:

Condition a) A acting on V is completely diagonalizable
and all eigenvalues are pure imaginary.

Condition b) A acting on V is diagonalizable, imaginary, and
the eigenvalues have distinct real parts.

These two conditions are in a sense at the opposite end of the spectrum
of possible sorts of stability behavior. With Condition (a) satisfied, the
one-parameter group $t \to \exp(tA)$ of transformations generated by A has
bounded norm as $t \to \pm\infty$. In particular, this group lies in a compact subset

of GL(V). By Cartan's conjugacy-of-maximal-compact-subgroup-theorem [5], $t \to \exp(tA)$ is conjugate to a fixed maximal compact subgroup of GL(V). Now, a maximal compact subgroup K of GL(V) may be constructed in the following way: a Hermitian-symmetric, positive definite inner product on V (i.e., a Hilbert space structure). Let K be the group of automorphisms of this Hermitian inner product (i.e., the unitary map). Now, any maximal compact subgroup of GL(V) is of this form; in particular, with Condition (a) there is such a Hermitian inner product with respect to which $t \to \exp(tA)$ is unitary, i.e., A is skew-Hermitian. Thus, the orbit $t \to \exp(tA)$ lies in a *maximal torus* of K, which more-or-less determines their stability properties.

<u>Remark</u>. This application of Cartan's theorem is rather trivial in the case A is time-independent. However, in the time dependent case,

$$\frac{dv}{dt} = A(t)v \quad , \tag{6.1}$$

it says a good deal more than is attainable with reasonably straightforward differential equation techniques. For, if all solutions of the *linear* time-dependent differential equations (6.1) are bounded as $t \to \pm\infty$, then the closure of the flow $t \to \phi_t$ on GL(V) it generates must also be in a maximal compact subgroup of GL(V). Cartan's theorem also "relativizes", i.e., if each $A(t)$ belongs to a fixed Lie subalgebra $\mathcal{G} \subset L(V)$, then conjugacy to a maximal compact subgroup holds to within Ad G, where G is the Lie subgroup of GL(V) generated by $\mathcal{G}$. This is especially useful, e.g., in the case where (6.1) is a *Hamiltonian* system, i.e., $G = Sp(n,R)$, the symplectic group.

Now turn to Condition (b). Let $v_1 \in G^m(V)$ be a zero point of the Riccati vector field A, i.e.,

$$A(V_1) \subset V_1 \quad .$$

As we have seen, $G^m(V)_{V_1}$, the tangent space to the Grassmann at V_1, is identified with

$$L(V_1, V_2) \quad ,$$

where V_2 is a linear subspace of V such that

$$V = V_1 \oplus V_2 \quad .$$

Since A is diagonalizable (hence, as a consequence of linear algebra, also completely reducible) we can choose V_2 so that also

LOCAL ANALYSIS NEAR A ZERO POINT

$$A(V_2) \subset V_2 \quad .$$

Now, $L(V_1, V_2)$ is also identified with

$$V_1^d \otimes V_2 \quad ,$$

where V_1^d is the dual space of V_1.

Exercise. With this identification of the tangent space to $G^m(V)$ at V with $V_1^d \otimes V_2$, show that the linearization of the Riccati vector field determined by A is the following linear map $V_1^d \otimes V_2 \to V_1^d \otimes V_2$:

$$\ell(A)(v_1^d \otimes v_2) = -A^d(v_1^d) \otimes v_2 + v_1^d \otimes A(v_2) \tag{6.2}$$

for $v_1^d \in V_1^d$; $v_2 \in V_2$.

Formula (6.2) makes it straightforward to compute the stability properties of the linearization of A, i.e., to compute the eigenvalues of $\ell(A)$. Since A acting on V is diagonalizable, V_1 and V_2 have bases consisting of eigenvalues of A. Suppose that

$$v_i \quad , \quad 1 \leq i \leq m$$

is such a basis for V_1 with

$$A(v_i) = \lambda_i v_i \quad .$$

Similarly, let w_a, $m+1 \leq a,b \leq m$, be a basis for V_2 consisting of eigenvectors, i.e.,

$$A(w_a) = \lambda_a w_a \quad .$$

Let v_i^d be the dual basis of v_1^d. Then, also

$$A^d(v_i^d) = \lambda_i v_i^d \quad . \tag{6.3}$$

Proof of (6.3).

$$A^d(v_i^d)(v_j) = v_i^d(Av_2) \quad \text{(by the definition of "dual map" } A^d \text{)}$$

$$= v_i^d \lambda_j v_j = \lambda_j \delta_{ij} \quad ,$$

which is precisely relation (6.3).

Hence,

$$\ell(A)(v_i^d \otimes w_a) = (\lambda_a - \lambda_i) v_i^d \otimes w_a \qquad (6.4)$$

If we assume that all the eigenvalues of A are distinct and have nonzero real parts, we deduce that the eigenvalues of $\ell(A)$ always have *nonzero real parts*. Hence, each fixed point has a "stable" and "unstable" manifold

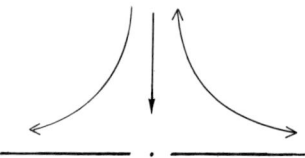

We shall see in a later chapter that there is a Lie group-theoretic way to compute these stable and unstable manifolds; in fact, they often appear as the *orbits* of certain groups.

Exercise. Extend the analysis sketched here to the case where V is a real vector space. This can either be done by a separate analysis, adapted to the real situation (instead of "eigenvectors", one studies two-dimensional linear subspaces V' such that

$$A(V') \subset V' \quad ,$$

and analyzes how a linear subspace $V_1 \subset V$ which is invariant under A may be split up as the direct sum of such two-dimensional subspaces) or by complexifying V and A, analyzing the resulting complex situation, then going back down to the reals.

Bibliography

1. P. Hartman, *Ordinary Differential Equations*, Wiley, New York, 1964.

2. S. Smale, "Differential Dynamical Systems", *Bull. Am. Math. Soc.* 73 (1967), 747-817.

3. C.R. Schneider, "Global Aspects of the Matrix Riccati Equation", *Mathematical Systems Theory* 7 (1973), 281-286.

4. A. Sagle and R. Walde, *Introduction to Lie Groups and Lie Algebras*, Academic Press, New York, 1973.

5. S. Helgason, *Differential Geometry and Symmetric Spaces*, Academic Press, New York, 1962.

Chapter 5

THE STATIONARY SOLUTIONS OF RICCATI EQUATIONS
AS THE ORBITS OF CENTRALIZERS

1. INTRODUCTION

Consider a time-dependent system of ordinary differential equations:

$$\frac{dx}{dt} = A(x,t), \qquad x \in R^n. \qquad (1.1)$$

It determines (modulo global complications, i.e., "finite escape time", which we shall, for the most part, ignore) a *flow* on R^n, i.e., a one-parameter family

$$t \to \phi_t$$

of transformations on R^n; the *orbits* of the flow, i.e., the curves of the form

$$t \to \phi_t(x_0)$$

are precisely the solution curves of (1.1). One says that the differential equation is the *infinitesimal generator* of the flow. (This is a familiar concept in fluid dynamics; if the flow ϕ_t is that corresponding to a fluid motion, n = 3, the time-dependent vector field $f(x,t)$ on R^3 is the Eulerian velocity field of the flow.)

Let G be a group of transformations on R^n. One says that the system (1.1) is a *Lie system associated to* G if, for each t, the transformation ϕ_t belongs to G. One can readily find an "infinitesimal" version of the definition. Let $\mathcal{G}$ be the Lie algebra of G. It acts on R^n as a Lie algebra of vector fields. The condition then amounts to saying that, for each t, the vector field

$$x \to A(x,t)$$

on R^n belongs to $\mathcal{G}$. Thus, a Lie system is determined by a curve in a Lie algebra $\mathcal{G}$ and an action of $\mathcal{G}$ on R^n (or, more generally, a differential manifold).

These systems were introduced by Sophus Lie himself [1], who proved many of the fundamental theorems about them. He showed that one could often derive useful information about the system (1.1) as *differential equations* knowing only that it was a Lie system associated to a certain type of Lie group. For example, the Lie systems associated to linear groups acting on R^n are the *linear homogeneous systems* of differential equations. The Lie systems associated with the affine groups on R^n are the *linear* inhomogeneous systems. Many of the well-known properties of these familiar systems can be given a group-theoretic interpretation in Lie's theory.

After Lie, the general theory of such systems lay dormant. (The exception is work by E. Vessiot [2], which also is unknown today.) Even with the revival of Lie group theory in recent times, there has been little attention by mathematicians to this topic. However, in the last fifteen years, various topics have arisen in Control and Systems Theory that cry out for unification within the context of Lie systems. For example, the matrix Riccati equations--which are basic to the practical applications--are beautiful examples of Lie systems [3,4].

2. NOTATION OF MANIFOLD THEORY

For the sake of efficiency, we shall work within the context of the theory of differentiable manifolds [3], although most everything can be readily reformulated in terms of the usual control theorist's notation for matrices and vectors. X will denote a manifold. $F(X)$ denotes the real valued C^∞ functions on X. A *vector field* A is a linear map: $F(X) \to F(X)$ such that:

$$FA(f_1 f_2) = A(f_1)f_2 + f_1 A(f_2) \ .$$

The collection of such vector fields is denoted by $V(X)$. Two such vector fields A and B can be added, and their *Jacobi bracket*

$$[A,B] = AB - BA$$

can be formed. Thus equipped, $V(X)$ becomes a Lie algebra over the real numbers.

Let (x^i), $1 \leq i,j \leq n$ denote a coordinate system for X. Each vector field A is then of the form of a first order linear differential operator:

$$A(f) = A^i \frac{\partial f}{\partial x^i} \ , \qquad (2.1)$$

where A^i are functions on X. This association $A \to (A^i)$ is basic to *tensor analysis*, i.e., the vector fields are identical to the one-contravariant tensor fields.

If $X = R^n$, with (x^i) the Cartesian coordinates, then each A determines a map

$$x \to (A^1(x), \ldots, A^n(x))$$

of $R^n \to R^n$, i.e., a "vector field" in the sense that that term is used in physics (e.g., for n = 3). Of course, we should keep in mind that many of the objects that physicists call "vector fields" (e.g., the electromagnetic field) are "really" the dual objects, *differential forms*.

Let t be a "time" coordinate varying, say, over $0 \leq t \leq \infty$. Consider a one-parameter family $t \to A_t$ of vector fields. The *orbits* of the family are the curves $t \to x(t)$ in X such that:

$$\frac{d}{dt} f(x(t)) = A_t(f)(x(t)) \tag{2.2}$$

for all $f \in F(X)$.

In terms of the coordinates (x^i), if

$$A_t = A^i(x,t) \frac{\partial}{\partial x^i} , \tag{2.3}$$

then the solutions of (2.2) are written in the more familiar form:

$$\frac{dx^i}{dt} = A^i(x,t) , \tag{2.4}$$

which is, when the indices are left off, the system (1.1). The curve $t \to A_t$ in V(X) is called the *infinitesimal generator* of the system of equations (2.2).

Definition. Let $\mathcal{G}$ be a Lie subalgebra of V(X). The system (2.2) (or (2.4)) is said to be a *Lie system* associated with $\mathcal{G}$ if $A_t \in \mathcal{G}$ for all t. If G is a Lie transformation group on X, (2.2) is a Lie system associated with G if it is associated with its Lie algebra $\mathcal{G}$ as a Lie algebra of vector fields on X (which amounts to saying, given the relation between G and $\mathcal{G}$, that the *generated* $t \to \phi_t$ belongs to G for all t.)

Example a) *Linear homogeneous system:*

X is a vector space

$$\frac{dx}{dt} = A(t)x .$$

Their solutions can be written as

$$x(t) = M(t)(x(0)) ,$$

where $t \to M(t)$ is a curve in GL(X). It is a Lie system with respect to the action of GL(X) on X.

Example b) *Linear homogeneous systems:*

$$\frac{dx}{dt} = A(t) + a(t) \quad .$$

The solutions can be written—with the "variation of parameters trick"—as

$$x(t) = M(t)y(t) \quad ,$$

where

$$\frac{dM}{dt} M^{-1} = A \quad ,$$

$$y(t) = y(0) + \int_0^t M(t)^{-1} a(t) \, dt \quad .$$

Thus,

$$x(t) = M(t)y(0) + \int_0^t M(t)M(s)^{-1} a(s) \, ds$$

$$= M(t)M(0)^{-1}x(0) + \int_0^t M(t)M(s)^{-1} a(s) \, ds$$

We see that this formula exhibits the solutions of (2.4) as orbits of curves in

$$G = \text{group of affine maps } X \to X \quad .$$

The Matrix Riccati equations are—as described in [3,4]—Lie systems, when one takes X as the Grassmann manifold of linear subspaces of another vector space Y, and G as the group $GL(Y)$ or one of its subgroups. (For example, the Matrix Riccati equation which occurs in linear Optimal Control problems is obtained by taking Y as R^{2n}, and G as $Sp(n,R)$.) We shall present more details about this most important special case later on.

Now, in the study of differential equations (2.2), there are two important qualitative questions: Find the *stationary*, i.e., *time-independent* solutions, and find the asymptotic behavior

$$\lim_{t \to \infty} x(t)$$

of solutions. We shall see that the special geometric nature of Lie systems gives useful information about these questions.

3. STATIONARY SOLUTIONS OF LIE SYSTEMS

Let X be a manifold and let $t \to A_t$ be a one-parameter family of vector fields on X. The *orbits* are the solutions of

$$\frac{dx}{dt} = A_t(x(t)) \quad . \tag{3.1}$$

A solution of (3.1) is *stationary* if it is constant, i.e., time-independent. Of course, it is determined by a point $x_0 \in X$ such that:

$$A_t(x_0) = 0 \tag{3.2}$$

for <u>all</u> t .

Geometrically, condition (3.2) means that x_0 is a *singular* or *fixed point* for the vector field A_t; the one-parameter group of diffeomorphisms generated by A_t leaves x_0 fixed.

Often in optimal control theory (and in many other parts of pure and applied mathematics) one is concerned with studying the set of $x_0 \in X$ which satisfy (3.2) for one value of t. *In general*, there is not much one can say: locally, a vector field is just a map

$$R^n \to R^n \quad ,$$

and one is simply asking for the fiber of this map above the point zero. Of course, one can apply the general techniques of differential topology—but we will not go into that at the moment!

However, for Lie systems, A_t *belongs to a given Lie algebra of vector fields on* X. Thus, one can hope to study the set of fixed points by means of Lie group and algebra theory. We shall, in fact, see that this is feasible, particularly in the case where the *Lie algebra acts transitively on* X.

4. SOME GENERAL FACTS ABOUT FIXED POINTS OF VECTOR FIELDS BELONGING TO FINITE DIMENSIONAL TRANSITIVE LIE ALGEBRAS. THE CENTRALIZER FOLIATION

Let X be a manifold, and let $\underset{\sim}{G}$ be a Lie algebra of vector fields on X. For $x \in X$, set:

$$\underset{\sim}{G}(x) = \{A(x) : A \in \underset{\sim}{G}\} \quad , \tag{4.1}$$

i.e., $\underset{\sim}{G}(x)$ is a linear subspace of X_x, the *tangent space to* X *at* x, and consists of the values at x of the vector fields in $\underset{\sim}{G}$.

<u>Definition</u>. $\underset{\sim}{G}$ is said to act *transitively on* X if the following condition is satisfied:

$$\underline{G}(x) = X_x \, , \tag{4.2}$$

for all $x \in X$.

Suppose that $\underline{G}$ acts transitively on X but that (for the moment) $\underline{G}$ is *not necessarily finite dimensional*. Let A be a fixed element of $\underline{G}$. Set:

$$Y = \{x \in X : A(x) = 0\} \, . \tag{4.3}$$

Our goal is to say as much as possible about the fixed set Y of A, especially with a view toward obtaining information about the space of all stationary solutions of Lie systems.

Now, set:

$$\underline{H} = \{B \in \underline{G} : [A,B] = 0\} \, . \tag{4.4}$$

$\underline{H}$ is the *centralizer of* A in $\underline{G}$. It is a Lie subalgebra of $\underline{G}$.

Consider $\underline{H}$ as a Lie algebra of vector fields on X. It defines a foliation--i.e., an equivalence relation on X such that the equivalence classes are connected submanifolds--as follows:

> Let us say that a curve in X is an *orbit curve* of $\underline{H}$ if it is continuous, piecewise C^∞, and each piece is an orbit curve of a vector field on $\underline{H}$. Say that two points $x, x' \in X$ are *equivalent* if they can be joined by an orbit curve.

It now follows from a theorem of Sussman [5]--which is a generalization of a theory of Chevalley [6] and Hermann [7,8] if $\underline{H}$ is finite dimensional--that the equivalence classes of this equivalence relation are indeed submanifolds of X. Let us call these submanifolds the *leaves* of the *centralizer foliation* determined by A.

For a general $\underline{H}$, Sussman shows that the tangent spaces to the leaves fill up a vector field system $\underline{H}'$ which contains $\underline{H}$. ($\underline{H}$ contains also the translates of $\underline{H}$ under one-parameter groups generated by elements of $\underline{H}$.) However, in this case--with $\underline{H}$ defined by (4.4)--it is obvious that $\underline{H}' = \underline{H}$. Let us sum up as follows.

Theorem 4.1. Let A be an element of a Lie algebra $\underline{G}$ of vector fields on a manifold X. Let $\underline{H}$ be the Lie subalgebra of $\underline{G}$ defined as the centralizer of A. Then, there is a foliation of X whose tangent space at each point x is precisely the values $\underline{H}(x)$ of $\underline{H}$ at x.

Definition. The leaf of this foliation passing through x is called the orbit of $\underline{H}$ through x.

We can now say something nontrivial about the fixed points of A. Let $x \in X$ be one. *Suppose* that the following condition is satisfied:

There is a linear subspace

$$\underset{\sim}{P} \subset \underset{\sim}{G}$$

such that:

$$[A, \underset{\sim}{P}] \subset \underset{\sim}{P} \tag{4.5}$$

$$X_x = \underset{\sim}{P}(x) \oplus \underset{\sim}{H}(x) \tag{4.6}$$

Every nonzero element of $\underset{\sim}{P}$ is nonzero at x. (4.7)

(We shall soon see how these conditions may be satisfied.) Here is a consequence of this assumption.

Theorem 4.2. Suppose that x is a fixed point of the vector field A and that a linear subspace $\underset{\sim}{P}$ of $\underset{\sim}{G}$ exists satisfying (4.5)-(4.7). Then, there is an open set O containing x such that the following condition is satisfied:

O ∩ (fixed set of A) = orbit of $\underset{\sim}{H}$ restricted to O

Proof. Let V be the orbit of $\underset{\sim}{H}$ through x. We can map

$$N \times \underset{\sim}{H} \to X$$

as follows:

The image of $(y, B) \in N \times \underset{\sim}{H}$ is the value at $t = 1$ of the orbit curve $t \to x(t)$ of B such that: $x(0) = y$. (4.8)

It is readily seen that (as a consequence of the hypotheses (4.5)-(4.7)) this map is a *local diffeomorphism*.

Now, consider the action of the one-parameter group

$$t \to \exp(tA) = g(t)$$

of diffeomorphisms generated by A. The map constructed by (4.8) is readily seen to intertwine the action of this group where $g(t)$ acts on $N \times \underset{\sim}{H}$ as follows:

$$g(t)(y,B) = (y, \exp(\mathrm{Ad}(tA))(B)) \ . \tag{4.9}$$

We see, using formula (4.9), that for t sufficiently small, the only fixed points of $g(t)$ on $N \times \underset{\sim}{H}$ are (in view of hypothesis (4.6), and the definition of $\underset{\sim}{H}$ as the centralizer of A) the points

$$(y,0) \ .$$

This proves Theorem 4.2.

5. THE FIXED POINT SET OF A VECTOR FIELD ARISING FROM A SEMISIMPLE ELEMENT OF A FINITE DIMENSIONAL LIE AGLEBRA. THE CARTAN DECOMPOSITION OF A SEMISIMPLE LIE ALGEBRA

Let us now specialize. Suppose $\mathscr{G}$ is a transitive *finite dimensional* Lie algebra of vector fields on a manifold X. Let A be one element of $\mathscr{G}$. Ad A denotes the linear transformation

$$B \to [A,B] \equiv \mathrm{Ad}\ A(B)$$

of $\mathscr{G} \to \mathscr{G}$.

<u>Definition</u>. A is a semisimple element of $\mathscr{G}$ if Ad A acting on $\mathscr{G}$ is completely reducible, i.e., if Ad A is the matrix which represents Ad A with respect to any basis of $\mathscr{G}$ is *diagonalizable*.

<u>Remark</u>. One must take care in using this terminology, since semisimple Lie algebras can have elements which are not semisimple. For example, if $\mathscr{G}$ = Lie algebra of $n \times n$ matrices, an element A is semisimple if and only if it is diagonalizable in the usual sense. An alternate and useful characterization is to say that A has the following "complete reducibility" property: If V is a linear subspace of $\mathscr{G}$ such that Ad A(V) $\subset$ V, there is a linear subspace V' $\subset \mathscr{G}$ such that $\mathscr{G} = V \oplus V'$ and Ad A(V') $\subset$ V'.

Now, consider A as a vector field in the manifold X. Set:

$$\mathrm{FS}_A = \{x \in X:\ X(x) = 0\}$$

$$\mathscr{C}_A = \{B \in \mathscr{G}:\ [A,B] = 0\} \ .$$

Since Ad A($\mathscr{C}_A$) $\subset \mathscr{C}_A$, by the complete reducibility property, there is a linear subspace $\mathscr{H} \subset \mathscr{G}$ such that:

$$[A, \mathcal{H}] \subset \mathcal{H}$$

$$\mathcal{G} = \mathcal{C}_A \oplus \mathcal{H} \quad .$$

As we have seen, this implies that the leaves of the foliation $\mathcal{C}_A$ are the connected components of the fixed set FS_A.

In case $\mathcal{G}$ arises from the global action of a Lie group, this can usefully be put another way. Let X be the coset space G/L, where G is a connected Lie group whose Lie algebra is $\mathcal{G}$, and where L is the isotropy subgroup of G at one point, which we call x_0. Let C_A denote the connected Lie subgroup of G generated by the Lie subalgebra $\mathcal{C}_A$, i.e.,

$$C_A = \text{maximal connected subgroup of the set of } g \in G \quad (5.1)$$
$$\text{such that } \text{Ad } g(A) = A \quad .$$

Thus we have proved:

<u>Theorem 5.1</u>. Let X be a coset space quotient G/L of a Lie group G by a closed subgroup L. Let A be a semisimple element of $\mathcal{G}$. Let C_A be the maximal connected subgroup of $g \in G$ which commute with A. Then, $FS_A \subset X$, the fixed point set of the vector field A, is a submanifold of X. Its connected components are orbits of the subgroup C_A of G.

6. MATRIX RICCATI EQUATIONS AS LIE SYSTEMS ON GRASSMANNIANS

Let V be an n-dimensional vector space over a scalar field K. (K = real or complex numbers.) For each integer m, $0 < m < n$, let $G^m(V)$ denote the space whose *points* are m-dimensional linear subspaces of V. It is called the *Grassmann manifold*.

Let GL(V) denote the Lie group of invertible linear maps $g: V \to V$. GL(V) acts on $G^m(V)$ in an obvious way.

If $\gamma \subset V$ is a linear subspace of V, then $g = \{gv: v \in \gamma\}$
is the image subspace under g.

As a transformation group on $G^m(V)$, GL(V) acts transitively. The isotropy subgroup is a closed subgroup of GL(V), hence the $G^m(V)$ is identified with a coset space, and a manifold structure.

A coordinate system can be constructed for $G^m(V)$ in the following way. Let

$$V = V_1 \oplus V_2 \quad (6.1)$$

be a fixed direct sum decomposition of V with $\dim V_1 = m$, i.e., $V_1 \in G^m(V)$. Let us say that a $\gamma \in G^m(V)$ is in *general position* with respect to the decomposition (5.1) if:

$$\gamma \cap V_2 = (0) . \qquad (6.2)$$

Let $G^m(V,V_2)$ denote the set of such γ's. Let $L(V_1,V_2)$ denote the vector space of linear maps $V_1 \to V_2$. γ determines uniquely a $\alpha(\gamma) \in L(V_1,V_2)$ so that:

$$\gamma = \{v_1 + \alpha(v_1) : v_1 \in V_1\} . \qquad (6.3)$$

The assignment $\gamma \to \alpha(\gamma)$ then defines a diffeomorphism between $G^m(V,V_2)$ -- which is an open subset of $G^m(V,V_2)$ --and the vector space $L(V_1,V_2)$. Then α sets up a *coordinate system* for the manifold $G^m(V)$.

Let $g \in GL(V)$. We shall compute the action of g in these coordinates. Write $v = v_1 \oplus v_2$, $v_1 \in V_1$, $v_2 \in V_2$, a column vector:

$$v = \begin{pmatrix} v_1 \\ v_2 \end{pmatrix}$$

Then,

$$gv = \begin{pmatrix} g_{11} & g_{12} \\ g_{21} & g_{22} \end{pmatrix} \begin{pmatrix} v_1 \\ v_2 \end{pmatrix} \qquad (6.2)$$

where $g_{11} \in L(V_1,V_1)$, $g_{12} \in L(V_2,V_1)$, $g_{21} \in L(V_1,V_2)$, $g_{22} \in L(V_2,V_2)$. (This is part of the linear algebra version of the classical idea of "partitioned matrix".)

Suppose that

$$g\gamma = \gamma' ,$$

$$\gamma = \left\{ \begin{pmatrix} v_1 \\ \alpha v_1 \end{pmatrix} : v_1 \in V_1 \right\}$$

$$\gamma' = \left\{ \begin{pmatrix} v_1' \\ \alpha' v_1' \end{pmatrix} : v_1' \in V_1 \right\} .$$

Then,

$$\begin{pmatrix} g_{11} v_1 + g_{12} \alpha v_1 \\ g_{21} v_1 + g_{22} \alpha v_1 \end{pmatrix} = \begin{pmatrix} v_1' \\ \alpha' v_1' \end{pmatrix}$$

or

$$\alpha'(g_{11}v_1 + g_{12}\alpha v_1) = g_{21}v_1 + g_{22}\alpha v_1 \quad ,$$

or

$$g(\alpha) = \alpha' = (g_{21} + g_{22}\alpha)(g_{11} + g_{12}\alpha)^{-1} \qquad (6.3)$$

The right hand side of (6.3) is traditionally called a *linear fractional transformation*. This formula shows that it is "really" just the coordinate-representation of the "natural" action of $GL(V)$ in the Grassmann manifold.

Now, consider a *flow* on $GL(V)$, i.e., a one-parameter family

$$t \to g(t)$$

of linear maps such that

$$g(0) = 1 \quad .$$

Set:

$$A(t) = \frac{dg}{dt} g^{-1} \quad .$$

The curve $t \to A(t)$ (a curve in the Lie algebra of $GL(V)$) is called the *infinitesimal generator* of the flow. Suppose $A(t)$ is partitioned in a way similar to g:

$$A(t) = \begin{pmatrix} A_{11} & A_{12} \\ A_{21} & A_{22} \end{pmatrix} \quad .$$

Then, a short calculation shows that:

$$\frac{d(g(t)(\alpha))}{dt} = (A_{21} + A_{22}g(t)(\alpha)) - g(t)\alpha(A_{11} + A_{12}g(t)\alpha) \quad .$$

This is a *matrix Riccati differential equation* for $t \to (g(t)\alpha)$, and shows explicitly how the most general such equation is a "Lie system" associated with the action of $GL(V)$ on $G^m(V)$.

We shall call a vector field on $G^m(V)$ which takes this form a *Riccati vector field*, i.e., a vector field which is the infinitesimal generator of a one-parameter group of linear transformations acting on the Grassmann manifold. Our strategy shall be to study the useful general properties of matrix Riccati equations using the tools of linear algebra and Lie theory, *not* the traditional "coordinate" representations.

7. THE FIXED POINT SET OF RICCATI VECTOR FIELDS CORRESPONDING TO DIAGONALIZABLE LINEAR TRANSFORMATIONS

Keep the notation of Section 6. Let the ground field K be $\mathbb{C}$, the complex numbers. Let A be a linear map: $V \to V$ which is diagonalizable. Let $\lambda_1, \ldots, \lambda_r$ be the distinct eigenvalues of A. Suppose $V_1, \ldots, V_r$ are the corresponding eigenvectors, i.e.,

$$A(v_1) = \lambda_1 v_1, \quad \text{for } v_1 \in V_1$$

$$A(v_2) = \lambda_2 v_2, \quad \text{for } v_2 \in V_2, \quad \text{etc.}$$

The assumption that A be diagonalizable means that V is the direct sum $V_1 + \cdots + V_r$. $C(A)$ consists of the linear maps $g \in GL(V)$ such that

$$g(V_1) = V_1$$
$$\vdots$$
$$g(V_r) = V_r$$

Let $\gamma \in G^m(V)$ be an m-dimensional linear subspace of V. The Riccati vector field determined by A vanishes at v if and only if

$$A(\gamma) \subset \gamma \quad .$$

This condition implies that:

$$\gamma = (\gamma \cap V_1) \oplus \cdots \oplus (\gamma \cap V_r)$$

Set:

$$d_1 = \dim(\gamma \cap V_1), \ldots, d_r = \dim(\gamma \cap V_r) \quad . \tag{7.1}$$

These integers $(d_1, \ldots, d_r)$ completely characterize which orbit of $C(A)$ the fixed point of A lies on.

Theorem 7.1. Two fixed points γ, γ' of the one-parameter group of linear fractional transformations generated by $A \in L(V)$ lie on the same orbit of $\mathscr{C}(A)$ if and only if the integers $(d_1, \ldots, d_r)$, $(d_1', \ldots, d_r')$ defined by (7.1) are equal.

Remark. Notice that this result, as trivial as it is from the point of view of linear algebra and Lie theory, says something significant from the applied point of view. Given an $L \in L(V)$, partitioned as

$$A = \begin{pmatrix} A_{11} & A_{12} \\ A_{21} & A_{22} \end{pmatrix} \quad ,$$

the set of $\alpha \in L(V_1, V_2)$ such that

$$(A_{21} + A_{22}\alpha) - (A_{11} + A_{12}\alpha)\alpha = 0 \quad , \tag{7.2}$$

(which is called the *algebraic Riccati* equation in the control theory literature, and is a quadratic equation for α) forms a discrete collection of manifolds (parameterized by the non-negative integers $(d_1, \ldots, d_r)$ such that $d_1 + \cdots + d_r = m$), each of which is an orbit of the Lie group $C(A)$. $C(A)$ itself is the direct product

$$GL(d, \mathbb{C}) \times \cdots \times GL(d_r, \mathbb{C}) \quad .$$

The main case of current interest in control theory literature is that where the matrix Riccati equation arises as a special solution of the Hamilton-Jacobi equation of a quadratic variational problem. (This first (?) arose in the classical work of Radon [9] on the Lagrange variational problem, which in turn is but a more classical name for the control theorist's problem.) We shall now investigate this case in more detail.

8. THE SYMPLECTIC MATRIX RICCATI EQUATION

Now, let V be a real, 2n-dimensional vector space with a non-degenerate, skew-symmetric bilinear form

$$\omega: V \times V \to R \quad .$$

Let $\mathscr{G}$ = set of linear maps $A: V \to V$ such that:

$$\omega(Av_1, v_2) + \omega(v_1, Av_2) = 0 \tag{8.1}$$

for $v_1, v_2 \in V$.

(A satisfying (8.1) is called *infinitesimally symplectic*.) It is readily verified that $\mathscr{G}$ forms a Lie algebra under commutation. It is called the *symplectic Lie algebra*. (It is the Lie algebra of the Lie group $Sp(n, R)$.)

Given $A \in \mathscr{G}$, notice that the bilinear form

$$(v_1, v_2) \to \omega(Av_1, v_2)$$

is *symmetric*. Set

$$h_A(v) = \omega(Av,v) \tag{8.2}$$

for $v \in V$.

h_A is a quadratic form on V called the *Hamiltonian* of A.

Here is the standard way to form such a V. Start off with a real vector space X of dimension n. Denote the dual vector space to X as X^d. Set:

$$Y = X^d \tag{8.3}$$

$$V = X \oplus Y. \tag{8.4}$$

Set:

$$\omega(x \oplus y, x' \oplus y') = y(x') - y'(x) \tag{8.5}$$

for $x, x' \in X$; $y, y' \in Y \equiv X^d \equiv L(X, \mathbb{R})$.

Write:

$$A \begin{pmatrix} x \\ y \end{pmatrix} = \begin{pmatrix} A_{11} & A_{12} \\ A_{21} & A_{22} \end{pmatrix} \begin{pmatrix} x \\ y \end{pmatrix}.$$

Suppose α is a linear map: $X \to T$ which is *symmetric*, i.e.,

$$\alpha(x)(x') = \alpha(x')(x) \tag{8.6}$$

for $x, x' \in X$.

Then

$$h_A(x, \alpha(x)) = \omega(A(x \oplus \alpha(x)), x \oplus \alpha(x)).$$

Now,

$$A \begin{pmatrix} x \\ \alpha(x) \end{pmatrix} = \begin{pmatrix} A_{11} & A_{12} \\ A_{21} & A_{22} \end{pmatrix} \begin{pmatrix} x \\ \alpha(x) \end{pmatrix}$$

$$= \begin{pmatrix} A_{11}x + A_{12}\alpha(x) \\ A_{21}x + A_{22}\alpha(x) \end{pmatrix}$$

$$= (A_{11}x + A_{12}\alpha(x)) \oplus (A_{21}x + A_{22}\alpha(x)).$$

Hence,

$$h_A(x,\alpha(x)) = -\alpha(x)(A_{11}x + A_{12}\alpha x) + (A_{21}x + A_{22}\alpha x)(x)$$

$$= -(\alpha^d(A_{11}x + A_{12}\alpha x))(x) + (A_{21}x + A_{22}\alpha(x))x \quad (8.7)$$

where α^d is the dual linear map $Y \to X$. Now, set:

$A(\alpha)$ = symmetric map $X \to Y$ such that

$$h_A(x,\alpha(x)) = A(\alpha)(x)(x) \quad (8.8)$$

We can read off from (8.7) what $A(\alpha)$ must be:

$$A(\alpha) = (A_{22}\alpha - \alpha^d A_{11}) + A_{21} - \alpha^d A_{12} . \quad (8.9)$$

Thus, the ordinary differential equation

$$\frac{d\alpha}{dt} = A(\alpha) \quad (8.10)$$

is the usual matrix Riccati equation of optimal control theory [4,15]. We can also compare it with the work of Section 6 and see that it is the infinitesimal generator of the group generated by A acting on the Grassmann manifold $G^n(V)$.
Set:

$$LG(V) = \{\gamma \in G^n(V): \omega(v_1, v_2) = 0 \text{ for } v_1, v_2 \in \gamma\}$$

$LG(V)$ is called the *Lagrange-Grassmann submanifold of* $G^n(V)$. Since $t \to \exp(tA)$ is a group of automorphisms of ω, we see that

> The orbits of $\exp(tA)$ at points of $LG(V)$ lie completely on $LG(V)$, i.e., the Riccati vector field defined by A in $G^n(V)$ is tangent to the submanifold $GL(V)$.

To complete this brief description of this circle of ideas, note that (8.9) is essentially the *Hamilton-Jacobi equation* of the calculus of variations. We have then shown the equivalence of finding a "complete solution of the Hamilton-Jacobi equation for a linear, optimal control problem" with the problem of "finding the orbits of the flow generated by A acting on $LG(V)$". (In fact, this equivalence also holds for time-dependent infinitesimal generators $t \to A(t)$.)

See References [10-15] for further information about matrix Riccati equations, mainly in the context of control theory.

Bibliography

1. S. Lie, *Transformationsgruppen*, Chelsea Publishing Co., New York.

2. E. Vessiot, *Encyclopédie des sciences mathematiques* (1910), Tome II, Vol. 3, 58-170.

3. R. Hermann, *Differential Geometry and the Calculus of Variations*, Academic Press, 1968. (Second Edition in preparation; to be published by Math Sci Press).

4. R. Hermann, *Algebraic Topics Useful in Systems Theory* (Interdisciplinary Mathematics, Vol. III), Math Sci Press, Brookline, MA, 1973.

5. H. Sussman, "Orbits of Families of Vector Fields and Integrability of Systems with Singularities", *Bull. Am. Math. Soc.* $\underline{79}$ (1973), 197-199.

6. C. Chevalley, *Lie Groups*, Princeton University Press, 1946.

7. R. Hermann, "The Differential Geometry of Foliations, II", *J. Math. Mech.* $\underline{11}$ (1962), 307-316.

8. R. Hermann, "Cartan Connections and the Equivalence Problem", *Contributions to Differential Equations* $\underline{3}$ (1964), 199-248.

9. J. Radon, "Zum Probleme von Lagrange", *Hamburg Math. Einzebschiften*, No. 2, B.G. Teuberg, Leipzig (1928).

10. M.K.H. Zakher-Itken, "The Matrix Riccati Differential Equation and the Semigroup of Linear Fractional Transformations", *Russian Math. Survey*.

11. J.C. Williams, "Least Squares Stationary Optimal Control and the Algebraic Riccati Equation", *IEEE Trans. Auto. Control* AC-16 (1971), 621-634.

12. C.R. Schneider, "Global Aspects of the Matrix Riccati Equation", *Mathematical Systems Theory* $\underline{7}$ (1973), 281-286.

13. W.A. Coppel, "Matrix Quadratic Equations", *Bull. Austral. Math. Soc.* $\underline{10}$ (1974), 377-401.

14. J.E. Potter, "Matrix Quadratic Solutions", *J. Scan. Appl. Math.* $\underline{14}$ (1966), 496-501.

15. B.D.O. Anderson and J.B. Moore, *Linear Optimal Control*, Prentice-Hall, New York

Chapter 6

MATRIX RICCATI EQUATIONS DEFINED BY LINEAR
QUADRATIC DIFFERENTIAL GAMES

1. INTRODUCTION

The theory of "differential games" is an interesting marriage of control and game theory. It too leads to "matrix Riccati equations" in a natural way. In this chapter, we shall study certain relations between the differential games and the Lie theory of their Riccati equations. The best introduction to game theory with which we are familiar is the book by Intriligator, and we will roughly follow his development.

2. CARATHEODORY'S METHOD FOR DIFFERENTIAL GAMES

We consider a state space whose element is denoted by x. (For simplicity, think of this as R^n, although the ideas are capable of vast generalization, e.g., to infinite dimensional spaces and/or differentiable manifolds.) There are two control vectors, labelled

$$u^1, u^2 \quad .$$

(This corresponds to a game with two "players". Generalizations can readily be made to games with more "players".) The *input-state equations* are:

$$\frac{dx}{dt} = f(x, u^1, u^2) \quad . \tag{2.1}$$

There are two *payoff functions*:

$$\mathscr{L}^1 = \int_{t_0}^{t_1} L^1(x, u^1, u^2) \, dt \tag{2.2}$$

$$\mathscr{L}^2 = \int_{t_0}^{t_1} L^2(x, u^1, u^2) \, dt \quad . \tag{2.3}$$

Player one chooses an input $t \to u^1(t)$, and player two chooses input $u^2(t)$. (2.1) is solved to determine the state evolution, beginning at $x(0)$ for $t = 0$. $\mathscr{L}^1$ and $\mathscr{L}^2$ are then the *pay offs* to players one and two for the "strategy" u^1, u^2. The ideas of game theory then suggest various optimization problems.

The Caratheodory method attempts to reduce these to the corresponding *finite dimensional* optimization problem. To do this, add on to the right hand side of (2.2)-(2.3) terms *which only depend on the end-times*.

$$\mathscr{L}^1 = \int_{t_0}^{t_1} (L^1(x,u^1,u^2) + S_x^1 f + S_t^1) \, dt \qquad (2.4)$$

$$\mathscr{L}^2 = \int_{t_0}^{t_1} (L^2(x,u^1,u^2) + S_x^2 f + S_t^2) \, dt \qquad (2.5)$$

S^1, S^2 are functions of x and t. Introduce another n-vector y, the "costate variable". Introduce Hamiltonian functions:

$$h^1(x,u^1,u^2,y) = L^1 + y'f \qquad (2.6)$$

$$h^2(x,u^1,u^2,y) = L^2 + y'f \qquad (2.7)$$

Now, hold x and y fixed, and regard h^1, h^2 are "payoff functions" for a "static" game, whose "strategy space" is u^1/u^2. Determine an "optimal strategy" (assuming a unique one exists) by one of the various schemes used in game theory (we shall examine some below), obtaining u^1/u^2 as *functions of* x, y:

$$u^1(x,y)$$

$$u^1 = \phi^1(x,y)$$

$$u^2 = \phi^2(x,y) \ .$$

These are the *feedback strategies*.

We shall restrict attention here to the simplest sort of game-theoretic possibility; namely, *zero sum*, i.e.,

$$\mathscr{L} \equiv \mathscr{L}^1 = -\mathscr{L}^2$$

and player one tries to maximize $\mathscr{L}^1$, player two to minimize it. Thus, we determine $\phi^1(x,y), \phi^2(x,y)$ by the conditions that:

$$h(x,u^1,\phi^2(x,y),y) \leq h(x,\phi^1(x,y),\phi^2(x,y),y)) \leq h(x,\phi^1(x,y),u^2,y) \qquad (2.8)$$

Condition (2.8) is a typical *saddle point* condition. For example, the function

$$u^1, u^2 \to (u^2)^2 - (u^1)^2 \quad \text{at} \quad (0,0) \quad .$$

In particular, notice that (2.8) implies the following "infinitesimal" conditions:

$$0 = h_{u^1}(x, \phi^1(x,y), \phi^2(x,y), y)$$
$$= h_{u^2}(x, \phi^1(x,y), \phi^2(x,y), y) \tag{2.9}$$

(Subscripts denote partial devivatives.)

$$h_{u^1 u^1} < 0 \tag{2.10}$$

$$h_{u^2 u^2} > 0 \tag{2.11}$$

Rather than continuing to work at this level of "nonlinear" generality, we shall now restrict to the case which only involves *quadratic forms*.

3. LINEAR QUADRATIC, TWO-PERSON, ZERO-SUM, DIFFERENTIAL GAMES

Suppose the state-input equations are:

$$\frac{dx}{dt} = Ax + B_1 u_1 + B_2 u_2 \quad . \tag{3.1}$$

The "payoff" function (to player one) is now:

$$\mathscr{L} = \int_{t_0}^{t_1} \frac{1}{2} (x'Qx - u_1'R_1 u_1 + u_2'R_2 u_2) \, dt \tag{3.2}$$

We now choose notation as follows: $x \in R^n$, $u_1 \in R^{m_1}$, $u_2 \in R^{m_2}$. The Hamiltonian function is now:

$$h = \frac{1}{2}(x'Qx + u_1'R_1 u_1 - u_2'R_2 u_2) + y'(Ax + B_1 u_1 + B_2 u_2) \tag{3.3}$$

$$h_{u_1'} = -R_1 u_1 + B_1' y \tag{3.4}$$

$$h_{u_2'} = R_2 u_2 + B_2' y \tag{3.5}$$

$$h_{u_1 u_1} = -2R_1 \tag{3.6}$$

$$h_{u_2 u_2} = 2R_2 \qquad (3.7)$$

Conditions (2.10) and (2.11) (hence, also (2.8)) can be satisfied if

$$R_1 > 0 ; \qquad R_2 > 0 \quad . \qquad (3.8)$$

We shall, in fact, assume that (3.8) holds.

The *Hamilton equations* are now constructed as follows:

$$\frac{dx}{dt} = h_{y'} = Ax + B_1 u_1 + B_2 u_2 \qquad (3.9)$$

$$\frac{dy}{dt} = -h'_x = -Qx - A'x \qquad (3.10)$$

Now, set the right hand side of (3.4)-(3.5) equal to zero:

$$\begin{aligned} u_1 &= R_1^{-1} B_1' y \\ u_2 &= -R_2^{-1} B_2' y \end{aligned} \qquad (3.11)$$

After these substitutions, the Hamilton equations (3.9)-(3.10) can be written in partitioned-matrix form as:

$$\frac{d}{dt} \begin{pmatrix} x \\ y \end{pmatrix} = \alpha \begin{pmatrix} x \\ y \end{pmatrix} \qquad (3.12)$$

with

$$\alpha = \begin{pmatrix} A, & B_1 R_1^{-1} B_1' - B_2 R_2^{-1} B_2' \\ -Q, & -A' \end{pmatrix} \qquad (3.13)$$

We see now why the "game" situation is more complicated than the "optimal control" situation from the view point of Lie theory. α is an element of $\mathcal{G}$, the Lie algebra of $Sp(n,R)$. The α's corresponding to the optimal control problems form a *cone* in $\mathcal{G}$, i.e., those of the form

$$\alpha = \begin{pmatrix} \alpha_{11} & \alpha_{12} \\ \alpha_{21} & -\alpha'_{11} \end{pmatrix}$$

with $\alpha'_{12} = \alpha_{12}$; $\alpha'_{21} = \alpha_{21}$,

$$\boxed{\alpha_{12} > 0, \quad \alpha_{21} > 0}$$

However, those of type (3.13) may be outside of the cone, i.e.,

α_{12} *is not necessarily positive definite.*

However, a natural question from the Lie point of view is:

When can α, defined by (3.13), belong to certain specified subspaces of $\mathscr{G}$?

I now leave this circle of ideas in this admittedly incomplete form. My aim has been to traverse the foothills in order to arrive in position for a later push up the main range--the mastery of matric Riccati equations with Lie theoretic tools.

GROUP II

HOLOMORPHIC VECTOR BUNDLES AND LINEAR SYSTEMS

INTRODUCTION

The dominant mathematical technology in the electrical engineering-control theory literature of the 1920's and 1930's was one-complex variable theory; it appeared in what engineers called *transfer functions*, which were usually the Laplace-transform of certain data involved in electrical circuits, transmission lines, and so on. In the 1950's and 1960's this technique--which was by its mathematical nature restricted to *linear* situation--was replaced with "state space" techniques; indeed, this caused a revolution in engineering education. The fundamental reason for this shift was *not* because of mathematics, but the greater compatibility of the state-space methods with computers. It is actually easier to solve a matrix differential equation on a computer than it is to make a "spectral factorization" of a rational function!

Several years ago, Clyde Martin and I undertook a broad probram of development of geometric methods in *linear* systems theory. We felt that some of the modern work in the theory of Lie and algebraic groups, group representation theory, etc. should be utilizable. The motivation from the practical side (which came via our work with the Ames Research Center of NASA and our links with Brian Doolin) was that we felt that the traditional methods based on 19th century matrix theory were becoming increasingly cumbersome and difficult as engineers considered more and more complicated systems. Also certain problems of a non-traditional type (such as the study of dependence of systems--*parameters*, and the modification of systems by failures) were arising for which the traditional methods were inadequate. We knew from our mathematical training that many of these classical techniques could be replaced (without any loss of effectiveness and intuitive appeal) with the more modern ideas, and *sometimes* spectacular simplifications and reasonably more powerful techniques developed. Of course, since we are *also* "applied mathematicians", we realize that it is not a simple matter of replacing "old-fashioned" techniques A,B,C,... with the "modern" techniques α,β,γ,... but required a much more subtle and delicate "fine-tuning" of the mathematics to the application.

In systems theory, I believe we have made an encouraging dent in two areas--improvement of the "pole placement via feedback" results and the general algebraic study of the feedback structure of linear, finite dimensional, time-invariant systems. The former work is presented in "Interdisciplinary

Mathematics", Volume 13, and the latter in the paper, "Applications of Algebraic Geometry to Systems Theory: The McMillan Degree and Kronecker Indices of Transfer Functions as Topological and Holomorphic System Invariants", SIAM J. Control $\underline{16}$ (1978), 743-755. There is also related material in the "Lie Groups" series, Volume 7 (The Ames 1976 Control Theory Conference). In this work, what we actually did was to rethink the *traditional* "transfer function" in modern terms. "Put it on the Grassmann manifold" was our motto.

Our work was essentially completed in 1976. Since then we have spasmodically attempted to improve it, both from the expository and research point of view, and extend it in various directions. Since we are no longer at the same place, it has been difficult to carry out such a sustained effort, and we have only done bits and pieces. This work is presented here. The most noteworthy *new* material is that in Chapters 10 and 11, concerned respectively with extensions to systems with *more independent variables* and *time-varying in one variable*. For the latter, note the exciting possibility of a connection between Systems Theory and the recent work by the Russian school of nonlinear wave-algebraic geometry.

Chapter 7

VECTOR BUNDLES DEFINED BY FINITE DIMENSIONAL,
LINEAR, TIME-INDEPENDENT SYSTEMS

1. INTRODUCTION

Work by C. Martin and myself [1-4], offers a new approach to certain problems of systems theory. Mathematically, this material is closely related to modern developments in complex variable theory, algebraic geometry, topology, etc., while from the point of view of engineering it has close links with the "older" style of control theory, which was in full development and application in the period 1930-1960. It is interesting to speculate on the role that fashion in mathematics played here; in this earlier period the theory of one complex variable was in its heyday, while by 1960 "differential equations" and "dynamical systems" were again in development by pure mathematicians, and formed the intellectual framework for the "modern, state-space" approach. Thus, it may seem perverse that Martin and I want to go back to the Bad Old Days before "state-spaces", but we believe that there are definite advantages to be gained and new insights to be developed into broad classes of system-theoretic problems that do not fit readily into the now standard framework.

As general remarks, we may note that present day linear systems theory makes most extensive use of the classical "matrix algebra" developed, say, in the treatises by Gantmacher [5] and Macduffee [6]. The theories of infinite dimensional linear systems and 2-D digital filters (to name two areas which are important to present day technology) present problems which transcend this traditional framework. What we have done (and I will continue this development here) is to recast the material in such a form that more "modern" and (sometimes) more powerful tools (of algebraic and differential geometry, Lie theory, several complex variables, etc.) may be used. Our central idea is to associate with linear systems various *vector bundles*. In the simplest case, these vector bundles will be non-singular finite dimensional and have the Riemannian spheres as base, but more complicated possibilities are interesting. It turns out [3] that basic systems theoretic properties about linear, time-invariant, finite dimensional systems become translated into basic algebro-geometric properties of these bundles.

In the more specialized treatises on the applications of Gantmacher-Macduffee type of matrix algebra to systems theory (e.g., [8,9]) a distinguished tool is the "coprime factorization" of matrices of rational functions. Now, as beautiful as this is, it is something that will definitely not generalize in the directions indicated above (i.e., infinite dimensional and 2-D). Our method replaces this algebraic tool and gives new algebraic and Lie-theoretic insights to even the well known material. For example in [2] we work out what "pole placement" means for *Hamiltonian* systems, where the standard approach *does not even make sense*. (A Hamiltonian matrix will *always* have eigenvalues which lie in the right half plane.) This is the prototype for a development of new algebraic tools for discussion of the "structure" of systems. Finally, I might mention that C. Byrnes [7] has used some of our methods (plus those of his own!) to dig significantly deeper into the theory of *delay* systems.

2. LINEAR INPUT-OUTPUT SYSTEMS

It is well known that a nifty general framework for much of what engineers call "systems theory" is a gadget of the form:

$$\begin{aligned} \frac{dx}{dt} &= Ax + Bu \\ y &= Cx \end{aligned} \qquad (2.1)$$

x, u, y are elements of vector spaces X, U, Y, say, with the complex numbers as field of scalars called the *state*, *input*, and *output spaces*. (See [10-11].) The differential equations (2.1) define a set of *input-output relations*. A curve $t \to (u(t), y(t))$ of curves in $U \times Y$ is called an *input-output pair* if there is a curve $t \to x(t)$ in X such that $t \to (x(t), u(t), y(t))$ is a solution of (2.1). The main object of the theory--especially in contrast to the traditional view of "dynamical systems" derived historically from the study of Newton's laws, is to *study these input-output relations*. A special goal is to design *systems so that they have desired input-output behavior*. As Brockett has put it in [12], control theory is the typical *prescriptive* science, while the disciplines derived from traditional physics (especially mechanics) is *descriptive*. This broad viewpoint is not as well-known in the scientific world as it might be. Most mathematicians confuse "control" with "optimal control", i.e., a generalized version of the classical calculus of variations, while on the other side, many scientists who potentially could use the ideas (physicists, biologists, economists,...) do not understand their full subtlety and complexity.

3. THE "FREQUENCY DOMAIN" DISCRIPTION VIA A VECTOR BUNDLE

In order to algebraicize (2.1), let us Laplace-transform it, with zero initial conditions in x:

$$\hat{x}(s) = \int_0^\infty e^{-st} x(t)\, dt$$

$$\hat{u}(s) = \int_0^\infty e^{-st} u(t)\, dt$$

$$\hat{y}(s) = \int_0^\infty e^{-st} y(t)\, dt$$

(2.1) then goes over to the following equations:

$$\boxed{\begin{aligned} s\hat{x} &= A\hat{x} + B\hat{u} \\ \hat{y} &= C\hat{x} \end{aligned}} \qquad (3.1)$$

We can then study the input-output relation (2.1) in the more "algebraic" form (3.1). It is convenient to do so by constructing objects that mathematicians call *vector bundles*. In fact, there are at least three such bundles which may be constructed and which seem to be related to the topics engineers and systems theorists find interesting.

Let s denote a complex number. We will denote the set of complex numbers by $\mathbb{C}$. Add the "point at infinity" to compactify $\mathbb{C}$, and obtain

$$S^2 \equiv \mathbb{C} \cup (\infty) \ .$$

(S^2 = Riemann sphere = two-dimensional sphere in R^3.) To make sense of (3.1) and $s = \infty$, we can substitute $s = 1/\tau$, and then let $\tau \to 0$. The result is the following set of equations:

$$\hat{x} = \tau(A\hat{x} + B\hat{u})$$

$$\hat{y} = C\hat{x}$$

or, as $\tau \to 0$,

$$\boxed{\begin{aligned} \hat{x} &= 0 \\ \hat{y} &= 0 \end{aligned}} \qquad (3.2)$$

Thus, at $s = \infty$, (3.2) determines the system relations.

Define

$$E = \text{set of all ordered quadruples } (s,\hat{x},\hat{u},\hat{y}) \text{ satisfying (3.1) if } s \in \mathbb{C}; \text{ set of all quadruples of the form } (\infty,0,\hat{u},0) \text{ if } s = \infty. \quad (3.3)$$

$$E' = \text{set of all ordered triples } (s,\hat{u},\hat{y}), \text{ for } s \in \mathbb{C}, \text{ such that there } exists \; \hat{x} \in X \text{ with Equations (3.1) satisfied. If } s = \infty, \text{ all } (\infty,\hat{u},0), \hat{u} \in U. \quad (3.4)$$

$$E'' = \text{set of all ordered triples } (s,\hat{x},\hat{u}) \text{ such that, if } s \in \mathbb{C}, s\hat{x} = A\hat{x} + B\hat{u}, \text{ and such that } \hat{x} = 0, \hat{u} \text{ arbitrary if } s = \infty. \quad (3.5)$$

Note that

$$\begin{aligned} E &\subset S^2 \times X \times U \times Y \\ E' &\subset S^2 \times U \times Y \\ E'' &\subset S^2 \times X \times U \end{aligned} \quad (3.6)$$

This identification of these "bundles" with subsets of sets already given to us, provides ways (if and when they are needed) of defining topologies, manifold structures, etc. on these bundles.

In all three cases, one can define maps $\pi: E \to S^2$, $\pi': E' \to S^2$; $\pi'': E'' \to S^2$, by projecting the Cartesian products in the first factor. Then

$$\pi(x,s,\hat{x},\hat{u},\hat{y}) = s$$

$$E(s) = \pi^{-1}(s); \quad E'(s) = {\pi'}^{-1}(s); \quad E''(s) = {\pi''}^{-1}(s)$$

are the fibers of these maps. Notice that they are *vector spaces*. For example, $E(s)$ is the set of all $(\hat{x},\hat{u},\hat{y})$ which satisfy the *linear* equations (3.1) *at this fixed value of* s. It is a linear subspace of the vector space $X \oplus U \oplus Y$, which is identified (as a *set*) with the Cartesian product $X \times U \times Y$.

This property--map $\pi: E \to S^2$ with the fibers given as *vector spaces*--defines these spaces as *vector bundles* [14,15]. Notice that the s-dependent solutions of Equations (3.1), i.e., essentially the "input-output relations" as the engineer knows them, are the *cross-sections* of these vector bundles.

Obviously, we have taken advantage of a very general construction-- representing a parametrized (by s) set of linear equations--like (3.1)--as a *vector bundle*, whose base space is the parameter. This is, so far, just a different "language". However, it turns out to be a powerful reformulation, because it makes contact with an extremely important part of "modern" mathematics. As we shall now see, we can attach to these vector bundles *maps*

$$\phi: S^2 \to G(V)$$

where $G(V)$ is the *Grassmann space*. It turns out that the properties of these maps are closely related to the "interesting" systems-theoretic properties of the original set of equations (3.1).

4. GRASSMANN SPACES AND THEIR CANONICAL VECTOR BUNDLES

Let V be a vector space. (Say that it has the complex numbers as a field of scalars and that it is finite dimensional. Generalizations in various directions are feasible.) A *linear subspace* of V is a certain type of subset. Construct a "new" space, called the *Grassmann space* of V denoted by $G(V)$; a "point" of $G(V)$ is a *linear subspace* of V.

Let $GL(V)$ denote the group of linear automorphisms of V. $GL(V)$ acts as a transformation group on $G(V)$:

> The transform of a linear subspace $U \subset V$ by the linear automorphism $A: V \to V$ is the linear subspace $A(U)$.

The orbits of $GL(V)$ on $G(V)$ are called the *Grassmann manifolds* of V. These orbits are parametrized by an integer m.

$$0 \leq m \leq \dim V \quad,$$

i.e., the orbit $G^m(V)$ consists of *the linear subspaces of* V *of dimension* m. (It follows from elementary linear algebra that two linear subspaces of dimension m can be transformed into each other by a "change of basis in V", i.e., by a transformation in $GL(V)$.)

The $G^m(V)$ are manifolds in the technical sense (since they are coset spaces of Lie groups), but $G(V)$ is not. The best one can do is to put a topology on $G(V)$ --even this has rather peculiar properties. We will not go into this at the moment.

There is a vector bundle E defined, with $G(V)$ as base space, in a natural way. We call it the *canonical bundle*. Here is how it is defined.

> A point of E is a pair (γ, v) of a linear subspace $\gamma \in G(V)$ and a vector $v \in \gamma$. The projection map $\pi: E \to G(V)$ is defined as $\pi(\gamma, v) = \gamma$. Thus, the fibers of E above the "point" γ in the base is γ itself *considered as a vector space*.

Let

$$\pi: X \to G(V)$$

be a map of any space X into $G(V)$. We can define a vector bundle E' with base X as follows:

A point of E' is a pair (x, v) with $x \in X$, $v \in \phi(v)$. E' is mapped into E as follows

$$(x, v) \to (\phi(x), v)$$

The corresponding diagram of maps

$$\begin{array}{ccc} E' & \longrightarrow & E \\ \downarrow & & \downarrow \\ X & \longrightarrow & G(V) \\ & \phi & \end{array}$$

is *commutative*. The vector bundle E' is said to be the *pull-back of the canonical bundle via the map* ϕ.

Thus, we have a method of replacing certain types of vector bundles with *maps* from their base space to a "standard" space, namely, the Grassmann space. It turns out that the "topological" and "analytic" properties of these maps are important for the study of the "invariants" of vector bundles. Applied to systems theory, Martin and I have shown that such basic linear systems-theoretic invariants as the *Macmillan degree* and the *state feedback invariants* are defined naturally in this way. We will now investigate a general setting for this construction.

5. VECTOR BUNDLES DEFINED BY LINEAR EQUATIONS

Let X be a space, and let V, W be vector spaces. Again, we shall assume that V and W are finite dimensional, although to a certain extent the ideas can be developed more generally. $L(V,W)$ denotes the vector space of linear maps $V \to W$. Let us suppose given a map

$$\alpha: S \to L(V,W) \quad .$$

Construct E as a subset of $S \times V$ as follows:

$$E = \text{set of } (s,v) \text{ such that } \alpha(s)(v) = 0 \tag{5.1}$$

Let $\pi: E \to S$ be the map π which sends (s,v) into s. Then, π defines E as a *vector bundle* with S as base.

This is the simplest and most direct way that vector bundles arise in practice. Here is a more general way. Suppose that S is the union of two subsets

$$S = S_1 \cup S_2 \quad .$$

Suppose also that

$$\alpha_1: S_1 \to L(V,W)$$

$$\alpha_2: S_2 \to L(V,W)$$

are mappings. Let

$$GL(V), \quad GL(W)$$

be the group of all linear automorphisms of the vector spaces V and W. Suppose given a map

$$\beta_{12}: S_1 \cap S_2 \to GL(V)$$

$$\beta'_{12}: S_1 \cap S_2 \to GL(W)$$

such that

$$\alpha_1(s) = \beta'_{12}(s)\alpha_2(s)\beta_{12}(s)^{-1} \quad \text{for all } s \in S_1 \cap S_2. \tag{5.2}$$

With this condition, we can construct vector bundles

$$E_1 \subset S_1 \times V = \text{set of } (s_1, v) \text{ such that } \alpha_1(s_1)(v) = 0$$

$$E_2 \subset S_2 \times V = \text{set of } (s_2, v) \text{ such that } \alpha_2(s_2)(v) = 0$$

Condition (5.2) guarantees that, for $s \in S_1 \cap S_2$,

$$E_1(s) = E_2(s) \quad .$$

Thus, we can define a vector bundle

$$E \subset S \times V$$

as the set of (s,v) such that

$$\alpha_1(s)(v) = 0, \quad \text{if } s \in S_1 ,$$

$$\alpha_2(s)(v) = 0, \quad \text{if } s \in S_2 ,$$

with the confidence that (because of condition (5.2)) E is *well defined* in the intersection $S_1 \cap S_2$, hence defines a vector bundle *globally* over S.

This procedure can obviously be generalized to the case where S is a space with a covering by an arbitrary number (even non-countable) of subsets $S_1, S_2, \ldots$. In the intersection $S_i \cap S_j$, we are given maps

$$\beta_{ij}, \beta'_{ij} : S_i \cap S_j \to GL(V), GL(W) \quad .$$

These then define *principal fiber bundles* in the sense of Steenrod [16]; the vector bundle we have constructed is essentially one of the "associated bundles". Thus, in a sense we are interested in a very special case of the general Steenrod construction. Now, we specialize even further to the situation of immediate interest for systems theory.

6. THE VECTOR BUNDLE ON THE RIEMANN SPHERE DEFINED BY AN INPUT SYSTEM. CONTROLLABILITY AS A BUNDLE PROPERTY

Here is a simple but useful example of the construction we have described in previous sections. Let X, U be vector spaces, $A: X \to X$, $B: U \to X$ linear maps. Consider the input-system:

$$\frac{dx}{dt} = Ax + Bu \tag{6.1}$$

Its Laplace transformed version is

$$s\hat{x} = A\hat{x} + B\hat{u} \quad . \tag{6.2}$$

VECTOR BUNDLES

This suggests the following construction. Regard s as a complex number, i.e., as an element of $\mathbb{C}$. Set

$$V = X \times U$$

$$W = X$$

$$\alpha(s)(x,u) = (A-s)x + Bu \qquad (6.3)$$

$s \to \alpha(s)$ is then a map

$$\mathbb{C} \to L(V,W)$$

Let $E \subset \mathbb{C} \times V$ be the subset of $(s,(x,u))$ such that

$$\alpha(s)(x,u) = 0 \quad ,$$

i.e., such that

$$Ax - sx + Bu = 0 \quad .$$

Set $s = 1/\tau$. Relation (6.2) becomes

$$\hat{x} = \tau A \hat{x} + \tau B \hat{u} \quad .$$

Let

$$\alpha'(\tau)(x,u) = (\tau A - 1)(x) + \tau Bu \qquad (6.4)$$

Then

$$\alpha'(\tau) = \tau \alpha\left(\frac{1}{\tau}\right) \qquad (6.5)$$

Define S^2, the Riemann sphere, in the following way. Take two copies of $\mathbb{C}$, call them $\mathbb{C}_1$, $\mathbb{C}_2$, coordinatized by s,τ. *Identify* a point of $\mathbb{C}_1 - (0)$ and a point of $\mathbb{C}_2 - (0)$ if their coordinates are *inverses* of each other. The quotient of $\mathbb{C}_1 \cup \mathbb{C}_2$ by this equivalence relation (or "pasting together") is S^2.

Let $V = X \times U$, the direct sum of the vector spaces X,U. The bundle is then a pull-back of the canonical bundle on

$$G(V) \quad ,$$

via the map

$$\gamma: S^2 \to G(V)$$

defined as follows:

$$\gamma(s) = \{(x,u): (A-s)x + Bu = 0\} \qquad (6.6)$$

Theorem 6.1. If the linear system (6.1) is *controllable*, then

$$\dim \gamma(s) = \dim U \tag{6.7}$$

for all $s \in \mathbb{C}$.

Proof. Recall that

$$V = X \oplus U,$$

$\alpha(s)$ is the linear map $X \oplus U \to X$ given by

$$\alpha(s)(x \oplus u) = Ax - sx + Bu \tag{6.8}$$

$\gamma(s)$ is the kernel of $\alpha(s)$. The elementary relation between the dimension of the kernel and range of a linear map gives the following relation:

$$\dim \alpha(s)(V) = \dim V - \dim \gamma(s)$$

$$= \dim X + \dim U - \dim \gamma(s).$$

Hence, condition (6.7) is *equivalent* to the condition

$$\dim \alpha(s)(V) = \dim X,$$

i.e.,

$$\boxed{\alpha(s) \text{ is onto.}} \tag{6.9}$$

Now, "controllability" of (6.1) means that

$$X = BU + ABU + \cdots \tag{6.10}$$

"Dually", it means that:

$$0 = \text{kernel } B^d \cap \text{kernel } (B^d A^d) \cap \cdots. \tag{6.11}$$

To prove that (6.9) follows from (6.11), suppose that $x^d \in X^d$, and

$$x^d(\alpha(s)(V)) = 0,$$

i.e., x^d is "perpendicular" to $\alpha(s)(V)$. Then,

$$\boxed{\begin{array}{l} x^d(Ax - sx + Bu) = 0 \\ \text{for all } x \in X,\ u \in U. \end{array}} \tag{6.12}$$

VECTOR BUNDLES

In particular, for $x = 0$, all $u \in U$,

$$0 = x^d Bu = (B^d x^d) u \ ,$$

i.e.,

$$B^d x^d = 0 \ .$$

Set $x = Bu/s$ in (6.9), obtaining

$$x^d \left(\frac{ABu}{s} - Bu + Bu \right) = 0 \ ,$$

or

$$x^d (ABu) = 0 \ ,$$

$$B^d A^d x^d u = 0$$

for $\underline{\text{all}}$ $u \in U \ ,$

i.e.,

$$B^d A^d x^d = 0 \ .$$

Continuing in this way, we see that

$$B^d (A^d)^j x^d = 0$$

for all integers j.

Condition (6.11) forces $x^d = 0$, which proves (6.9), hence completes the proof of Theorem 6.1.

<u>Exercise</u>. If (6.7) is satisfied, is the system (6.1) controllable?

<u>Exercise</u>. Let $E(\infty)$ be the fiber of the vector bundle at $s = \infty$. Show that

$$\dim E(\infty) = \dim U \ .$$

Theorem 6.1 is a basic result for our approach. For it says that a systems theoretic property--controllability--implies that the vector bundle we have constructed on S^2 is non-singular. A basic theorem of Grothendieck [17] "classifies" (i.e., gives all "invariants") such bundles. In the next chapter we shall show how Kronecker's algebraic theory of such "pencils" of linear maps (i.e., linear maps depending *linearly* on the parameters) determines exactly the "invariants" predicted by Grothendieck's general theorem.

7. CONTROLLABILITY AND OBSERVABILITY AS A VECTOR BUNDLE PROPERTY

Now, consider an input-output system:

$$\frac{dx}{dt} = Ax + Bu$$
$$y = Cx \quad . \tag{7.1}$$

Convert it into an algebraic form by means of Laplace transform:

$$s\hat{x} = A\hat{x} + B\hat{u}$$
$$\hat{y} = C\hat{x} \tag{7.2}$$

We shall now construct two vector bundles on S^2, denoted as E, E'. A point of S^2 is denoted by s; the fibers of E, E' by $E(s)$, $E'(s)$.

$$E(s) = \{(x,u): sx = Ax + Bu\}$$

$$E'(s) = \{(u,y): \exists\, x \in X \text{ such that (7.2) is satisfied}\}$$

(For typographical convenience, from now on leave off the symbol ^.) Let

$$C: E \to E'$$

be the linear bundle map defined as follows:

$$C((x,u)) = (u, Cx) \tag{7.3}$$

for $(x,u) \in E(s)$.

Theorem 7.1. If the input-output system (7.1) is observable *and* controllable, then the bundle map $C: E \to E'$ is a *bundle isomorphism*.

Proof. We have seen in Section 6 that *controllability* of (7.1) implies the bundle E is non-singular, i.e., all its fibers are of the same dimension. At the points of the base which are not eigenvalues of A, the fibers are obviously both of the same dimension as U, the input space. (Note that:

$$E'(s) = \{(u,y): y = T(s)u, \text{ where } T(s) = C(s-A)^{-1}B\}$$

$$E(s) = \{(x,u): x = (s-A)^{-1}Bu\} \quad .)$$

We shall now show that: For all $s \in \mathbb{C}$,

$$\boxed{C: E(s) \to E'(s) \text{ is one-one}} \tag{7.4}$$

VECTOR BUNDLES

Let $(x,u) \in E(s)$ be such that

$$C(x,u) = 0 ,$$

i.e.,

$$u = 0 = Cx .$$

Then,

$$sx = Ax ,$$

i.e., s is an eigenvector of A. Also,

$$CAx = sCx = 0$$

$$CA^2 x = 0 , \ldots$$

Observability now implies that $x = 0$, proving (7.4). C is now an isomorphism because--by construction-- C maps $E(s)$ onto $E'(s)$.

In summary, I have used the "system" (7.1) to construct the two vector bundle E and E' which are at the heart of the method Martin and I have developed to apply algebraic geometry to linear systems theory. It will turn out to be very significant that the linear map C between them is an isomorphism when the system is controllable and observable. This fact will turn out to be a key link between the "systems theoretic" nature of the problem and the "algebro-geometric".

Remark. In our original paper [3], Martin and I used a different method to construct E'. Consider the "transfer map"

$$T(s) = C(s-A)^{-1} B .$$

It is a standard technique of systems theory [8,9] to factor $T(s)$ as

$$T(s) = D(s)^{-1} N(s) ,$$

where $s \to D(s)$, $s \to N(s)$ are polynomial maps $(N(s) \in L(U,Y); D(s) \in L(Y,Y))$. Then,

$$E'(s) = \{(u,y): y = D(s)^{-1} N(s) u\}$$

$$\equiv \{(N(s)u, D(s)y)\}$$

This factorization is said to be *coprime* if $\dim E'(s)$ is then always (i.e., even at the eigenvalues of A) equal to the dimension of U. It turns out to be essentially unique.

However, I believe the technique developed here has great potentialities for working in situations where the "coprime factorization" would break down, e.g., for certain infinite dimensional systems. Such potentialities will be explored (in part) in later chapters. Much remains to be done.

Bibliography

1. R. Hermann and C. Martin, "Applications of Algebraic Geometry to Systems Theory", Part I, IEEE Trans. Aut. Control 22 (1977), 19-25.

2. C. Martin and R. Hermann, "Applications of Algebraic Geometry to Systems Theory, Part II: Feedback and Pole Placement for Linear Hamiltonian Systems", Proceedings IEEE 65 (1977), 841-848.

3. R. Hermann, "Applications of Algebraic Geometry to Systems Theory, Part III: The Macmillan Degree as Intersection Number; Part IV: Vector Bundles, Feedback Invariants and Kronecker Indices", preprint Harvard Univ., 1976, submitted to SIAM J. Control; Part V: "Ramifications of the Macmillan Degree"; Part VI: "Infinite Dimensional Linear Systems and Properties of Analytic Functions", Proceedings of the Ames Research Center (NASA) 1976 Conference on Geometric Control, Math Sci Press, Brookline, Mass.

4. C. Martin and R. Hermann, "A New Application of Algebraic Geometry to Systems Theory", Proceedings of IEEE Conf. on Decision and Control, 1976.

5. E. Gantmacher, Matrix Theory, Chelsea Pub. Co., N.Y.

6. C.C. Macduffee, The Theory of Matrices, Chelsea Pub. Co., N.Y.

7. C. Byrnes, "Realization and Stabilization of Certain Non-Linear Systems", Proceedings 1977 JACC Conference.

8. H. Rosenbrock, State Space and Multivariable Theory, Wiley, N.Y., 1970.

9. W. Wolovich, Linear Multivariable Systems, Springer-Verlag, N.Y., 1974.

10. R. Hermann, Interdisciplinary Mathematics, Vol. 3 (Algebraic Topics in Systems Theory); Vol. 8 (Linear Systems Theory and Introductory Algebraic Geometry); Vols. 9 and 11 (Geometric Structure of Systems-Control Theory and Physics, Parts A and B). Math Sci Press, Brookline, Mass.

11. R. Hermann and C. Martin, Algebro-Geometric and Lie-Theoretic Techniques in Systems Theory, Part A (Interdisciplinary Mathematics, Vol. 13), Math Sci Press, Brookline, Mass.

12. R. Brockett, "Control Theory and Analytical Dynamics", Proceedings of the Ames Research Center (NASA) 1976 Conference on Geometric Control (C. Martin and R. Hermann, eds.), Math Sci Press, Brookline, Mass.

13. C. Martin and R. Hermann, "Linear Systems with Structure Groups and their Feedback Equivalence", Proceedings 1977 JACC Conference.

14. J. Milnor and J. Stasheff, Characteristic Classes, Princeton Univ. Press, 197 .

15. R. Hermann, Vector Bundles in Mathematical Physics, Vols. I and II, W. A. Benjamin, Reading, Mass., 1970.

16. N. Steenrod, The Topology of Fiber Bundles, Princeton Univ. Press, 1951.

17. A. Grothendieck, Amer. J. Math. $\underline{79}$ (1957), 121-138.

18. D. Husemoller, Fiber Bundles, Second Edition, Springer-Verlag, 1974.

19. F. Hirzebruch, Topological Methods in Algebraic Geometry, Springer-Verlag, 1966.

Chapter 8

THE KRONECKER THEORY, VECTOR BUNDLES,
AND INPUT-OUTPUT SYSTEMS

1. INTRODUCTION

The Kronecker theory is basically the study of pairs (A_0, A_1) of linear maps $V \to W$ between finite dimensional vector spaces. It gives information about two types of algebraic objects:

The orbits of $GL(V) \times GL(W)$ on $L(V,W) \times L(V,W)$

The vector bundles $s \to E(s)$, $s \to E^d(s)$ whose fibers are the kernels of the maps $A_0 + sA_1$ and $A_0^d + sA_1^d$.

We shall now see how this information may be used in the study of input-output systems.

2. THE KRONECKER INVARIANTS OF INPUT SYSTEMS

If $A_0, A_1 \in L(V,W)$ are a pair of linear maps, the *Kronecker invariants* are numbers attached to the orbits of $GL(V) \times GL(W)$ acting on $L(V,W)$. We shall now associate a pair to an input system, and describe the relations between the Kronecker invariants and the "natural" systems theoretic invariants.

Let X, U be vector spaces with

$$\frac{dx}{dt} = Ax + Bu , \qquad (2.1)$$

$$A \in L(X,X), \quad B \in L(U,X)$$

a given input system. As usual, we shall work with its Laplace transformed version:

$$sx = Ax + Bu . \qquad (2.2)$$

Associate with (2.2) the following pairs of vector spaces

$$V = X \oplus U$$

$$W = X .$$

Denote a point of V by

$$\begin{pmatrix} x \\ u \end{pmatrix} .$$

Then (2.2) can be written as follows:

$$s(1,0)\begin{pmatrix} x \\ u \end{pmatrix} = (A,0)\begin{pmatrix} x \\ u \end{pmatrix} + (0,B)\begin{pmatrix} x \\ u \end{pmatrix} .$$

Rewrite these equations as:

$$sA_1\begin{pmatrix} x \\ u \end{pmatrix} + A_0\begin{pmatrix} x \\ u \end{pmatrix} ,$$

with:

$$A_0\begin{pmatrix} x \\ u \end{pmatrix} = -Ax - Bu$$

$$A_1\begin{pmatrix} x \\ u \end{pmatrix} = \begin{pmatrix} sx \\ 0 \end{pmatrix}$$

A more convenient procedure is to work with the usual formula of "partitioned matrices", defining

$$A(s) = (A - sI, B) \qquad (2.3)$$

(This is classically called a "pencil" of linear maps.) Suppose now that

$$(A',B') \qquad (2.4)$$

is another system. Associate with it another "pencil":

$$A'(s) = (A' - sI, B') .$$

The system (2.1) and (2.4) are *Kronecker equivalent* if there is a pair

$$\gamma \in GL(W)$$

$$\beta \in GL(V)$$

such that:

$$A(s) = A'(s) \qquad (2.5)$$

for all $s \in \mathbb{C}$. In order to see what (2.5) means, write β as a partitioned matrix:

$$\beta = \begin{pmatrix} \beta_{11} & \beta_{12} \\ \beta_{21} & \beta_{22} \end{pmatrix}$$

with

$$\beta_{11} \in L(X,X) \ ; \qquad \beta_{12} \in L(U,X)$$

$$\beta_{21} \in L(X,U) \ ; \qquad \beta_{22} \in L(U,U)$$

$$\gamma((A-sI),B) = (A'-sI,B') \begin{pmatrix} \beta_{11} & \beta_{12} \\ \beta_{21} & \beta_{22} \end{pmatrix}$$

$$= ((A'-sI)\beta_{11} + B'\beta_{21}, (A'-sI)\beta_{12} + B'\beta_{22})$$

or

$$\gamma(A-sI) = (A'-sI)\beta_{11} + B'\beta_{21}$$

$$\gamma B = (A'-sI)\beta_{12} + B'\beta_{22} \ . \tag{2.6}$$

Equate coefficients of s on both sides of (2.6):

$$\gamma = \beta_{11} \tag{2.7}$$

$$\beta_{12} = 0 \ . \tag{2.8}$$

Insert (2.7) and (2.8) back into (2.6)

$$\beta_{11} A = A'\beta_{11} + B'\beta_{21}$$

$$\beta_{11} B = B'\beta_{22} \ ,$$

or

$$A = \beta_{11}^{-1} A' \beta_{11} + \beta_{11}^{-1} B' \beta_{21}$$

$$B = \beta_{11}^{-1} B' \beta_{22} \ . \tag{2.9}$$

Conditions (2.9) can be interpreted in a group-theoretical way. Let Σ be the space of all systems (2.1), i.e., the space of all pairs (A,B). We then have three groups acting on Σ. The first is $GL(X)$:

$$(A,B) \to (gAg^{-1}, gB) \qquad (2.10)$$

for $g \in GL(X)$

This is the group of *algebraic state equivalence*. The second group is $GL(U)$:

$$(A,B) \to (A, Bg^{-1}) \qquad (2.11)$$

for $g \in GL(U)$.

The third is *state feedback*:

$$(A,B) \to (A - BF, B) \qquad (2.12)$$

for $F \in L(X,U)$.

Now, the action of the first two groups commutes with each other. Thus, they define an action of the direct product group

$$GL(X) \times GL(U)$$

on : $L(X,U) \times L(X,U)$

$$(g_1, g_2)(A,B) = (g_1 A g_1^{-1}, g_1 B g_2^{-1}) \qquad (2.13)$$

for $g_1 \in GL(X)$, $g_2 \in GL(U)$

Let us see how this fits in with the action of the feedback group. Write:

$$F(A,B) = \text{right hand side of } (2.12) \quad .$$

Then,

$$(g_1, g_2)(F)(A,B) = (g_1(A-BF)g_1^{-1}, g_1 B g_2^{-1})$$
$$= (g_1 A g_1^{-1} - g_1 BF g_1^{-1}, g_1 B g_2^{-1}) \qquad (2.14)$$

Set:

$$F' = g_2 F g_1^{-1} \quad . \qquad (2.15)$$

Then,

$$F'(g_1, g_2)(A,B) = F'(g_1 A g_1^{-1}, g_1 B g_2^{-1})$$
$$= (g_1 A g_1^{-1}, -g_1 B g_2^{-1} F', g_1 B g_2^{-1})$$
$$= (g_1 A g_1^{-1}, -g_1 BF g_1^{-1}, g_1 B g_2^{-1}) \quad .$$

KRONECKER THEORY

Hence:

$$(g_1, g_2)(F) = F'(g_1 g_2) \quad . \tag{2.16}$$

This proves:

Theorem 2.1. The composite G of the three transformation groups defined above acting on Σ is the semidirect sum of $GL(X) \times GL(U)$ and the abelian feedback group (2.12). Two systems are--as Kronecker pencils--strongly equivalent if and only if they lie on the same orbit under G.

Remark. This group action is of interest from the point of view of general transformation group theory. For it is an example of a group which has orbits which are parameterized with a combination of "discrete" and continuous parameters. We shall now investigate this point.

3. INPUT SYSTEM OF KRONECKER TYPE

Consider an input system of the usual type

$$sx = Ax + Bu \tag{3.1}$$

$$x \in X, \quad u \in U \quad .$$

Set:

$$V = X \oplus U ; \quad W = X$$

$$A_0(x \oplus u) = Ax + Bu$$

$$A_1(x \oplus u) = -x$$

Thus, (3.1) is written in the form

$$(A_0 + sA_1)(x \oplus u) = 0 \quad .$$

Definition. The input system (3.1) is said to be of *Kronecker type* if there is a map $\alpha : V \to V$ such that

$$A_1 = A_0 \alpha \tag{3.2}$$

$$\alpha^n = 0, \quad \text{for } n \text{ sufficiently large} \tag{3.3}$$

Remark. This type is suggested by the Kronecker theory, particularly Theorem 5.2 of the previous chapter. It might also be given the name "canonical" because it is related to the various "canonical forms" that are customarily used in algebraic systems theory.

Let us now see more explicitly what condition (3.2) means. Suppose that

$$\alpha(x \oplus u) = (\alpha_{11} x + \alpha_{12} u) \oplus (\alpha_{21} x + \alpha_{22} u) \tag{3.4}$$

where $\alpha_{11}, \ldots, \alpha_{22}$ are linear maps of the appropriate domain and range.

Remark. This notation is more understandable in terms of the more classical notions of "partitioned matrices":

$$\alpha \begin{pmatrix} x \\ u \end{pmatrix} = \begin{pmatrix} \alpha_{11} & \alpha_{12} \\ \alpha_{21} & \alpha_{22} \end{pmatrix} \begin{pmatrix} x \\ u \end{pmatrix} . \tag{3.5}$$

In view of the definition of A_0, A_1, we can rewrite (3.2) as follows:

$$-x = A_0 \alpha \begin{pmatrix} x \\ u \end{pmatrix}$$

$$= A(\alpha_{11} x + \alpha_{12} u) + B(\alpha_{21} x + \alpha_{22} u)$$

for all $x \in X$, $u \in U$.

This leads to the following relations between maps:

$$-1 = A\alpha_{11} + B\alpha_{21} \tag{3.6}$$

$$0 = A\alpha_{12} + B\alpha_{22} \tag{3.7}$$

4. THE INPUT VECTOR BUNDLE FOR KRONECKER INPUT SYSTEMS

Let

$$sx = Ax + Bu \tag{4.1}$$

be an input system (in algebraic form). For $s \in \mathbb{C}$, let

$$E(s) = \{(x,u) \in X \oplus U : Ax + Bu = sx\} \tag{4.2}$$

KRONECKER THEORY

The assignment $s \to E(s)$ of a vector space to point $s \in S^2$ defines a vector bundle over S^2, called the *input bundle*.

Set

$$V = X \oplus U ; \qquad W = X$$

$$A_0(x \oplus u) = Ax + Bu \tag{4.3}$$

$$A_1(x + u) = -x . \tag{4.4}$$

Thus, (4.1) can be written as:

$$(A_0 + sA_1)(x \oplus u) = 0 . \tag{4.5}$$

Let us *suppose* that the system are of *Kronecker type*, in the sense that there is a linear map

$$\alpha : V \to X$$

such that

$$A_1 = -A_0 \alpha ; \qquad \alpha^n = 0$$

As we have seen already, such an Ansatz enables us to construct the input bundle explicitly:

$v \in E(s)$ if and only if

$$(A_0 - sA_0 \alpha)(v) = 0$$

or

$$A_0(1 - s\alpha)(v) = 0 ,$$

or

$$(1 - s\alpha)^{-1}(v) \in \text{kernel } A_0 , \tag{4.6}$$

Hence:

Theorem 4.1. If the input system is of Kronecker type, the input bundle $s \to E(s)$ is determined by the following formula

$$E(s) = (1 + s\alpha + s^2 \alpha^2 + \cdots + s^{n-1} \alpha^{n-1})(\text{kernel } A_0) \tag{4.7}$$

Proof. Since α is nilpotent, $(1 - s\alpha)^{-1}$ is given by the geometric series $1 + s^2 \alpha + s^2 \alpha^2 + \cdots$ *cut off at a finite stage*. This, combined with (4.6), leads to (4.7).

Remark. Notice how nicely (4.7) gives a "formula" for the input bundle and exhibits its "algebraic" nature.

Theorem 4.2. If the system (4.1) is *controllable*, then the input bundle (assumed Kroneckerian) always has a fiber $E(s)$ whose dimension is equal to the dimension of U, *including the point* $s = \infty$.

Proof. Actually, a proof has already been given for this result in the previous chapter. However, it is instructive and convenient to prove it again with the methods being developed here.

First, look at the point $s = \infty$. In order to handle it, set

$$\tau = \frac{1}{s} .$$

Then,

$$E(s) = \text{set of } (x,u) \text{ such that } \frac{1}{\tau} x = Ax + Bu ,$$

or

$$x = \tau(Ax + Bu) .$$

Thus,

$$E(\infty) = E(s)_{\tau=0} = \{(x,u) : x = 0\}$$

$$= 0 \times U .$$

This shows that

$$\dim E(\infty) \equiv \dim U .$$

Formula (4.7) shows that:

$$\dim E(s) = \dim E(0)$$

for all $s \in \mathbb{C}$.

Hence, we must calculate $\dim E(0)$. Now,

$$E(0) = \{(x,u) : A_0(x,u) \equiv Ax + Bu = 0\}$$

$$\equiv \text{kernel } A_0 .$$

Now, the dual map A_0^d maps $X^d \to V^d \equiv X^d \oplus U^d$.

$$A_0^d(x^d)(x \oplus u) = x^d(A_0(x \oplus u))$$
$$= x^d(Ax + Bu)$$
$$= (A^d x^d)(x) + (B^d x^d)(u) \quad ,$$

i.e.,

$$\boxed{A_0^d(x^d) = (A^d x^d) \oplus (B^d x^d)} \qquad (4.8)$$

The condition

$$\dim(\text{kernel } A_0) = \dim U$$

is equivalent to

$$A_0(X \oplus U) = X \quad ,$$

or dualizing, that:

$$\boxed{A_0^d \text{ is one-one}} \qquad (4.9)$$

Let us prove (4.9) (with the controllability hypothesis) to finish the proof of Theorem 4.2.

Suppose there is a $x^d \in X^d$ such that $A_0^d(x^d) = 0$, i.e.,

$$A^d x^d = 0 = B^d x^d \quad .$$

Then,

$$B^d (A^d)^n (x^d) = 0$$

for all integers n.

This means that

$$x^d (BU + ABU + \cdots) = 0 \quad ;$$

controllability of (A,B) now implies that $x^d = 0$.

Chapter 9

THE KRONECKER THEORY

1. INTRODUCTION

As has been indicated already in Interdisciplinary Mathematics, Volumes 3 and 9, Kronecker's theory of "pencils" of matrices is the algebraic foundation of much of the standard theory of linear, time-invariant, finite dimensional systems modelled by ordinary differential equations. Unfortunately, the Kronecker theory has almost disappeared from contemporary algebra--the only complete reference is in Gantmacher's marvelous book. In fact, the recent theory of "quivers" due to P. Gabriel shows that the Kronecker theory is very isolated and exceptional from a certain general invariant-theoretic point of view.

In this chapter I want to develop the relations between the Kronecker theory, the theory of vector bundles over the Riemann sphere, and linear systems theory. We shall see that there are in fact close and interesting interrelations.

2. PAIRS OF LINEAR MAPS FROM THE KRONECKER POINT OF VIEW

Let V, W be finite dimensional vector spaces over a field of complex numbers as scalars. $L(V,W)$ denotes the space of linear maps: $V \to W$. $GL(V)$ and $GL(W)$ denote the group of linear automorphisms of V and W. The Kronecker theory involving a pair

$$(A_0, A_1) \in L(V,W) \times L(V,W)$$

of linear maps. Two such pairs $(A_0, A_1), (A_0', A_1')$ are said to be *equivalent* if there is a pair $(g_1, g_2) \in GL(W) \times GL(V)$ such that:

$$A_0' = g_1 A_0 g_2^{-1}$$
$$A_1' = g_1 A_1 g_2^{-1}$$
(1.1)

The formula (1.1) defines a group action of $GL(W) \times GL(V)$ on $L(V,W) \times L(V,W)$. (1.1) means that the two elements lie on the same orbit. The Kronecker theory describes these orbits.

Let A_0, A_1 be a pair of linear maps. Here are some basic concepts.

Definition. The pair (A_0, A_1) is said to be *regular* if *either* A_0 or $A_0 + sA_1$ for some $s \in \mathbb{C}$, is an isomorphism of $V \to W$.

Definition. The Kronecker pair (A_0, A_1) is said to be *irreducible* if there is no proper pair of subspaces $V' \subset V$, $W' \subset W$ such that

$$A_0(V') \subset W' \; ; \quad A_1(V') \subset W' \quad .$$

The Kronecker pair (A_0, A_1) is said to be *decomposable* if there are direct sum decompositions

$$V = V' \oplus V''$$

$$W = W' \oplus W''$$

such that

$$A_0(V') \subset W' \; ; \quad A_0(V'') \subset W''$$

$$A_1(V') \subset V' \; ; \quad A_1(V'') \subset W''$$

Given a Kronecker pair $(A_0, A_1) \in L(V,W) \times L(V,W)$, one can form the *dual pair*

$$(A_0, A_1)^d \equiv (A_0^d, A_1^d) \in L(W^d, V^d) \times L(W^d, V^d) \quad .$$

3. THE KRONECKER VECTOR BUNDLE DEFINED BY A PAIR OF LINEAR MAPS

Let

$$A_0, A_1 : V \to W$$

be a pair of linear maps between finite dimensional complex vector spaces. Let s denote a complex number. Let

$$E = \{(s,v) \in \mathbb{C} \times V : (A_0 + sA_1)(v) = 0\} \tag{3.1}$$

Map $E \to \mathbb{C}$ as $(s,v) \to s$. This defines E as a vector bundle over $\mathbb{C}$. Its fiber is $E(s) = \{v \in V : (A_0 + sA_1)(v) = 0\}$. Extend this to the Riemann sphere, S^2, i.e., to the point $s = \infty$, as follows

$$E(\infty) = \{v \in V : A_1(v) = 0\} \quad . \tag{3.2}$$

We will call this the *Kronecker vector bundle* defined by the pair (A_0, A_1).

This vector bundle may be singular, i.e., its fiber may not have constant dimension in s. One can also form the bundle associated with the dual Kronecker pair (A_0^d, A_1^d).

4. POLYNOMIAL CROSS-SECTIONS OF THE KRONECKER VECTOR BUNDLE

Let $A_0, A_1 \in L(V,W)$ be a pair of linear maps, and let E be the Kronecker vector bundle defined by the pair. It is a vector bundle with the Riemann sphere S^2 as base, coordinatized by s. For $s \in S^2$, the fiber $E(s)$ is the space of vectors $v \in V$ such that

$$(A_0 + sA_1)(v) = 0 \quad.$$

Let $\mathbb{C} \subset S^2$ be the subset of S^2 consisting of the points s which are at *finite* distance from the origin. (Thus, S^2 is $\mathbb{C}$ plus the point at infinity.) A *cross-section*

$$\gamma: \mathbb{C} \to E$$

is a map such that

$$\gamma(s) \in E(s)$$

for all $s \in \mathbb{C}$. Since E over $\mathbb{C}$ is a subset of $\mathbb{C} \times V$, we know what it means for such a γ to be a *polynomial map*. Explicitly, such a polynomial cross-section is determined by a map

$$s \to \underline{v}(s)$$

such that:

a) $\underline{v}(s)$ is of the form

$$\underline{v}(s) = v_n x^n + v_{n-1} s^{n-1} \cdots + v_0 \quad (4.1)$$

with $v_0, \ldots, v_n \in V$.

b) $(A_0 + sA_1)(\underline{v}(s)) = 0 \quad (4.2)$

for all $s \in \mathbb{C}$.

The integer n is called the *degree* of the polynomial cross-section.

<u>Theorem 4.1.</u> The Kronecker vector bundle has a nonzero polynomial cross-section if and only if the following condition is satisfied:

> The kernel of the map
>
> $$A(s) = A_0 + sA_1 : V \to W \quad (4.3)$$
>
> is nonzero for *each* $s \in \mathbb{C}$.

Proof. Suppose that

$$\underline{v}(s) = v_n s^n + \cdots + v_0 ,$$

with $v_0, \ldots, v_n \in V$

is such a polynomial cross-section. Let n be the minimal number that can appear in this way. We know (IM, Vol. 3, p. 63) that *the vectors $v_0, \ldots, v_n$ are linearly independent*. Hence, $\underline{v}(s)$ is nonzero for *each* $s \in \mathbb{C}$, and lies in the kernel of $A(s)$.

Conversely, suppose that

$$\text{kernel } A(s) \neq 0$$

for all values of s. For each $s_0 \in \mathbb{C}$, we have

$$\dim A(s)(V) \geq \dim A(s_0)(V) \quad (4.4)$$

for all $s \in \mathbb{C}$ which is sufficiently close to s_0.

(The proof of (4.4) goes as follows. Let $v_1', \ldots, v_m'$ be elements of V such that $A(s_0)(v_1'), \ldots, A(s_0)(v_m')$ are a basis for $A(s_0)(V)$. Then, by continuity, $A(s)(v_1'), \ldots, A(s)(v_m')$ are linearly independent for s sufficiently close to s_0', hence $\dim A(s)(V) \geq m$. (4.4) follows.)

Since $\dim V = \dim \ker A + \dim A(V)$, we have

$$\dim \ker A(s) \leq \dim \ker A(s_0) \quad (4.5)$$

for all $s \in \mathbb{C}$ sufficiently close to s_0.

Now, *choose* s_0 so that

$$\dim \ker A(s_0) \leq \dim \ker A(s) \quad (4.6)$$

for all $s \in \mathbb{C}$

(i.e., chose s_0 so that this integer valued function takes its minimum).

Combining with (4.6), we see that we have found a $s_0 \in \mathbb{C}$ such that:

$$\boxed{\begin{array}{l} \dim \text{kernel } A(s) = \dim \text{kernel } A(s_0) \\ \text{for all } s \text{ sufficiently close to } s_0. \end{array}} \qquad (4.7)$$

Now, choose a linear subspace V' of V so that:

$$V = \text{kernel } A(s_0) \oplus V' \quad .$$

Thus,

$$V = \text{kernel } A(s_0) \oplus V'$$

for all s sufficiently close to s.

Let $\pi(s): V \to V$ be the projection of V on $A(s)$.

Exercise. Show that $\pi(s)$ is a *rational* function of s. (Hint: Note that it is determined by solving linear equations with coefficients which are polynomials in s; namely $\pi(s)\pi(s) = \pi(s)$; $A(s)\pi(s) = 0$.) Hence, for $v_0 \in V$.

$$\underline{v}(s) = \pi(s)(v_0)$$

determines a map $s \to E(s)$, i.e., a cross-section of the Kronecker bundle. This rational map can be multiplied by a scalar polynomial to make it a polynomial map. Thus, we have finished the proof of Theorem 4.1.

5. THE LINE BUNDLES DETERMINED BY THE KRONECKER REDUCTION PROCESS

Let V and W continue as finite dimensional complex vector spaces and let

$$A_0, A_1: V \to W$$

be linear maps. Let s denote a complex variable, and let, for $s \in \mathbb{C}$,

$$E(s) = \{v \in V: (A_0 + sA_1)(v) = 0\}$$

i.e., $s \to E(s)$ is the *Kronecker bundle* of the pair. $E(\infty) = \{v \in V: A_1(v) = 0\}$. E so defined is a vector bundle over $S^2 = \mathbb{C} \cup (\infty)$.

Now, by the argument given in Section 4, *either* of the following mutually exclusive alternatives holds

$$E(s) = 0, \quad \text{for all but possibly a finite number of } s \qquad (5.1)$$

$$E(s) \neq 0, \quad \text{for all } s \in \mathbb{C} . \qquad (5.2)$$

(Exercise. Prove this statement.)

It is the case (5.2) in which we are interested here. Assuming (5.2), by Theorem 4.1 there is a polynomial cross-section

$$s \to \underline{v}(s) = s^n v_n + \cdots + v_0 \qquad (5.3)$$

with $v_0, \ldots, v_n \in V$, i.e.,

$$(A_0 + sA_1)(\underline{v}(s)) = 0 \qquad (5.4)$$

for all $s \in \mathbb{C}$.

Let us choose n as the minimal integer n that can appear in a relation like (5.4). We know that if n is chosen in this way, then

$$v_0, \ldots, v_n \text{ are } linearly\ independent \qquad (5.5)$$

Let us make (5.4) explicit:

$$0 = (A_0 + sA_1)(v_n s^n + \cdots + v_0)$$

$$= A_0 v_n s^n + \cdots + A_0 v_0 + A_1 v_n s^{n+1} + A_1 v_{n-1} s^n + \cdots + sA_1 v_0$$

Thus, (5.4) is equivalent to the following relations:

$$\boxed{\begin{aligned} A_1 v_n &= 0 = A_0 v_0 \\ A_1 v_{n-1} + A_0 v_n &= 0 \\ &\vdots \\ A_1 v_0 + A_0 v_1 &= 0 \end{aligned}} \qquad (5.6)$$

We can interpret these relations in the following "basis-independent" way: Let V_1 be the linear subspace spanned by the vectors $v_0, \ldots, v_n$. By (5.5), we know that:

$$\dim V_1 = n + 1 \qquad (5.7)$$

Define a linear map

$$\alpha_1 : V_1 \to V_1$$

by the following formula:

$$\begin{aligned} \alpha_1(v_0) &= -v_1 \\ \alpha_1(v_1) &= -v_2 \\ &\vdots \\ \alpha_1(v_{n-1}) &= -v_n \\ \alpha_1(v_n) &= 0 \end{aligned} \qquad (5.8)$$

Comparing (5.6) and (5.7), we then have:

$$A_1(v) = A_0 \alpha_1(v) \qquad (5.9)$$
$$\text{for all } v \in V_1$$
$$\alpha_1^n = 0 \qquad (5.10)$$

Also,

$$\text{kernel } \alpha_1 = (v_n) ,$$

i.e.,

$$\dim \alpha_1(V_1) = n . \qquad (5.11)$$

Set:

$$W_1 = A_0(V_1) \qquad (5.12)$$

Using (5.9) we also have:

$$W_1 \supset A_1 V_1 . \qquad (5.13)$$

Thus, $V_1 \subset V$, $W_1 \subset W$ are linear subspaces of V, W such that

$$A_0(V_1) \subset W_1$$
$$A_1(V_1) \subset W_1$$

i.e., the pair (V_1, W_1) decomposes the pair (A_0, A_1). A basic Kronecker theorem (see Gantmacher and Interdisciplinary Mathematics, Volume 9 for a further discussion) is that the process can be continued, i.e., there are linear subspaces $V_1' \subset V$, $W_1' \subset W$ such that

$$V = V_1 \oplus V_1'$$

$$W = W_1 \oplus W_1',$$

$$A_0(V_1') \subset W' \quad ; \quad A_1(V_1') \subset W_1'.$$

Let $s \to \underline{v}'(s)$ be a polynomial of minimal degree with coefficients in V_1' such that

$$(A_0 + sA_1)(\underline{v}'(s)) \equiv 0.$$

Again, we know that the coefficients of this polynomial span a linear subspace V_2 of V; there is a linear map

$$\alpha_2 : V_2 \to V$$

such that

$$\boxed{\begin{array}{l} A_1(v) = A_0 \alpha_2(v) \\ \text{for all } v \in V_2 \\ \alpha_2^{n'} = 0 \end{array}}$$

Continuing in this way, we decompose V and W into linear subspaces

$$V = V_1 \oplus \cdots \oplus V_m \oplus V_{m+1}$$

$$W = W_1 \oplus \cdots \oplus W_m \oplus W_{m+1}$$

such that:

$$\boxed{(A_0 + sA_1)(V_i) \subset W_i}$$

for $1 \leq i \leq m+1$; all s. There are linear maps

$$\alpha_i : V_i \to V_i, \qquad 1 \leq i \leq m,$$

$$(A_1 - A_0 \alpha_i)(V_i) = 0$$

for $1 \leq i \leq m$

$(A_0 + sA_1)$ restricted to V_{m+1} is one-one for all but a finite number of s's.

We can now use these basic results to decompose the Kronecker vector bundle. Recall that

$$E(s) = \{v \in V : (A_0 + sA_1)(v) = 0\}$$

$E: s \to E(s)$ is the *Kronecker vector bundle*. Set:

$$E_1'(s) = \{v \in V_1 : (A_0 + sA_1)v = 0\}$$

Thus, if $v \in E_1(s)$,

$$\begin{aligned} 0 &= (A_0 + sA_1)(v) \\ &= (A_0 + sA_0 \alpha_1)(v) \\ &= A_0(1 + s\alpha_1)(v) \quad, \end{aligned}$$

or

$$(1 + s\alpha_1)(v) \in \text{kernel } A_0 \quad.$$

Since α_1 is nilpotent, $1 + s\alpha_1$ is isomorphic for all s, i.e.,

$$\dim E_1(s) = \dim \text{kernel } A_0 = 1$$

Then, $s \to E_1'(s)$ is a *subbundle* of E whose fibers have dimension one. Such an object is called a *line bundle*. The integer n is called its *degree* or *Kronecker index*. (In a later chapter we shall describe the relation between this algebraic invariant and the "topological" object--the "characteristic class"--called the "degree". It is one of the basic results of the subject that the "topological" invariant equals the "algebraic" one!

In general, there is no á priori reason that this line subbundle "splits", i.e., that there should be another subbundle E" of E such that

$$E = E' \oplus E" \qquad (5.14)$$

((5.14) is the "Whitney sum" [1]. This means that $E(s) = E'(s) \oplus E"(s)$ for all $s \in \mathbb{C}$.) However, as proved by Grothendieck [2] it is a special property of (holomorphic or "algebraic") vector bundles over the Riemann sphere as base that this in fact happens. In our special situation--bundles defined by pairs of linear maps--such a splitting is in fact implied by Kronecker's original work. We can now sum up as follows.

<u>Theorem 5.1.</u> Let $A_0, A_1: V \to W$ be a pair of linear maps. For $s \in S_1^2$ let

$$E(s) = \{v \in V: (A_0 + sA_1)(v) = 0\} \qquad .$$

Then, the vector bundle $s \to E(s)$ is a direct sum of line bundles, plus a bundle whose fiber is nonzero at at most a finite number of points of S^2.

Here is another version of the Kronecker theory at the level of "pure" linear algebra.

<u>Theorem 5.2.</u> Let $A_0, A_1: V \to W$ be linear maps between finite dimensional vector spaces. (The scalar field may now be arbitrary.) Then, there is a direct sum decomposition:

$$V = V' \oplus V", \quad W = W' \oplus W"$$

with the following properties:

$$A_0(V') \subset W'$$

$$A_1(V') \subset W'$$

$$A_0(V") \subset W"$$

$$A_1(V") \subset W" \qquad .$$

$A_0 + sA_1$ maps $V" \to W"$ in a one-one way for all but possibly a finite number of s. There is a map $\alpha: V' \to V'$ such that

$$A_1 = A_0 \alpha \quad \text{restricted } V', \quad \alpha = 0 \text{ for an integer } n.$$

This way of interpreting the Kronecker theory makes it easy to deduce Theorem 5.1. A polynomial map $s \to \underline{v}'(s)$; $\mathbb{C} \to V'$, such that

$$(A_0 + sA_1)(\underline{v}'(s)) = 0$$

is determined as a solution of

$$(A_0 + sA_0\alpha)(\underline{v}'(s)) = 0 \quad,$$

or

$$A_0((1 + s\alpha)(\underline{v}'(s)) = 0 \quad,$$

or

$$\underline{v}'(s) \in (1 + s\alpha)^{-1}(\text{kernel } A_0) \quad.$$

Now,

$$(1 + s\alpha)^{-1} = 1 - s\alpha + s^2\alpha^2 - \cdots \pm s^{n-1}\alpha^{n-1}$$

if $\alpha^n = 0$. Let

$$E'(s) = \{v' \in V' : (A_0 + sA_1)(v') = 0\} \quad.$$

Then pick an arbitrary element

$$v' \in \text{kernel } A_0 \quad.$$

$$v'(s) = (1 - s\alpha + s^2\alpha^2 - \cdots)(v')$$

is a *formula* for the polynomial cross-section. The one-dimensional subbundles it generates then are those predicted by Grothendieck's structure theorem for holomorphic vector bundles on the Riemann sphere.

Bibliography

1. J. Milnor and J. Stasheff, <u>Characteristic Classes</u>, Princeton Univ. Press.
2. A. Grothendieck, Am. J. Math. <u>79</u> (1957), 121-138.

Chapter 10

TRANSFER FUNCTIONS AND HOLOMORPHIC VECTOR BUNDLES FOR
1-D AND n-D SYSTEMS

1. INTRODUCTION

The geometric methods introduced by Martin and me into System Theory depend on associating with a finite dimensional, time-invariant input-output system a *vector bundle* on the Riemann sphere. Note that the Riemann sphere (the complex numbers with its one-point compactification) is the only *Riemann surface of genus zero*. This point of view is not really so avante-garde as it might sound, since it is very much in the spirit of the one-complex variable methods used by electrical engineers in the 1920's and 1930's. (This is the material that was supplanted by "state space methods" in the 1950's.) We believe that there is considerable potential for applying these methods (enriched by the modern mathematical research) to more general systems. In this chapter I will develop material in this direction for systems described by ordinary *and* partial differential equations.

There is a vast field of engineering *practice* called the "theory of two-dimensional filters", which as yet has no systematization and mathematization (as the "state space" theory systematized and mathematized one-dimensional filters). I believe the ideas presented here will ultimately put us on the road to such a theory. Another interesting possibility is an application of complex manifold methods to the systems of *partial* differential equations which occur in physics. Here again, the reader will notice a theme which pervades my own work--the *unification of systems theory* and *physics*.

2. THE USUAL ONE-DIMENSIONAL TRANSFER FUNCTION AS A MEROMORPHIC DIFFERENTIAL IN THE RIEMANN SURFACE OF GENUS ZERO

Consider a finite dimensional, linear, time-invariant input-output system

$$\frac{dx}{dt} = Ax + Bu$$
$$y = Cx \quad .$$
(2.1)

Convert this into an algebraic equation using the Laplace transform

$$\hat{x}(s) = \int_0^\infty x(t) e^{-st} \, dt \quad ,$$

with zero initial conditions.

$$s\hat{x}(s) = A\hat{x} + B\hat{u}$$

$$\hat{y} = C\hat{x}$$

or

$$\hat{y} = (C(s-A)^{-1}B)\hat{u}$$

$$T(s) = C(s-A)^{-1}B \qquad (2.2)$$

is the *transfer* or *frequency response*. If U and Y are the linear input and output spaces, T is a natural mapping

$$\mathbb{C} \to L(U,Y) \quad .$$

A more appropriate gadget from the Riemann surface is the sequence

$$\theta_n = C(s-A)^{-1}Bs^n ds \qquad (2.3)$$

of differential forms. Let us see how they behave at $s = \infty$. Set:

$$u = \frac{1}{s}$$

$$du = -u^2 ds$$

$$\begin{aligned}
\theta_n &= -C\left(\frac{1}{u} - A\right)^{-1} B u^{-n} u^{-2} du \\
&= -C(1 - Au)^{-1} B u^{-(n+1)} du \\
&= -C(1 + Au + A^2 u^2 + \cdots) B u^{-(n+1)} du \qquad (2.4)
\end{aligned}$$

This shows that the θ_n are *mesomorphic differential forms* on the Riemann sphere. Their *residue* at $u = 0$, i.e., $s = \infty$, is especially important; it is the "Hankel data".

$$\boxed{\text{residue } \theta_n = -CA^n B} \qquad (2.5)$$

We are especially interested in the inverse Laplace transform

$$\int_{a-i\infty}^{a+i\infty} T(s) e^{st} ds = \sum_{n=0}^{\infty} \int \frac{\theta_n}{n!} t^n \quad . \qquad (2.6)$$

This formula describes the input-output relations in terms of objects which have an a priori meaning in terms of Riemann surfaces, the "Abelian Integrals".

$$\int \theta_n$$

It is a well-known and classical idea (in mathematics) that the appropriate way to generalize these ideas is to replace the Riemann sphere with other compact complex manifolds.

3. THE "KRONECKER PENCIL" FORM OF THE INPUT-OUTPUT RELATIONS

Let us rewrite (2.1) as

$$\frac{dx}{dt} - Ax + Bu = 0$$
$$y - Cx = 0 \quad .$$
(3.1)

These equations have the form

$$\alpha_1 \frac{d}{dt} \begin{pmatrix} x \\ u \\ y \end{pmatrix} + \alpha_0 \begin{pmatrix} x \\ u \\ y \end{pmatrix} = 0 \quad , \tag{3.2}$$

where α_1, α_0 are linear maps: $X \oplus U \oplus Y \to X \oplus Y$. The algebraic theory of such pairs (α_1, α_0) of linear maps (and their equivalence under the action of the linear automorphism group on their vector space domains and ranges) is covered by Kronecker's theory of pencils. (Gantmacher's *Theory of Matrices* is the standard reference.) I have given a more "modern" version in "Interdisciplinary Mathematics", Volume 9.

Thus, the *general* problem of the theory of linear, time-invariant, finite dimensional systems is covered by the following equations

$$\alpha_1 \frac{dv}{dt} + \alpha_0 v = 0 \quad , \tag{3.3}$$

where V, W are vector spaces, α_1, α_0 are linear maps $V \to W$. Associated with this is the *pencil*

$$\alpha(s) = \alpha_1 s + \alpha_0 \tag{3.4}$$

of linear maps. In later chapters we shall study such pencils in more detail.

We can assign to the pencil (3.4) a *vector bundle*

$$\pi: E \to P_1(\mathbb{C}) \quad .$$

(Assuming all vector spaces and maps are defined over the complex numbers as field of scalars, $P_1(\mathbb{C})$ is the Riemannian sphere.)

$$E(s) = \{(s,v): s \in \mathbb{C}, v \in V: \alpha(s)v = 0\}$$

It turns out that the basic structure theory for such bundles is closely related to the Kronecker pencil theory *and* to the basic algebra of linear system theory.

4. A GENERALIZATION TO PARTIAL DIFFERENTIAL EQUATIONS

Let $t_1,\ldots,t_n$ be independent variables; V,W are finite dimensional vector spaces.

$$\alpha_0, \alpha_1, \ldots, \alpha_n : V \to W$$

are linear maps. Consider the differential equations:

$$\alpha_1 \frac{\partial v}{\partial t_1} + \cdots + \alpha_n \frac{\partial v}{\partial t_n} + \alpha_0 v = 0 \quad . \tag{4.1}$$

Associated with this we have:

$$s = (s_1, \ldots, s_n) \in \mathbb{C}^n \quad . \tag{4.2}$$

$$\alpha(s) = \alpha_1 s_1 + \cdots + \alpha_n s_n \tag{4.3}$$

$$E = \{(s,v): \alpha(s)(v) = 0\} \tag{4.4}$$

$$\pi : E \to \mathbb{C}^n \quad .$$

$$\pi(s,v) = s \quad .$$

π should be *completed* to be a *vector bundle* over a compact complex analytic manifold X on which $\mathbb{C}^n$ is embedded as an open subset. How this is to be done *in general* will be left open for the moment. Instead, we shall consider some simple examples motivated by physics.

5. AN EXAMPLE FROM CLASSICAL MATHEMATICAL PHYSICS

Suppose given two independent variables t_1, t_2. (In the physical applications, they may be two space variables or one space, one time variable.) Denote partial derivatives of functions of these variables by subscripts. Consider the equation of "Helmholtz" type:

TRANSFER FUNCTIONS   131

$$x_{t_1 t_1} + x_{t_2 t_2} + Ex = u \qquad . \tag{5.1}$$

E is a constant, u scalar functions of t_1, t_2, u is the "input", x the "output". This can, of course, be converted to an equation of type (4.1) by introducing more dependent variables, but it will be (for the moment) more convenient to work directly with Equation (5.1). The "transfer function" is obviously the following rational function of two complex variables s_1, s_2:

$$T(s_1, s_2) = \frac{1}{s_1^2 + s_2^2 + E} \tag{5.2}$$

Thus,

$$x(t_1, t_2) = \iint_Y T(s_1, s_2)\, e^{(s_1 t_1 + s_2 t_2)} \, \hat{u}(s_1, s_2)\, ds_1\, ds_2 \tag{5.3}$$

taken over an appropriately chosen two-dimensional submanifold of $\mathbb{C}^2$, might be a solution of (5.1).

Note that (5.1) is invariant under the rotation group $SO(2,R)$ acting in R^2. This translates into T depending only on $s_1^2 + s_2^2$, as we know explicitly as shown from formula (5.3).

Introduce polar coordinates in the integral (5.3):

$$z^2 = s_1^2 + s_2^2$$

$$s_1 = z \cos\theta, \qquad s_2 = z \sin\theta$$

$$ds_1 \wedge ds_2 = (dz \cos\theta - z \sin\theta\, d\theta) \wedge (dz \sin\theta + z \cos\theta\, d\theta)$$

$$= z\, dz \wedge d\theta \qquad .$$

Set:

$$w = \cos\theta$$

$$dw = -\sin\theta\, d\theta ,$$

$$d\theta = -\frac{1}{\sqrt{1-w^2}}\, dw$$

$$\boxed{ds_1 \wedge ds_2 = \frac{z}{\sqrt{1-w^2}}\, dw \wedge dz} \tag{5.4}$$

Let us now suppose that u *is a function of* z *alone*. Then (5.3) can be rewrite as:

$$x(t_1, t_2) = \iint e^{z(t_1 w + t_2 \sqrt{1-w^2})} \frac{\hat{u}(z)}{z^2 + E} \frac{z}{\sqrt{1-w^2}} \, dw \wedge dz$$

$$= \int \left(\int e^{z(t_1 w + t_2 \sqrt{1-w^2})} \frac{dw}{\sqrt{1-w^2}} \right) \frac{z\hat{u}(z)}{z^2 + E} \, dz \quad (5.5)$$

This formula exhibits the potential "algebro-geometric" nature of the situation. We have the *algebraic correspondence*

$$(s_1, s_2) \to z = \sqrt{s_1^2 + s_2^2} \quad (5.6)$$

of $\mathbb{C}^2 \to \mathbb{C}^1$. (Note that it is *not* a "rational map".) The fibers are the orbits of the orthogonal group $SO(2, \mathbb{C})$. They are the *quadrics*

$$s_1^2 + s_2^2 = z_1^2$$

which are one-dimensional "algebraic curves" of *genus zero*.

In order to get a better idea of the algebro-geometric nature of the integrals (5.5), let us use a power series expansion for the exponential function. Set:

$$x_{n,m} = \iint (zw)^n (z \sqrt{1-w^2})^m \frac{dw}{\sqrt{1-w^2}} \frac{z}{z^2 + E} \hat{u}(z) \, dz \quad (5.7)$$

i.e.,

$$x(t_1, t_2) = \sum_{n,m=0}^{\infty} x_{n_1 n_2} \frac{t_1^n t_2^m}{n! m!} \quad (5.8)$$

Thus, it *might* be appropriate to consider the differential forms

$$\theta_{n,m} = (zw)^n (z\sqrt{1-w^2})^m \frac{dw}{\sqrt{1-w^2}} \frac{z}{z^2 + E} \, dz \quad (5.9)$$

$$= \frac{s_1^n s_2^m}{s_1^2 + s_2^2 + E} \, ds_1 ds_2 \quad (5.10)$$

as the characteristic "geometric objects" attached to the input-output system (5.1).

TRANSFER FUNCTIONS   133

This suggests that we consider separately the differential forms

$$\alpha_{n,m} = (zw)^n \frac{(z\sqrt{1-w^2})^2 \, dw}{\sqrt{1-w^2}} \quad , \tag{5.11}$$

with z considered as a *parameter*. They are essentially differential forms on the Riemann surface whose local variable is w, i.e., the algebraic curve

$$w_1^2 + w_2^2 = 1 \quad .$$

Of course, another point of view is to regard

$$e^{z(t_1 w + t_2 \sqrt{1-w^2})} \frac{dw}{1-w^2}$$

as a differential form on this Riemann surface. (Its indefinite integral can be written down explicitly in terms of Bessel functions.)

This simple example clearly gives us much new material to consider for a *general* theory of systems from a complex manifold-algebraic geometry point of view.

6. VECTOR BUNDLES ON ORBIT SPACES

We can immediately see a general pattern to the example treated in the previous section.

Let X, Y be spaces,

$$\phi : X \to Y$$

a map. Suppose *given* a vector bundle

$$\pi : E \to Y \quad .$$

For $y \in Y$, the fiber $E(y) = \pi^{-1}(y)$ is a vector space. It defines a vector bundle

$$\phi^{-1}(E)$$

on X, called the *pull-back bundle*

$$\phi^{-1}(E) = \{(x,v) : v \in E(\phi(x))\} \quad .$$

There is a commutative diagram

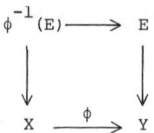

Consider a given bundle E' on X and a map $\phi: X \to Y$, we can *ask whether there is a bundle* E *on* Y *such that*

$$E' = \phi^{-1}(E) \quad .$$

One can also ask whether there is an *equivalent vector bundle* E'' and a bundle E such that

$$E'' = \phi^{-1}(E) \quad .$$

These questions are especially interesting for Systems Theorem (and physicists!) if the following additional structure is put on:

> G is a transformation group on X. $Y = G\backslash X$ is the orbit space, $\phi: X \to G\backslash X = Y$ the map which sends $x \in X$ into the orbit Gx on which it lies. G acts as a group of automorphisms of the vector bundle E'.

7. VECTOR BUNDLES DEFINED BY SYSTEMS OF PARTIAL DIFFERENTIAL EQUATIONS AND LINEAR GROUPS OF SYMMETRIES

Let V, W be complex vector spaces, n an integer, and let $\alpha_0, \alpha_1, \ldots, \alpha_n : V \to W$ be linear maps. We can then construct the system of *linear constant coefficient* partial equations:

$$\alpha_1 \frac{\partial v}{\partial t_1} + \cdots + \alpha_n \frac{\partial v}{\partial t_n} - \alpha_0 v = 0 \tag{7.1}$$

to be solved for a function

$$t = (t_1, \ldots, t_n) \to v(t) \quad ,$$

i.e., a map $R^n \to V$.

We can then associate with this system the n-complex variable "pencil"

$$\alpha(s) = \alpha_1 s_1 + \cdots + \alpha_n s_n + \alpha_0 \tag{7.2}$$

of linear maps: $V \to W$, and the *holomorphic vector bundles*

$$E = \{(s, v): s = (s_1, \ldots, s_n) \in \mathbb{C}^n; \ \alpha(s)v = 0\} \tag{7.3}$$

TRANSFER FUNCTIONS   135

Now, let G be a group. Suppose given *three* linear actions on $\mathbb{C}^n$, V and W.

Definition. G acts as a *symmetry group* of the pencil $\alpha(s)$ if the following condition is satisfied

$$g\alpha(s)g^{-1} = \alpha(g(s)) \tag{7.4}$$

for all $g \in G$.

such an action determines an action of G on the vector bundle E:

$$g(s,v) = (gs, gv) . \tag{7.5}$$

(If $\alpha(s)v = 0$, note that using (7.4)

$$\alpha(g(s))(gv) = g\alpha(s)(v) = 0 ,$$

so that the action (7.5) really does map E onto itself.)

Thus, we can form the orbit space

$$G\backslash\mathbb{C}^n ,$$

and ask whether the bundle E comes from a bundle on this orbit space. This is clearly an interesting system-theoretic way of defining vector bundles!

8. $SO(3,\mathbb{C})$ -BUNDLES

If G has a known Lie-theoretic structure, we can use Lie group representation theory to analyze the bundles constructed in Section 7. One of the simplest cases (and the most important for physicists) is that where G is the three-dimensional rotation group. For algebraic reasons, we complexify everything.

We are supposing that $n = 3$. Refer to my *Lectures on Mathematical Physics*, Volume II or W. Miller's *Symmetry Groups and their Applications* for the notation and ideas used here.

Thus we suppose that $\alpha_0, \alpha_1, \alpha_2, \alpha_3$ are linear maps: $V \to W$. If

$$s = \begin{pmatrix} s_1 \\ s_2 \\ s_3 \end{pmatrix} ,$$

$$g = \begin{pmatrix} g_{11} & g_{12} & g_{13} \\ g_{21} & g_{22} & g_{23} \\ g_{31} & g_{32} & g_{33} \end{pmatrix}$$

$$g(s) = gs \equiv \text{matrix multiplication}$$

It is convenient to write

$$\alpha = (\alpha_1 \alpha_2 \alpha_3)$$

$$\alpha(s) = (\alpha_1 \alpha_2 \alpha_3)\begin{pmatrix} s_1 \\ s_2 \\ s_3 \end{pmatrix} + \alpha_0 \quad .$$

Suppose g is an orthogonal 3×3 complex matrix, i.e.,

$$g' = g^{-1} \quad .$$

(' denotes transpose of a matrix.) Let

$$\sigma_1 : G \to L(V) \quad ,$$

$$\sigma_2 : G \to L(W)$$

be the given representations of G by linear transformations on V and W.

$$\sigma_2(g)\alpha(s)\sigma_1(g^{-1}) = \alpha(g(s))$$

or

$$\sigma_2(g)(\alpha_1\alpha_2\alpha_3)\sigma_1(g^{-1})\begin{pmatrix} s_1 \\ s_2 \\ s_3 \end{pmatrix} + \sigma_2(g)\alpha_0\sigma_1(g^{-1}) = (\alpha_1\alpha_2\alpha_3)g\begin{pmatrix} s_1 \\ s_2 \\ s_3 \end{pmatrix} + \alpha_0$$

$$= (\alpha_1\alpha_2\alpha_3)(g)\begin{pmatrix} s_1 \\ s_2 \\ s_3 \end{pmatrix} + \alpha_0$$

or

$$\sigma_2(g)\alpha_0\sigma_1(g^{-1}) = \alpha_0 \tag{8.1}$$

$$\sigma_2(g)(\alpha_1\alpha_2\alpha_3)\sigma_1(g^{-1}) = (\alpha_1\alpha_2\alpha_3)g \tag{8.2}$$

Now,

$$L(V,W) \equiv V \otimes W^d$$

(W^d = dual space to W. In fact, all finite dimensional representations of

TRANSFER FUNCTIONS 137

G are self-dual.) The "Clebsch-Gordan" rules for decomposition of tensor product of representations of SO(3) tell how many such independent α's there are.

The *irreducible* (finite dimensional) representations of G are parameterized by an integer $j \geq 0$ ($\equiv spin$). The vector space for the spin j-representation has dimension $(2j+1)$. Thus, V and W can be split up into a direct sum of vector spaces V_j, W_j in each of which G acts via a direct sum of spin j-representation. Using "Clebsch-Gordan" we see that

$$\alpha(s)(V_j) \subset W_{j+1} + W_j + W_{j-1} \qquad (8.3)$$

$$\alpha_0(V_j) \subset W_j \quad ,$$

$$j = 0,1,2,\ldots \quad .$$

Example. Consider the Helmholtz operator with input u:

$$s^2 x + Ex - u = 0 \quad .$$

Set:

$$x_i = s_i x \, , \qquad i = 1,2,3 \quad .$$

Thus,

$$\sum_{i=1}^{3} s_i x_i = -Ex + u$$

or

$$\boxed{\begin{array}{l} s_i x - x_i = 0 \, , \qquad i = 1,2,3 \\[1em] \sum_i s_i x_i + Ex - u = 0 \end{array}} \qquad (8.4)$$

$$V = R^5$$

$$= \left\{ \begin{pmatrix} x_1 \\ x_2 \\ x_3 \\ x \\ u \end{pmatrix} \right\}$$

$$W = R^4 \quad .$$

$$\alpha(s) \begin{pmatrix} x_1 \\ x_2 \\ x_3 \\ x \\ u \end{pmatrix} = \begin{pmatrix} s_1 x - x_1 \\ s_2 x - x_2 \\ s_3 x - x_3 \\ \sum_i s_i x_i + Ex - u \end{pmatrix} \qquad (8.5)$$

Thus we see that V is a direct sum of a one subspace S_1 and two zero subspaces. W is the direct sum of a S_1 and an S_0. Now, Clebsch-Gordan gives:

$$S_1 \otimes S_1 = S_2 \oplus S_1 \oplus S_0$$

$$S_0 \otimes S_1 = S_1$$

i.e., there are intertwining maps

$$S_1 \otimes S_1 \to S_0$$

$$S_0 \otimes S_1 \to S_1 \quad .$$

It is these that are obviously present in (8.4).

9. THE COMPACTIFIED VECTOR BUNDLES DETERMINED BY THE HELMHOLTZ EQUATION

Continue with the situation of Section 8, i.e., the linear differential equations which are invariant under the action of $SO(3,\mathbb{C})$ on $\mathbb{C}^3$. (These equations cover most of those of interest in classical mathematical physics!) For $s \in \mathbb{C}^3$ let

$$E(s) = \{v \in V: \alpha(s)(v) = 0\} \quad . \qquad (9.1)$$

As s varies over $\mathbb{C}^3$, $E(s)$ defines a *vector bundle* whose basis is $\mathbb{C}^3$. $G \equiv SO(3,\mathbb{C})$ acts linearly on E.

We want to examine the algebro-geometric structure of this bundle, and any possible bundles with orbit spaces

$$G \backslash \mathbb{C}^3 \quad .$$

The first step is to "homogenize" Equations (8.2). Introduce "homogeneous coordinates"

$$\tau_0, \tau_1, \tau_2, \tau_3 \quad ,$$

with

TRANSFER FUNCTIONS

$$s_i = \frac{\tau_i}{\tau_0}, \qquad i = 1, 2, 3 \quad .$$

Equations (8.2) take the form

$$\boxed{\begin{array}{c} \tau_i x - x_i \tau_0 = 0 \\ \\ \sum_{i=1}^{3} \tau_i x_i + (Ex - u)\tau_0 = 0 \end{array}} \tag{9.2}$$

These equations *determine a vector bundle* E *whose base is* $P_3(\mathbb{C})$, *which restricts to the given bundle when* $\mathbb{C}^3$ *is embedded on the "affine" subspaces of* $P_3(\mathbb{C})$.

Let us work out the fibers of this bundle. If $\tau_0 \neq 0$, these equations are equivalent to (8.2), and the solutions E( ) obviously form a one-dimensional vector space. The "input-output" relations are

$$\boxed{x = \frac{1}{s^2 + E} u}$$

This gives a "geometric", *systems-theoretic* meaning to $1/(s^2 + E)$ as the *"transfer-function-fundamental solution-Green's function"* for the Helmholtz equation.

For

$$\tau_0 = 0 \quad ,$$

Equations (9.2) take the following form:

$$\begin{array}{c} \tau_i x = 0 \\ \\ \sum_{i=1}^{3} \tau_i x_i = 0 \end{array} \tag{9.3}$$

Since one of the τ_i is nonzero, (9.3) forces $x = 0$. Then, (9.3) reduces to

$$\boxed{\begin{array}{c} x = 0 \\ \\ \sum_{i=1}^{3} \tau_i x_i = 0 \end{array}} \tag{9.4}$$

$E(\tau)$ consists of the (x,u,x_i) such that *only* (9.3) is satisfied. Thus,

$$\boxed{\dim E(\tau) = 3}$$

The vector bundle E does *not* have constant dimension as τ ranges over $P_3(\mathbb{C})$. We see that there is here a distinct difference from the one-dimensional situation!

However, we can construct an *equivalent* bundle E' on $\mathbb{C}^3$ which does have a non-singular extension to $P_3(\mathbb{C})$. Namely,

$$E' = \left\{(x,u,s): x = \frac{1}{s^2 + E} u\right\}$$

Again, put

$$s_i = \frac{\tau_i}{\tau_0} .$$

Thus,

$$E'(\tau) = (x,u): x = \frac{1}{\sum \frac{\tau_i \tau_i}{\tau_0^2} + E} u$$

$$= (x,u): x = \frac{\tau_0^2}{\sum \tau_i \tau_i + \tau_0^2 E} u$$

Thus, if $\tau_0 = 0$,

$$E'(\tau) = \{(x,u): x = 0\} .$$

We see that for *all* $\tau \in P_3(\mathbb{C})$,

$$\boxed{\dim E'(\tau) = 1}$$

We can also examine how E' projects down to the orbit space $SO(3,\mathbb{C}) \setminus \mathbb{C}^3$. Map

$$\phi: \mathbb{C}^3 \to \mathbb{C}$$

TRANSFER FUNCTIONS   141

as follows

$$\phi(s) = s^2 = z \quad.$$

The fibers are the orbits of $SO(3,\mathbb{C})$. We would now like to "projectify" this situation.

$$s_i = \frac{\tau_i}{\tau_0}$$

$$z = \frac{z_1}{z_0}$$

$$\phi(\tau) = \left(\tau_0^2, \sum_{i=1}^{3} \tau_i \tau_i\right) \tag{9.5}$$

Note this map sends one-dimensional linear subspaces of $\mathbb{C}^4$ into one-dimensional linear subspaces of $\mathbb{C}^2$, hence it passes to the equivalent to define a map

$$\phi: P_3(\mathbb{C}) \to P_1(\mathbb{C}) \quad. \tag{9.6}$$

The orbits of $SO(3,\mathbb{C})$ *lie in the fibers of* ϕ.

ϕ and the vector bundle E' are very compatible.

$$E(\tau) = \left\{(x,u): x = \frac{\tau_0^2}{\sum_i \tau_i \tau_i + E\tau_0^2} u\right\}$$

Set:

$$E''((z_0,z_1)) = \left\{(x,u): x = \frac{z_0}{z_1 + Ez_0} u\right\}$$

This formula defines a *non-singular* vector bundle on $P_1(\mathbb{C})$.

$$E' = \phi^{-1}(E'') \quad,$$

i.e., E' is the *pull-back* of the bundle on $P_1(\mathbb{C})$.

Remark. These constructions obviously have a great deal to do with the classical "separation of variables" methods of classical mathematical physics. These methods also have their extensions and ramifications with *relativistic quantum field theory*, which I intend to go into in more detail later on. Here I will do several illustrative examples.

10. THE VECTOR BUNDLES DEFINED BY MAXWELL'S EQUATIONS

Adopt the notation of relativistic physics.

$$x = (x^\mu)$$

denotes a point of R^4, $0 \leq \mu,\nu \leq 3$. Adopt the summation convention on these indices.

$$A_\mu(x)$$

is the *electromagnetic potential*

$$\partial_\mu = \frac{\partial}{\partial x^\mu} .$$

$$F_{\mu\nu} = \partial_\mu(A_\nu) - \partial_\nu(A_\mu) \tag{10.1}$$

is the *electromagnetic field*.

$$g_{\mu\nu} = \begin{cases} c^2 & \text{if } \mu = \nu = 0 \\ 0 & \text{if } \mu \neq \nu \\ -1 & \text{if } 1 \leq \mu = \nu \leq e \end{cases}$$

is the *Minkowski metric tensor*.

$$g^{\mu\nu} = \text{inverse matrix to } g_{\mu\nu}$$

$$F^\mu_\nu = g^{\mu\nu'} F_{\nu\nu'} .$$

$$\boxed{\partial_\nu(F^\nu_\mu) = J_\mu} \tag{10.2}$$

is Maxwell's equation. J_μ is the *current*.

Let us now think of these equations as analogous to the input equations

$$\frac{dx}{dt} = Ax + Bu \tag{10.3}$$

of linear systems theory. Here, (J_μ) is the input. Just as we convert (10.3) into an algebraic form

$$\hat{x}s = A\hat{x} + B\hat{u} \tag{10.4}$$

with s a complex variable, let us do the same with (10.1) - (10.2), introducing four complex variables $(s_\mu) = s$.

$$\hat{F}_{\mu\nu} = s_\mu \hat{A}_\nu - s_\nu \hat{A}_\mu \tag{10.5}$$

$$s_\nu \hat{F}^\nu_\mu = \hat{J}_\mu \tag{10.6}$$

or

$$\hat{J}_\mu = s_\nu g^{\nu\nu'} \hat{F}_{\mu\nu'}$$

$$= s_\nu g^{\nu\nu'} (s_\mu \hat{A}_{\nu'} - s_{\nu'} \hat{A}_\mu)$$

$$= g^{\nu\nu'} s_\nu s_\mu \hat{A}_{\nu'} - s^2 \hat{A}_\mu$$

$$= (\hat{A} \cdot s) s_\mu - s^2 \hat{A}_\mu \tag{10.7}$$

with

$$s^2 = g^{\nu\nu'} s_\nu s_{\nu'} \tag{10.8}$$

$$\hat{A} \cdot s = g^{\nu\nu'} s_\nu \hat{A}_{\nu'} \tag{10.9}$$

We can now treat A_μ, $\hat{J}_\mu$ as components of four-vectors, and rewrite (10.7) as:

$$\boxed{\hat{J} = (\hat{A} \cdot s) s - s^2 \hat{A}} \tag{10.10}$$

We similarly write (10.5) - (10.6) in four-vector notation:

$$\boxed{\begin{array}{c} \hat{F} = s \wedge \hat{A} \\ s \lrcorner\, g(\hat{F}) = \hat{J} \end{array}} \tag{10.11}$$

In (10.11) g is the isomorphism that Minkowski symmetric bilinear form defines between the $\mathbb{C}^4 \wedge \mathbb{C}^4$ and its *dual* $\mathbb{C}^{4d} \wedge \mathbb{C}^{4d}$. $\lrcorner$ is the *contraction*.

The system theorist might now say that (10.11) is the "state space version" of (10.12), *since it is of first degree in the variables* s.

We can now define two vector bundles E,E' over $\mathbb{C}^4$. The fibers above the point $s \in \mathbb{C}^4$ are denoted as E(s), E'(s).

$$E(s) = \{(\hat{A}, \hat{J}): (10.10) \text{ is satisfied}\} \tag{10.12}$$

$$\subset \mathbb{C}^4 \oplus \mathbb{C}^4$$

$$E'(s) = \{(\hat{A},\hat{F},\hat{J}): \text{Equation (10.11) are satisfied}\} \tag{10.13}$$
$$\subset \mathbb{C}^4 \oplus (\mathbb{C}^4 \quad \mathbb{C}^4) \oplus \mathbb{C}^4 \quad.$$

Note that these bundles are *isomorphic*; the isomorphism is the projection

$$(\hat{A},\hat{F},\hat{J}) \to (\hat{A},\hat{J}) \quad.$$

Notice a special feature here: The $\hat{J}$ satisfying (10.10) is not an arbitrary element of $\mathbb{C}^4$, but must satisfy the following condition:

$$\hat{J} \cdot s = 0 \quad. \tag{10.14}$$

Systems theoretically, this means that the "transfer function" relating the input-output data is more complicated algebraically than it is in the one-dimensional case (i.e., $T(s) = C(s-A)^{-1}B$). The "poles" are the values of s for which there are conditions on $\hat{J}$ in addition to (10.14). In fact, if $s^2 \neq 0$, note that the set of A satisfying (10.10) with $\hat{J} = 0$ is one-dimensional, namely those lying in the one-dimensional linear subspace spanned by s. If $s^2 = 0$, they are three-dimensional. Thus,

> The "poles" of the system are the elements $s \in \mathbb{C}^4$ such that $s^2 = 0$. (10.15)

We can now "homogenize" the equations and attempt to extend the vector bundles to $P_4(\mathbb{C})$. Set:

$$s = \frac{\tau}{\tau_0} \quad, \tag{10.16}$$

with $\tau \in \mathbb{C}^4$, $\tau_0 \in \mathbb{C}$. Regard (τ,τ_0) as "homogeneous coordinates" of a point of $P_4(\mathbb{C})$. In this way,

$$s \to (s,1)$$

embedds $\mathbb{C}^4$ as an open subset of $P_4(\mathbb{C})$. The subset

$$\{\tau,0\}$$

is the "hyperplane at ∞"; it is the complement of $\mathbb{C}^4$ in $P_4(\mathbb{C})$.

Insert (10.16) into (10.10) and (10.11), in order to see how the vector bundles E, E' extend to $P_4(\mathbb{C})$.

$$\hat{J} = \left(\hat{A} - \frac{\tau}{\tau_0}\right)\frac{\tau}{\tau_0} - \frac{\tau^2}{\tau_0^2}\hat{A} \quad ,$$

or

$$\tau_0^2 \hat{J} = (\hat{A}\cdot\tau)\tau - \tau^2 \hat{A} \tag{10.17}$$

Now, set $\tau_0 = 0$; resulting in

$$\boxed{(\hat{A}\cdot\tau)\tau - \tau^2 \hat{A} = 0} \tag{10.18}$$

Thus, $E(\tau,0)$ consists of the pairs $(\hat{J},\hat{A})$ with $\hat{J}$ *arbitrary*, but with $\hat{A}$ constrained by the linear relation (10.18). It is of greater dimension than $E(\tau,\tau_0)$, with $\tau_0 \neq 0$. Thus, we see that the vector bundle E can be defined over all of $P_4(\mathbb{C})$, but *its fiber has greater dimension on the hyperplane at infinity*; it is "singular" there.

We can deal in a similar way with E', using the Equations (10.11), "homogenized" to the following form:

$$\boxed{\begin{aligned} \tau_0 \hat{F} &= \tau \wedge \hat{A} \\ \tau \lrcorner\, g(\hat{F}) &= \tau_0 J \end{aligned}} \tag{10.19}$$

Now, for $\tau_0 = 0$, these equations become the following ones:

$$\boxed{\begin{aligned} 0 &= \tau \wedge \hat{A} \\ 0 &= \tau \lrcorner\, g(\hat{F}) \end{aligned}} \tag{10.20}$$

Let us compute the dimension of the fiber. If $\tau_0 \neq 0$, $\hat{F}$ and $\hat{J}$ are determined by $\hat{A}$, hence;

$$\boxed{\text{dimension fiber} = 4}$$

Equation (10.20) is equivalent to

$\hat{A}$ is a linear multiple of τ.

$g(\hat{F})$ lies on $\mathbb{C}^4/(\tau) \approx \mathbb{C}^3$. There are no constraints in $\hat{J}$. Now,

dimension fiber for $\tau_0 = 0$ is

$$1 + \frac{3 \cdot 2}{2} + 2 = 8 \quad .$$

It is clear that a theory of holomorphic vector bundles on $P_4(\mathbb{C})$ with algebraic singularities on hyperplanes would be very important for application to this sort of "systems theory" (and a fortiori for "relativistic physics").

Chapter 11

THE RIEMANN SURFACE AND TRANSFER FUNCTION VECTOR BUNDLE
OF A LINEAR, TIME-VARYING SYSTEM DEFINED IN TERMS OF
LINEAR DIFFERENTIAL OPERATOR SYMMETRIES

1. INTRODUCTION

Modern systems theory has its roots in the electrical engineering of the 1920's and 30's. In turn, this subject was, in its mathematical orientation, heavily influenced by Maxwell-Heaviside electromagnetic theory *and* by the dominance of one-variable complex function theory in the mathematics of the day. (In fact, this is an interesting example of how mathematics can be "relevant" and still first-rate *and* interesting to even the "pure" mind.)

The generation of electrical engineers which matured in the 1950's and 60's then had to fight the usual battle to introduce *"state space"* ideas, replacing the "transfer function" of the earlier generation. Of course there is still much of value in the older methods, and spasmodic attempts have been made in the recent control theory literature to adapt them to more recent concerns.

These older methods were the inspiration for the work by Clyde Martin and myself [3] on algebro-geometric methods in systems theory. We showed that one natural setting for the theory of transfer functions with *rational* functions as coefficients was the theory of *holomorphic vector bundles on the Riemann sphere*. This suggested to us that the theory of holomorphic vector bundles on other complex manifolds should play a role in the study of systems. The aim of this chapter is to describe how holomorphic vector bundles on arbitrary "Riemann surfaces" (i.e., complex one-dimensional manifolds) can be used to describe the *input-output behavior* of linear, *time-varying* systems. This work ties in with recent important work on nonlinear wave equations and their associated linear "inverse scattering" equations.

In order to define these holomorphic bundles, I introduce a new concept--*linear differential operator symmetries of time-varying linear systems*.

2. LINEAR DIFFERENTIAL OPERATOR SYMMETRIES OF LINEAR, TIME-VARYING SYSTEMS

Consider a linear, time-varying system of the usual form:

$$\frac{dx}{dt} = A(t)x + B(t)u$$

$$y = C(t)x \quad .$$

(2.1)

Denote a curve in the vector spaces U, Y, X as $\underline{u}$, $\underline{y}$, $\underline{x}$. Let

$$IO = \{(\underline{u},\underline{y}): \exists \underline{x} \text{ such that } \underline{u},\underline{y},\underline{x} \text{ satisfies (2.1)}\} \quad (2.2)$$

In words, IO is the set of *input-output pairs* of the system (2.1). $\underline{U},\underline{Y}$ denotes the space of $\underline{u},\underline{y}$. It is a vector space. Thus,

$\underline{U} \oplus \underline{Y}$ is also a vector space

IO is a linear subspace of this vector space.

Definition. A *linear, differential operator symmetry* of the system (2.1) is a linear differential operator

$$\alpha: \underline{U} \oplus \underline{Y} \to \underline{U} \oplus \underline{Y}$$

such that

$$\boxed{\alpha(IO) \subset IO} \quad (2.3)$$

3. DIFFERENTIAL OPERATOR SYMMETRIES OF SCALAR INPUT OUTPUT SYSTEMS OF CLASSICAL TYPE

Now, suppose

$$U = Y = R \quad,$$

so that $\underline{U},\underline{Y}$ consists of the (C^∞) real-valued functions of the time variable t. Suppose that (Δ_1, Δ_2) are a pair of linear differential operators such that the input-output relations are of the following form:

$$IO = \{(\underline{u},\underline{y}): \Delta_1 \underline{y} = \Delta_2 \underline{u}\} \quad (3.1)$$

Let

$$\alpha_1: \underline{Y} \to \underline{Y} \quad, \qquad \alpha_2: \underline{U} \to \underline{U}$$

be differential operators. Define $\alpha: \underline{U} \oplus \underline{Y} \to \underline{U} \oplus \underline{Y}$ as follows:

$$\alpha(u \oplus y) = \alpha_2 u \oplus \alpha_1 y \quad. \quad (3.2)$$

Let us look for the conditions that α be a symmetry, i.e., the

$$\alpha A(IO) \subset IO \quad. \quad (3.3)$$

TIME-VARYING SYSTEMS

Theorem 3.1. If α_1, α_2 satisfy the following conditions:

$$\alpha_1 \Delta_1 = \Delta_1 \alpha_1 \qquad (3.4)$$

$$\Delta_2 \alpha_2 = \alpha_1 \Delta_2 \qquad (3.5)$$

then α defined by (3.2) is an input-output symmetry.

Proof. Let $(\underline{u}, \underline{y}) \in IO$, i.e.,

$$\Delta_1 \underline{y} = \Delta_2 \underline{u} \quad . \qquad (3.6)$$

Then

$$\Delta_1(\alpha_1 \underline{y}) - \Delta_2(\alpha_2 \underline{u}) = \quad , \text{ using } (3.4)-(3.5),$$

$$\alpha_1 \Delta_1 \underline{y} - \alpha_1 \Delta_2 \underline{u}$$

$$= 0 \quad , \text{ using } (3.6) \quad .$$

The simplest situation where such a symmetry exists is that where the system is time-independent, i.e., Δ_1, Δ_2 are constant coefficient linear differential operators. Then,

$$\alpha_1 = \alpha_2 = \frac{d}{dt}$$

will do the trick.

Another simple situation is that where:

$$\Delta_2 = \text{multiplication by a constant} \quad .$$

Then, α_1 and α_2 can be chosen to be *equal*, provided that:

$$\boxed{\alpha_1 \text{ commutes with } \Delta_1 \quad .}$$

4. OLD AND NEW FACTS ABOUT COMMUTING SCALAR LINEAR DIFFERENTIAL OPERATORS. RELATIONS WITH ALGEBRAIC GEOMETRY

Continue with scalar input-output systems of classical type. As we have just seen, the problem of "symmetries" leads to the following, more classical, problem of:

> Given a linear, ordinary, scalar differential operator Δ, find all differential operators Δ' such that
>
> $$[\Delta, \Delta'] = 0$$

(4.1)

Of course, (4.1) can be regarded as a *differential equation for* Δ'. Approached directly in this way, it is rather complicated and has never been completely treated. However, there is now available considerable partial information.

This development goes back to work of Burchnall and Chaundy [1]. Apparently, this work was completely forgotten, and is essentially duplicated in recent work in the theory of nonlinear waves. See [2], for a recent survey of this work and [5] for further ramifications.

There seem to be two essential facts:

a) The set of all such Δ' commute among themselves, if Δ is "generic",

b) There is a polynomial $F(x,y)$ in two complex variables such that the following relation follows from (4.1):

$$F(\Delta, \Delta') = 0 \qquad (4.2)$$

This *algebraic* relation between differential operators is the key fact which introduces *algebraic geometry* into the picture. It should be clear that the properties of the "algebraic curve"

$$\{(x,y) \in \mathbb{C}^2 : F(x,y) = 0\}$$

are destined to play a key role in the study of the properties of the differential operator Δ'. Later on, I will use this relation to generate vector bundles on algebraic curves (i.e., Riemann surfaces) for time-dependent linear systems, generalizing work done by Martin and myself [3] in the time-independent case.

Note that property (a) is very reminiscent of an important fact about a finite dimensional semi-simple Lie algebra $\mathcal{G}$: If $A \in \mathcal{G}$ is a generic element, then the commutator of A in $\mathcal{G}$ is *abelian*, and its dimension is equal to the rank of $\mathcal{G}$. This commutator Lie algebra plays a key role in semisimple Lie algebra theory, and is called the *Cartan subalgebra*. I suspect that many other facts about the *infinite dimensional* Lie algebra of linear differential operators are related, perhaps only in a "weak" way, to those for finite dimensional semisimple Lie algebras. Another important relation with Lie theory is via the Picard-Vessiot theory.

TIME-VARYING SYSTEMS   151

An immediate relation between the algebraic curve

$$F(x,y) = 0$$

and the pair of operators (Δ,Δ') can be established via *eigenvalue problems*.
If $\lambda \in \mathbb{C}$, let

$$\mathscr{F}_\lambda = \text{set of functions } t \to f(t) \text{ such that } \Delta f = \lambda f$$

If Δ' commutes with Δ, then, of course,

$$\Delta'(\mathscr{F}_\lambda) \subset \mathscr{F}_\lambda \quad.$$

Since $\mathscr{F}_\lambda$ is finite dimensional, there exists an $f \in \mathscr{F}_\lambda$ which is also an eigenvalue of Δ', i.e.,

$$\Delta'(f) = \Delta' f$$
$$\Delta(f) = f \quad.$$
(4.3)

Substituting f into the relation (4.2) gives:

$$\boxed{F(\lambda,\lambda') = 0} \qquad (4.4)$$

The trivial observation indicates very strongly that the "spectrum" of Δ will be very closely related to the algebraic curve determined by Δ. In [4] Martin and I have presented some preliminary work about such spectral matters for general systems. More is presented elsewhere in this volume. Of course, the key notion of "isospectral deformation" is also related.

All of this becomes "trivial"--but enormously significant--if Δ is a *constant coefficient* differential operator. Then,

$$\Delta' = \frac{d}{dt} \quad.$$

Thus,

$$\Delta' = p\left(\frac{d}{dt}\right) \quad,$$

where $p(\ )$ is a polynomial in one variable.

$$F(x,y) = p(x) - y \quad. \qquad (4.5)$$

The algebraic curve $F(x,y) = 0$ is then of *genus zero*.

5. THE VECTOR BUNDLE DEFINED BY A DIFFERENTIAL OPERATOR SYMMETRY OF A LINEAR TIME-VARYING SYSTEM

Return to the general situation

$$\frac{dx}{dt} = A(t)x + B(t)u \quad (5.1)$$

$$y = C(t)x \quad .$$

Here, x,u,y are vectors of finite dimensional complex vector spaces X,U,Y. t is a real variable. $\underline{Y},\underline{U},\underline{Y}$ denote the space of all curves $t \to \underline{x}(t)$, $t \to \underline{u}(t)$, $t \to \underline{y}(t)$. IO is the set of all curves $(u,y) \in U \times Y$ such that there exists an $(x,u,y) \in X \times U \times Y$ which solves the differential equation (5.1).

A *linear differential operator symmetry* is a linear differential operator

$$\alpha: \underline{U} \times \underline{Y} \to \underline{U} \times \underline{Y}$$

such that

$$\alpha(IO) \subset IO \quad . \quad (5.2)$$

$\underline{U} \times \underline{Y}$ is a vector space (i.e., the direct sum of the vector spaces $\underline{U}$ and $\underline{Y}$). Hence α *is an ordinary differential operator, and the kernel of* α *is finite dimensional*

Construct a vector bundle over $\mathbb{C}$ as follows. For $s \in \mathbb{C}$, its fiber $V(s)$ is a vector space, namely:

$$V(s) = \{v \in \underline{U} \times \underline{Y}: \alpha(v) = sv\} \quad (5.3)$$

i.e., $V(s)$ is the space of eigenfunctions of the differential operator α. Then

$$V = \{(s,v): s \in \mathbb{C}, v \in V(s)\} \quad (5.4)$$

Of course, V defined in this way may have singularities, i.e., points s in the neighborhood of which the fibers $s \to V(s)$ may not have constant dimension. Often, one can *desingularize* the situation in the following way:

> Find a discrete subset $D \subset \mathbb{C}$ such that the bundle $V_{\mathbb{C}-D}$ over $\mathbb{C}-D$ can be completed to be a non-singular bundle on a *compact* Riemann surface M in which $\mathbb{C}-D$ is an open, dense subset.

TIME-VARYING SYSTEMS

If this desingularization were possible for a reasonably broad class of systems, it would be clear that the theory of *holomorphic non-singular vector bundles on compact Riemann surfaces* is applicable to the study of the structure of linear, time-varying systems. Now, there is one case, the *time-invariant system*, for which we already know the story [3]. I will now review this familiar material from this new point of view.

6. THE INPUT-OUTPUT BUNDLE FOR TIME-INVARIANT SYSTEMS

Suppose the input-output system is

$$\frac{dx}{dt} = Ax + Bu$$
$$y = Cx \tag{6.1}$$

with $A \in L(X,X)$; $B \in L(U,X)$; $C \in L(X,Y)$. Let IO be the set of input-output pairs $(\underline{u},\underline{y})$.

$$\alpha = \frac{d}{dt} \tag{6.2}$$

Since (6.1) is constant coefficient, obviously

$$\alpha(IO) \subset IO ,$$

α is a *symmetry*.

For $s \in \mathbb{C}$,

$$V(s) = \left\{ v = (\underline{u},\underline{y}) : \frac{dv}{dt} = sv \right\} ,$$

Thus, if $v = (\underline{u},\underline{y}) \in V(s)$, then

$$\frac{d\underline{u}}{dt} = s\underline{u} , \quad \frac{d\underline{y}}{dt} = s\underline{y} ,$$

or

$$\underline{u}(t) = e^{st} \underline{u}_0 ; \quad \underline{y}(t) = e^{st} \underline{y}_0 . \tag{6.3}$$

The condition that $(\underline{u},\underline{y}) \in IO$ means that there exists a curve $t \to x(t)$ in X such that

$$(\underline{u}: t \to e^{st}\underline{u}_0; \quad \underline{x}: t \to x(t); \quad \underline{y}: t \to e^{st}\underline{y}_0)$$

satisfies (6.1). Hence,

$$\frac{dx}{dt} = Ax + e^{st} Bu_0$$

$$e^{st} y_0 = Cx \quad .$$

(6.4)

We can now differentiate (6.4) n times (n = dim X) in order to eliminate x using Cayley-Hamilton:

$$se^{st} y_0 = C \frac{dx}{dt}$$

$$= C(Ax + e^{st} Bu_0)$$

etc.

Hence,

$$p(s) e^{st} y_0 = e^{st} (a_1 CB + a_2 CAB + \cdots) u_0 \quad ,$$

where

$$p(s) = a_0 + a_1 s + \cdots$$

is the *characteristic polynomial* of the matrix A. We see that

$$y_0 = T(s) u_0 \quad , \tag{6.5}$$

with

$$T(s) = C(s - A)^{-1} B \tag{6.6}$$

the *transfer function* (or frequency response) of the system (6.1). Thus, we reach the starting point of [3]. In this case, the bundle V over $\mathbb{C}$ is already non-singular. It can be compactified to give a bundle over $P_1(\mathbb{C})$, where "invariants" in the sense of algebraic-holomorphic geometry turn out to be precisely the natural systems-theoretic invariants.

7. THE VECTOR BUNDLES DEFINED BY THE BURCHNALL-CHAUNDY THEORY FOR CERTAIN SCALAR INPUT-OUTPUT SYSTEMS

Consider now a time-varying system

$$\frac{dx}{dt} = A(t) x + B(t) u$$

$$y = C(t) x \quad , \tag{7.1}$$

with U = Y = R, i.e., *scalar inputs and outputs*. Suppose

$$= a_n(t) \frac{d^n}{dt^n} + \cdots \tag{7.2}$$

is a scalar, linear differential operator such that the input-output relations of the system (7.1) are of the form

$$\Delta \underline{y} = \underline{u} . \tag{7.3}$$

By the Burchnall-Chaundy theory [1,2,5], there is another scalar differential operator α and a polynomial $F(s,\tau)$ in two complex variables s,τ such that:

$$[\alpha, \Delta] = 0 \tag{7.4}$$

$$F(\alpha, \Delta) = 0$$

Because we are assuming that the input-output relations have the simple form (7.3), α is an input-output symmetry.

In the way that F is defined by Burchnall and Chaundy in [3], F has the following property:

$$\boxed{\begin{array}{c} F(s,\tau) = 0 \\ \text{for } (s,\tau) \in \mathbb{C}^2 \\ \text{if and only if in the differential equations} \\ \Delta - \tau = 0 \\ \alpha - s = 0 \\ \text{have a solution in } common. \end{array}} \tag{7.5}$$

For $s \in \mathbb{C}$, set

$$V(s) = v \in IO: v = sv ,$$

i.e., $V(s)$ is the fiber of the *input-output vector bundle above the point* $s \in \mathbb{C}$. In this case,

$$V(s) = \{\underline{u}: t \to u(t); \underline{y}: t \to y(t): \alpha \underline{u} = s\underline{u}, \alpha \underline{y} = s\underline{y}, \\ \alpha \underline{y} = \underline{u}\} \tag{7.6}$$

Let $W(s)$ be the space of all scalar functions $\underline{w}$ such that

$$\alpha \underline{w} = s\underline{w} .$$

Since Δ commutes with α, we have:

$$\Delta(W(s)) \subset W(s) \qquad (7.8)$$

The flow $V(s)$ is then essentially just the *graph* of Δ acting in $W(s)$. In particular:

Theorem 7.1. For each $s \in \mathbb{C}$, the dimension of the fiber $V(s)$ of the input-output bundle is equal to the dimension of $W(s)$.

8. THE BURCHNALL-CHAUNDY VECTOR BUNDLE

Continue with the notation of Section 7. We have defined the *input-output vector bundle* V, whose base is $\mathbb{C}$ and whose fiber above the point $s \in \mathbb{C}$ is the set of all input-output pairs $(\underline{u},\underline{y})$ which are eigenvectors of α, with eigenvalue s. Using Δ, α, and the polynomial $F(s,\tau)$ in two complex variables having the property described in (7.6), we will now describe another vector bundle. Let

$$B = \{(s,\tau) \in \mathbb{C}^2 : F(s,\tau) = 0\} . \qquad (8.1)$$

We know that for $(s,\tau) \in B$, there is a solution $\underline{w}$ of the eigenvalue problem

$$\begin{aligned} \Delta \underline{w} &= s\underline{w} \\ \alpha \underline{w} &= \tau \underline{w} \end{aligned} \qquad (8.2)$$

Let $E(s,\tau)$ be the vector space of all solutions $\underline{w}: t \to w(t)$ of Eq. (8.2). As (s,t) varies over B, $E(s,t)$ is the fiber of a *vector bundle*, which we will call the B-C *bundle*.

B is an affine algebraic set. It can be extended to a projective one:

$$B' = \left\{(s_0,s,\tau) \in \mathbb{C}^2 : F\left(\frac{s}{s_0}, \frac{\tau}{s_0}\right) = 0\right\}$$

The fiber of the B-C bundle becomes the set of scalar valued functions $\underline{w}$ such that

$$\Delta \underline{w} = \frac{s}{s_0} \underline{w}$$

$$\alpha \underline{w} = \frac{\tau}{s_0} \underline{w} .$$

The fiber at infinity, i.e., $s_0 = 0$, then must be defined by a study of the asymptotics of this eigenvalue problem. At the moment I do not see how to do very much in *general* about this.

Bibliography

1. J.L. Burchenall and T.W. Chaundy, "Commutative ordinary differential operators", Proc. London Math. Soc. 21 (1923), 420-440.

2. M. Adler and J. Moser, "On a class of polynomials connected with the Korteweg-de Vries equation", Comm. Math. Phys. 61 (1978), 1-30.

3. R. Hermann and C. Martin, "Algebro-geometric methods in systems theory: The Macmillan degree and Kronecker indices" (to appear, SIAM J. Control).

4. C. Martin and R. Hermann, "Algebraic-geometric aspects of systems theory", Parts 5-6, Proc. 1976 Ames Conf. on Geometric Control, Math Sci Press, 1977.

5. I.M. Kricever, "Algebraic-geometric construction of the Zaharov-Sabat equations and their periodic solutions", Soviet Math. Dokl. 17 (1976), 394-397.

GROUP III

ISOSPECTRAL DEFORMATION OF INPUT-OUTPUT SYSTEMS

INTRODUCTION

The "isospectral deformation" concept is basic to the theory of nonlinear waves, although it refers, at first sight, to *linear* differential equations. This group of chapters develops these ideas in the context of *linear input-output systems*. I believe this provides certain insights even into the theory of nonlinear waves, as well as suggesting new areas of research into systems theory.

Chapter 12

ISOSPECTRAL DEFORMATIONS OF LINEAR, TIME-VARYING SYSTEMS

1. INTRODUCTION

One of the landmarks of contemporary mathematics and physics is the "Inverse Scattering-Isospectral Deformation" method of Gardner, Greene, Kruskal, and Miura [1]. Although many striking and marvelous mathematical concepts and insights have already fallen out of this work, the most interesting differential equations for applications lie tantalizingly just beyond our grasp. This has motivated a search for methods of generalization and reformulation. At this point, I will mention four:
 a) The "isospectral deformation" method of Lax [2]
 b) The "generalized Fourier transform" of Ablowitz, Kaup, Newell, and Segur [3].
 c) The exterior differential systems-prolongation-connections in fiber bundles-with structure-group method of Estabrook, Wahlquist and myself.
 d) The hybrid but very effective and convenient methods in the paper by Wadati, Sonuki and Konno, which I have already quoted extensively.

(c) and (d) are closely related, as I have already described in "Interdisciplinary Mathematics", Volume 18. In this chapter I want to pursue the relations between these approaches and the theory of *time-varying linear systems*, as described, for example, in the book by Roger Brockett [4].

2. ISOSPECTRAL DEFORMATION OF FINITE DIMENSIONAL LINEAR MAPS

First, some generalities. Let V be a finite dimensional vector space, $L(V)$ the space of linear maps: $V \to V$. $GL(V)$ is the group of invertible linear maps. Let $A \in L(V)$. The *spectrum*, $\sigma(A)$ of A are the complex numbers λ such that

$$\det (A - \lambda) = 0 \ . \tag{2.1}$$

A *deformation* of A is a curve

$$\tau \to A(\tau)$$

in $L(V)$. The deformation is *isospectral* if

$$\sigma(A(\tau))$$

is independent of τ.

One way of achieving the property is to allow all the $A(\tau)$ to be similar, i.e., to suppose there is a curve $\tau \to g(\tau)$ in $GL(V)$ such that:

$$A(\tau) = g(\tau) A(0) g(\tau)^{-1} \quad . \tag{2.2}$$

To find another version of this condition, differentiate (2.2):

$$\frac{dA}{d\tau} = \frac{dg}{d\tau} A(0) g(\tau)^{-1} - g A(0) g(\tau)^{-1} \frac{dg}{d\tau} g^{-1}$$

$$= \left(\frac{dg}{d\tau} g^{-1}\right) A(\tau) - A(\tau) \left(\frac{dg}{d\tau} g^{-1}\right) \quad .$$

Or,

$$\frac{dA}{d\tau} = [B(\tau), A(\tau)] \tag{2.3}$$

where

$$\tau \to B(\tau) = \frac{dg}{d\tau} g(\tau)^{-1} \tag{2.4}$$

Conversely, if (2.3) holds, for *some* curve $\tau \to B(\tau)$ in $L(V)$, then the steps are reversible to show that the deformation $\tau \to A(\tau)$ is isospectral.

Here is another way of thinking of relation (2.3). Consider the system of *partial* differential equations

$$\frac{\partial v}{\partial \tau} = A(\tau) v$$

$$\frac{\partial v}{\partial \tau} = B(\tau) v \tag{2.5}$$

to be solved for *a surface* $(t, \tau) \to v(t, \tau)$ in V.

Let us investigate *complete integrability* (in the differential-geometric Mayer-Frobenius sense). This requires that the compatibility conditions resulting from cross-differentiating (2.5) be satisfied *identically*.

$$\frac{\partial^2 v}{\partial \tau} = \frac{\partial A}{\partial \tau} v + A \frac{\partial v}{\partial \tau}$$

$$= \frac{\partial A}{\partial \tau} v + ABv$$

$$= \frac{\partial}{\partial \tau} (B(\tau) v)$$

$$= B(\tau) Av \quad ,$$

$$\boxed{\frac{\partial A}{\partial \tau} + [A(\tau), B(\tau)] = 0} \qquad (2.6)$$

which is just the condition (2.3).

It will be convenient to have another way of interpreting this condition which does not involve adding a "new" variable t. Consider the "mixed" algebraic-differential system of equations:

$$\boxed{\begin{aligned} A(\tau)(v) + \lambda v &= u \\ \frac{\partial v}{\partial \tau} &= B(\tau) v \end{aligned}} \qquad (2.7)$$

The "unknowns" are curves $t \to (v(t), u(t))$ in $V \times V$. λ is an arbitrary parameter. Let us investigate "complete integrability" of this mixed system by differentiating with respect to τ:

$$\frac{\partial A}{\partial \tau}(\tau) v + A(\tau) \frac{\partial v}{\partial \tau} + \lambda \frac{\partial v}{\partial \tau} = \frac{\partial u}{\partial \tau} \;;$$

Using relations (2.6) we have:

$$\frac{\partial A}{\partial \tau}(\tau) v + A(\tau) B(\tau) v + \lambda B(\tau) v = \frac{\partial u}{\partial \tau}$$

$$= \frac{\partial A}{\partial \tau} v + A(\tau) B(\tau) v + B(u - Av) \;.$$

This finally gives us the identity

$$\boxed{\frac{\partial A}{\partial \tau} v + [A, B] v = \frac{\partial u}{\partial \tau} + bu} \qquad (2.8)$$

Now, "complete integrability" requires that the compatibility equations be satisfied *identically*. This means that the following identities should be satisfied:

$$\boxed{\begin{aligned} \frac{\partial Av}{\partial \tau} + [A, B] v &= 0 \\ \text{for all} \quad v \end{aligned}} \qquad (2.9)$$

$$\boxed{\frac{\partial u}{\partial \tau} + Bu = 0 \quad \text{for all } u} \qquad (2.10)$$

Now, (2.9) is an *algebraic equation*, which forces

$$\boxed{\frac{\partial A}{\partial \tau} + [A,B] = 0} \qquad (2.11)$$

Condition (2.9) is a *differential equation*, which imposes no further compatibility conditions. Thus we see that, pursuing the "complete integrability" of the system leads us right back to the condition (2.11).

3. THE KRUSKAL-LAX "ISOSPECTRAL DEFORMATION" CONDITIONS FOR LINEAR DIFFERENTIAL OPERATORS

The material of Section 2 is obviously valid quite generally, for general vector spaces V and linear maps: $V \to V$. I will now look at this from a more *classical* point of view where:

V = space of vector-valued functions on R^n

$A: V \to V$ is a linear differential operator.

The independent variables will be $x = (x^1, \ldots, x^n) \in R^n$. We use the following notation. If

$$a = (a_1, \ldots, a_n) \in Z_+^n$$

is a sequence of non-negative integers, ∂_a is the differential operator

$$\partial_a = \frac{\partial^{a_1}}{\partial x_1^{a_1}} \cdots \frac{\partial^{a_n}}{\partial x_n^{a_n}} \quad . \qquad (3.1)$$

Thus, if

$$|a| = a_1 + \cdots + a_n \quad , \qquad (3.2)$$

then $|a|$ is the *order* of the differential operator ∂_a. A *linear differential operator* is of the form:

$$\Delta = \sum_a \alpha^a \partial_a \quad ,$$

ISOSPECTRAL DEFORMATION

where the sum runs over a *set of* subsets of Z_+^n, and each α^a is an $m \times m$ matrix function of x. Δ acts on m-vector functions

$$x \to \psi(x) = \begin{pmatrix} \psi_1(x) \\ \vdots \\ \psi_m(x) \end{pmatrix}$$

in R^n.

$$\Delta(\psi) = \sum_a \alpha^a \partial_a (\psi) \qquad (3.4)$$

Let $D(m,m)$ be the set of differential operators of this type. Consider curves in $D(m,m)$ of the form

$$\tau \to \Delta(\tau) , \qquad (3.5)$$

with τ a real variable. Thus, if

$$\Delta(\tau) = \sum_a \alpha^a(x,\tau) \partial_a ,$$

then

$$\frac{d\Delta}{d\tau} = \sum_a \frac{\partial}{\partial \tau} (\alpha^a) \partial_a . \qquad (3.6)$$

<u>Definition</u>. Such a curve in $D(m,m)$ is an *isospectral deformation* (in the sense of *Kruskal-Lax*) if there is a curve $\tau \to \beta(\tau)$ in $D(m,m)$ such that:

$$\frac{d\Delta}{d\tau} = [\beta(\tau), \Delta] . \qquad (3.7)$$

We are particularly interested in such deformations which are generated in the following way. Let S be a subset of $D(m,m)$. For each $\Delta \in S$, let

S_Δ be the subset $D(m,m)$ consisting of the "tangent vectors" to curves $D(m,m)$ beginning at Δ, i.e., those of the form

$$\frac{d}{d\tau} \phi(\Delta(\tau))\big|_{\tau=0} , \qquad \Delta(0) = \Delta .$$

Let

$$\phi: S \to D(m,m)$$

such that the following condition is satisfied:

$$[\Delta, \phi(\Delta)] \in S_\Delta$$

for all $\Delta \in S$.

The *Kruskal-Lax equation* is then:

$$\frac{d\Delta}{dt} = [\phi(\Delta), \Delta] \quad . \tag{3.8}$$

Notice that this is a *nonlinear partial differential equation* for the coefficients of Δ. It is known that a few partial differential equations can be written in this form (Korteweg-de Vries, Boussinesq,...)

This is just a general version of the well-known material in Lax's paper [2]. I now want to put this in another way which is more conducive to the systems-theoretic view point.

4. DEFORMATIONS OF LINEAR SYSTEMS

In engineering, a *linear system* is a set of differential equations of the form

$$\frac{dx}{dt} = Ax + Bu$$
$$y = Cx \quad . \tag{4.1}$$

u,x,y are vectors, the *input*, *state* and *output*. The A,B,C may be functions of t. In this section I will only consider the case where

> B,C are independent of t, but
> $t \to A(t)$ is time-varying.

Let Σ denote the set of systems of this type. Denote an element Σ by σ.

A *deformation* of linear systems is a curve

$$\tau \to \sigma_\tau = (A_\tau(t), B_\tau, C_\tau)$$

Here I will only consider deformations of the following type:

> B_τ, C_τ do not depend on t. There is a matrix $D(t,)$ such that:
>
> $$\frac{\partial A}{\partial \tau} - \frac{\partial D}{\partial t} = [D, A] \tag{4.2}$$

ISOSPECTRAL DEFORMATION

Conditions (4.2) may be generated in the following way. Suppose given the matrix-valued function

$$(t,\tau) \to D(t,\tau)$$

in R^2. Consider the following set of equations:

$$\boxed{\begin{aligned} \frac{\partial x}{\partial t} &= A(x,t)x + Bu \\ \frac{\partial x}{\partial \tau} &= Dx \end{aligned}} \qquad (4.3)$$

Let us investigate "complete integrability" of these equations:

$$\begin{aligned} \frac{\partial^2 x}{\partial t \partial \tau} &= \frac{\partial A}{\partial \tau} x + A \frac{\partial x}{\partial \tau} + B \frac{\partial u}{\partial \tau} \\ &= \frac{\partial A}{\partial \tau} x + ADx + B \frac{\partial u}{\partial \tau} \\ &= \frac{\partial D}{\partial t} x + d \frac{\partial x}{\partial t} \\ &= \frac{\partial D}{\partial t} x + D(Ax + Bu) \quad , \end{aligned}$$

hence:

$$\frac{\partial A}{\partial \tau} x - \frac{\partial D}{\partial t} x = [D,A]x + DBu - B \frac{\partial u}{\partial \tau} \quad .$$

The "integrability conditions" are then conditions (4.2) *plus*

$$\boxed{B \frac{\partial u}{\partial \tau} = DBu}$$

Now, suppose that we require that A *and* D *be independent of* τ. (4.2) reduces to

$$\frac{\partial A}{\partial \tau} = [D,A] \quad ,$$

which, as we have seen, reduces to "isospectral deformation" condition of the single operator A in finite dimensional vector spaces. Thus, we obtain interesting *nonlinear* differential equations (of the Euler rotating rigid body-Toda lattice type) by postulating $D = \phi(A)$ as a function of A. This corresponds to the "finite number of degrees of freedom" completely integrable (in the sense of Hamiltonian mechanics) physical system". As we shall see (and as I have already discussed in Volumes 12, 14, 15, and 18 of "Interdisciplinary Mathematics") the Equations (4.2) are those involved in "Nonlinear Wave Theory". For example, D could be a function

$$D = \phi(A, A_t, A_{tt}, \ldots)$$

of A and its t-derivatives. (Physically, this corresponds to "x", i.e., "space".) (4.2) then becomes a *partial* differential equation:

$$\frac{\partial A}{\partial \tau} - \phi_A A_t - \phi_{A_t} A_{tt} \cdots = [\phi(A, A_t, \ldots), A] \qquad (4.4)$$

5. LINEAR DIFFERENTIAL OPERATORS IN ONE INDEPENDENT VARIABLE THAT ARE THE INPUT-OUTPUT RELATIONS OF LINEAR SYSTEMS. SPECIALIZATION TO STURM-LIOUVILLE

As I have just mentioned, there is a notational mis-match between systems theory and nonlinear waves. Where the former think of "time", t, the latter thinks of "space". Hence, I will now shift gears in order to keep closer to the "nonlinear wave" literature.

Consider a linear system of the form

$$\begin{aligned}\frac{dz}{dx} &= A(x)z + Bu \\ y &= Cz\end{aligned} \qquad (5.1)$$

Here, B.C are constant matrices, A is a matrix-valued function of the real variable x. (Thus, "x" now replaces "t", "z" replaces "x".)

<u>Definition</u>. Let Δ_1, Δ_2 be linear differential operators on the variables x. The pair (Δ_1, Δ_2) is said to define the *input-output relations* of the system (5.1) if, for *every* solution

$$x \to (u(x), z(x), y(x))$$

of (5.1), the following differential equations hold between y and u:

$$\Delta_1 y = \Delta_2 u$$

Example. *Sturm-Liouville*

$$A(x) = \begin{pmatrix} \lambda, & q(x) \\ 1, & -\lambda \end{pmatrix}$$

$$z = \begin{pmatrix} z_1 \\ z_2 \end{pmatrix} \in R^2$$

t   u(t) is a scalar

q(x)   is an arbitrary function of x

$$Bu = \begin{pmatrix} b_1 u \\ b_2 u \end{pmatrix}$$

$$C \begin{pmatrix} z_1 \\ z_2 \end{pmatrix} = z_2$$

$$z(x) = \begin{pmatrix} z_1(x) \\ z_2(x) \end{pmatrix}$$

$$\begin{pmatrix} \dfrac{dz_1}{dx} \\ \dfrac{dz_2}{dx} \end{pmatrix} = \begin{pmatrix} \lambda, & q(x) \\ 1, & -\lambda \end{pmatrix} \begin{pmatrix} z_1 \\ z_2 \end{pmatrix} + Bu$$

$$= \begin{pmatrix} \lambda z_1 + q z_2 \\ z_1 - \lambda z_2 \end{pmatrix} + \begin{pmatrix} b_1 u \\ b_2 u \end{pmatrix}$$

$$\boxed{\begin{aligned} \frac{dz_1}{dx} &= \lambda z_1 + q z_2 + b_1 u \\[6pt] \frac{dz_2}{dx} &= z_1 - \lambda z_2 + b_2 u \\[6pt] y &= z_2 \end{aligned}}$$

(5.3)

(5.4)

Let us eliminate z_1 from these relations:

$$\frac{d}{dx}\left(\frac{dz_2}{dx} + \lambda z_2 + b_2 u\right) = \lambda\left(\frac{dz_2}{dx} + \lambda z_2 + b_2 u\right) + qz_2 + b_i u$$

or

$$\frac{d^2 y}{dx^2} + (-q - \lambda^2) y = \lambda b_2 u + b_1 u - \frac{d}{dx}(b_2 u)$$

This clearly exhibits the input-output relations for this system in the form (5.2).

6. ISOSPECTRAL DEFORMATION OF LINEAR SYSTEMS AND ISOSPECTRAL LINEAR DIFFERENTIAL OPERATORS

Consider (as in Section 5) a linear system of the form:

$$\frac{dz}{dx} = A(x) z + Bu$$
$$y = Cz \quad .$$
(6.1)

For simplicity, we restrict attention to the case where B and C are *constant* matrices. Denote the system as follows

$$\sigma = (A(x), B, C) \quad .$$

Let τ be a deformation parameter. We then consider a one-parameter family of systems

$$\sigma_\tau = (A(\tau, x), B, C)$$

parameterized by τ. (We assume for simplicitly that B and C do not vary.)

Definition. The deformation σ_τ is *isospectral* if there is a matrix-valued function $D(x, \tau)$ such that the system of partial differential equations

$$\frac{\partial z}{\partial x} = A(\tau, x) z + Bu$$

$$\frac{\partial z}{\partial \tau} = D(\tau, x) z$$

is *completely integrable* in the Mayer-Frobenius sense.

ISOSPECTRAL DEFORMATION

Remark. I keep on making this pedantic distinction between different types of "complete integrability" in order to remind the reader that the term "complete integrability" in the nonlinear wave literature usually refers to another concept, "solubility in the sense of Hamiltonian mechanics".

Let us work out the complete integrability conditions of the system (6.2) once again.

$$\frac{\partial^2 z}{\partial x \partial \tau} = \frac{\partial A}{\partial \tau} z + A \frac{\partial z}{\partial \tau} + B \frac{\partial u}{\partial \tau}$$

$$= \frac{\partial D}{\partial x} z + D \frac{\partial z}{\partial x}$$

$$= \frac{\partial A}{\partial \tau} z + ADz + B \frac{\partial u}{\partial \tau}$$

$$= \frac{\partial D}{\partial x} z + D(Az + Bu)$$

hence:

$$\boxed{\frac{\partial A}{\partial \tau} - \frac{\partial D}{\partial x} = [A, D]} \qquad (6.3)$$

Let us now *also* suppose that the solutions to (6.1) satisfy the "input-output" relations:

$$\boxed{\Delta_1^\tau y = \Delta_2^\tau u} \qquad (6.4)$$

where Δ_1, Δ_2 are linear differential operators in the variable x, with τ as *parameter*. Let us differentiate (6.4) with respect to τ and see what happens:

$$\frac{\partial}{\partial \tau}(\Delta_1^\tau) y + \Delta_1^\tau \frac{\partial y}{\partial \tau} = \frac{\partial \Delta_2^\tau}{\partial \tau} u + \Delta_2^\tau \frac{\partial u}{\partial \tau}$$

$$\frac{\partial}{\partial \tau}(\Delta_1^\tau) y + \Delta_1^\tau C \frac{\partial x}{\partial \tau} = \frac{\partial \Delta_2^\tau}{\partial \tau} u + \Delta_2^\tau \frac{\partial u}{\partial \tau}$$

$$= \frac{\partial}{\partial \tau}(\Delta_1^\tau) y + \Delta_1^\tau (CDx) \qquad . \qquad (6.5)$$

Now, recall that

$\tau \to \Delta_1^\tau$ defines a *isospectral deformation in the sense of differential operations* (in the Kruskal-Lax sense) if there is a τ-dependent linear differential operator x, β_τ, such that

$$\frac{\partial}{\partial \tau}(\Delta_1^\tau) = \beta_\tau \Delta_1^\tau - \Delta_1^\tau \beta_\tau \qquad (6.6)$$

With the aid of (6.6), (6.5) takes the following form:

$$\frac{\partial \Delta_2^\tau}{\partial \tau} u + \Delta_2^\tau \frac{\partial u}{\partial \tau} = \beta_\tau \Delta_1^\tau Cz - \Delta_1^\tau \beta_\tau (Cz) + \Delta_1^\tau (CDz)$$

$$= \beta_\tau \Delta_2^\tau u - \Delta_1^\tau \beta_\tau (Cz) + \Delta_1^\tau (CDz) \quad .$$

This seems to lead to the following relations:

$$\beta_\tau Cz = CDz \quad , \qquad (6.7)$$

which enables one to determine β_τ in terms of D.

From now on, I will shift emphasis from the Kruskal-Lax differential operator condition for the "isospectral deformation" to the related "systems" version. This latter version is more closely related to the Estabrook-Wahlquist "connection" approach developed in Volumes 12, 14, 15, and 18.

7. ISOSPECTRAL DEFORMATION AND INPUT-OUTPUT RELATIONS OF THE SYSTEMS-THEORETIC ISOSPECTRAL DEFORMATION

Carry over the following equation:

$$\frac{\partial z}{\partial x} = A(x,\tau)z + Bu$$

$$\frac{\partial z}{\partial \tau} = D(x,\tau)z \quad , \qquad (7.1)$$

with the "isospectral integrability" conditions

$$\frac{\partial A}{\partial \tau} - \frac{\partial D}{\partial x} = [A, D] \qquad (7.2)$$

Let $\mathcal{G}$ be the Lie algebra of matrices generated by the $A(x,\tau)$, $D(x,\tau)$. Let G be the Lie group of matrices whose Lie algebra is $\mathcal{G}$. Thus, there is a unique map

ISOSPECTRAL DEFORMATION

$$(x,\tau) \to g(x,\tau)$$

$$R^2 \to G \quad ,$$

such that:

$$\boxed{\begin{aligned} \frac{\partial g}{\partial x} &= A(x,\tau)g \\[4pt] \frac{\partial g}{\partial \tau} &= D(x,\tau)g \\[4pt] g(0,0) &= 1 \end{aligned}}$$

Thus, we can write solutions to (7.1) in the following form:

$$z(x,\tau) = g(x,\tau)w(x,\tau)$$

where $(x,\tau) \to w(x,\tau)$ is a vector-valued function. Then,

$$\begin{aligned}
\frac{\partial z}{\partial x} &= \frac{\partial g}{\partial x} w + g \frac{\partial w}{\partial x} \\[4pt]
&= Agw + g \frac{\partial w}{\partial x} \\[4pt]
&= Az + g \frac{\partial w}{\partial x} \\[4pt]
&= Az + Bu \quad ,
\end{aligned}$$

$$\boxed{\frac{\partial w}{\partial x} = g^{-1}Bu} \qquad (7.3)$$

Also,

$$\begin{aligned}
\frac{\partial z}{\partial \tau} &= \frac{\partial g}{\partial \tau} w + g \frac{\partial w}{\partial \tau} \\[4pt]
&= Dz + g \frac{\partial w}{\partial \tau} \\[4pt]
&= Dz \quad ,
\end{aligned}$$

whence:

whence:

$$\boxed{\frac{\partial w}{\partial \tau} = 0} \qquad (7.4)$$

Thus, we are looking for a vector-valued function $w(x)$ of x *alone* such that:

$$\boxed{\frac{dw}{dx}(x) = g(x,\tau)^{-1} Bu(x,\tau)} \qquad (7.5)$$

This condition forces the input $u(x,\tau)$ to satisfy additional compatibility conditions

$$0 = \frac{\partial}{\partial \tau}(g^{-1} Bu)$$

$$= -g^{-1} \frac{\partial g}{\partial \tau} g^{-1} Bu + g^{-1} B \frac{\partial u}{\partial \tau}$$

$$= -g^{-1} DBu + g^{-1} B \frac{\partial u}{\partial \tau},$$

or

$$\boxed{D(x,\tau) Bu(x,\tau) = B \frac{\partial u}{\partial \tau}(x,\tau)} \qquad (7.6)$$

Thus, with (7.3) satisfied, we can write $w(x)$ as follows:

$$w(x) = \int_{x_0}^{x} g(x',0)^{-1} Bu(x',0)\, dx' + g(x_0,0)^{-1} z(x_0)$$

$$z(x,\tau) = \int_{x_0}^{x} g(x,\tau) g(x',0)^{-1} Bu(x',0)\, dx' + g(x,\tau) g(x_0,0)^{-1} z(x_0)$$

Thus, if

ISOSPECTRAL DEFORMATION

$$y = Cz$$

is the read-out map, we then have the following input-output relations:

$$y(t) = \int_{x_0}^{x} Cg(x,\tau)g(x',\tau')Bu(x',\tau') \, dx' + Cg(x,\tau)g(x_0,\tau')^{-1}x(x_0)$$

(7.7)

In the next chapter I will look at these relations in more detail.

Bibliography

1. C.S. Gardner, J.M. Greene, M.D. Kruskal, and R.M. Miura, Method for solving the Korteweg-de Vries equation, Phys. Rev. Lett. $\underline{19}$ (1967), 1095-1097.

2. P.D. Lax, Integrals of nonlinear equations of evolution and solitary waves, Comm. Pure Appl. Math. $\underline{21}$ (1968), 647-690.

3. M.J. Ablowitz, D.J. Kaup, A.C. Newell, and H. Segur, The inverse scattering transform--Fourier analyses for nonlinear problems, Studies in App. Math. $\underline{53}$ (1974), 249-315.

4. R. Brockett, Finite Dimensional Linear Systems, Wiley, New York, 1965.

Chapter 13

RELATIONS BETWEEN ISOSPECTRAL DEFORMATIONS OF SYSTEMS AND DIFFERENTIAL OPERATORS

1. INTRODUCTION

In the last chapter I have discussed general features of "isospectral deformation". I now want to go into more detail about the systems-theoretic version I have presented above and the original Kruskal-Lax definition.

2. LINEAR SYSTEMS ASSOCIATED WITH DIFFERENTIAL OPERATORS

As before, consider a linear system:

$$\boxed{\begin{aligned} \frac{dz}{dx} &= A(x)z + Bu \\ y &= Cz \end{aligned}} \quad (2.1)$$

with B and C constant matrices, and $x \to A(x)$ varying with the independent variable x. Suppose, for simplicity, that the systems *scalar input-output*, i.e., $x \to u(x)$, $y(x)$ are real (or complex) functions of a variable x. Let $\underline{U}$ be the set of input functions

$$x \to u(x), \quad -\infty < x < \infty,$$

and let $\underline{Y}$ be the set of outputs

$$x \to y(x).$$

Suppose

$$\Delta: \underline{Y} \to \underline{U}$$

is a linear differential operator.

<u>Definition</u>. The system (2.1) is *associated with the differential operator* Δ *if*

$$\Delta y = u \quad (2.2)$$

for each set (u,z,y) of solutions of (2.1).

We can also write this in terms of the *weighting pattern* of (2.1), i.e., the function

$$(x,x') \to W(x,x')$$

such that: for each solution (u,z,y) of (2.1), each real number x_0,

$$y(x) = \int_{x_0}^{x} W(x,x')u(x')\,dx' + Cz(x_0) \quad . \tag{2.3}$$

In turn, W can be written as:

$$W(x,x') = Cg(x)g(x')^{-1}B \quad , \tag{2.4}$$

where $x \to g(x)$ is the *fundamental matrix solution*, i.e.,

$$\frac{dg}{dx} = Ag \tag{2.5}$$

$$g(x_0) = 1 \quad .$$

3. CONDITION FOR STURM-LIOUVILLE TO BE ASSOCIATED WITH A 2×2 SYSTEM

Suppose the system (2.1) is 2×2, Δ is Sturm-Liouville, and it is associated with the system (2.1).

$$\Delta = \frac{d^2}{dx^2} + q(x) \quad . \tag{3.1}$$

Then,

$$\frac{dy}{dx} = C(Az + Bu)$$

$$\frac{d^2y}{dx^2} = C\frac{dA}{dx}B + CA^2z + CABu + CB\frac{du}{dx} \tag{3.2}$$

Since A is 2×2, it satisfies its own characteristic equation.

$$A^2 - \text{trace}(A)A + (\det A) = 0 \tag{3.3}$$

Hence, (3.2) becomes:

$$\frac{d^2y}{dx^2} = C\frac{dA}{dx}z + C(\text{trace } A)Az + CABu - C(\det A)z + CB\frac{du}{dx}$$

$$= C\frac{dA}{dx}z + (\text{trace } A)CAz - C(\det A)z + CABu + CB\frac{du}{dz} \tag{3.4}$$

SYSTEMS AND OPERATORS   179

Thus, if Δ is associated with the system (2.1), the right hand side of (3.4) should equal

$$u - qCz \quad .\tag{3.5}$$

Now, this equality should hold *whatever* u. Start off with

$$u \equiv 0 \quad .$$

We must then have:

$$C \frac{dA}{dx} z + (\text{trace } A)CAz - C(\det A)z = -qCz$$

whenever

$$\frac{dz}{dx} = Ax \quad .$$

Since z(x) can have an arbitrary value subject to this differential equation, we must have:

$$\boxed{C \frac{dA}{dx} + (\text{trace } A)CA - C(\det A) = -qC}\tag{3.6}$$

Substitute (3.6) into (3.4) for *arbitrary* u, and equate to (3.5): We obtain:

$$CABu + CB \frac{du}{dx} = u$$

$$\text{for all } u \in \underline{U} \quad ,$$

which forces the following relations:

$$\boxed{\begin{aligned} CB &= 0 \\ CAB &= 1 \end{aligned}}\tag{3.7}$$

For example, here is one way of satisfying relations (3.6):

$$\boxed{\begin{aligned} \text{trace } A &= 0 \\ \det A &= q \\ C \frac{dA}{dx} &= 0 \end{aligned}}\tag{3.8}$$

Conditions (3.7)-(3.8) can be realized as follows:

$$A = \begin{pmatrix} \lambda & q+\lambda^2 \\ -1 & -\lambda \end{pmatrix}$$

for arbitrary constant λ.

$$C = (0,-1)$$

$$B = \begin{pmatrix} 1 \\ 0 \end{pmatrix}$$

This is, in fact, the realization we have already used.

This analysis is obviously widely generalizable; it would lead to a *realization theory* for time-varying systems.

4. INPUT-OUTPUT SYSTEMS AND COVARIANT DIFFERENTIATION

Now, consider a general system:

$$\frac{dz}{dx} = Az + Bu \qquad (4.1)$$

$$y = Cz \quad .$$

However, for simplicity, we continue to suppose that this is *scalar* input and output, i.e., C is $1 \times n$, B is $n \times 1$. Let us attempt to find two scalar linear differential operators

$$\Delta_1 = a_0 + a_1 \frac{d}{dx} + \cdots + \frac{d}{dx^n} \qquad (4.2)$$

$$\Delta_2 = b_0 + \cdots + b_m \frac{d}{dx^m} , \qquad m < n , \qquad (4.3)$$

such that for every solution (u,z,y) of (4.1)

$$\Delta_1 y = \Delta_1 u \quad . \qquad (4.4)$$

Let us rewrite (4.1) in terms of another first order differential operator which I have called *covariant differentiation* [3], and which plays a basic role in the study of observability and controllability of systems of this type

$$\nabla z = \frac{dz}{dx} - Az \quad . \qquad (4.5)$$

Formally, we can then rewrite (4.1) as

$$z = \nabla^{-1} Bu \qquad (4.6)$$

$$y = Cz = C\nabla^{-1} Bu$$

Thus, if we could write

$$C\nabla^{-1} = C\Delta_1^{-1} D \quad , \qquad (4.7)$$

where Δ_1 is a *vector scalar* differential operator, D a matrix differential operator, we would have:

$$\Delta_1 y = CDBu \equiv \Delta_2 u \quad , \qquad (4.8)$$

which is the relation (4.4).

Let us now try to satisfy (4.7) as:

$$C\Delta_1 = CD\nabla \quad . \qquad (4.9)$$

In order to satisfy (4.9), suppose that

$$\delta = \frac{d}{dx} \quad .$$

$$D = \delta^{n-1} + B_{n-2}\delta^{n-2} + \cdots + B_0 \quad , \qquad (4.10)$$

where $B_0, \ldots, B_{n-2}$ are $n \times n$ matrices (n = dimension of the state space, i.e., the number of components in z.) Thus, satisfying (4.9) requires that

$$CD\nabla = C \times \text{scalar differential operator} \quad . \qquad (4.11)$$

This is a differential equation for the coefficient $B_0, \ldots, B_{n-2}$.

Example. $n = 2$.

Look for D of the following form:

$$D = \delta + B \quad ,$$

with B an $n \times n$ matrix function of x.

$$\begin{aligned}
D\nabla &= (\delta + B)(\delta - A) \\
&= \delta^2 + B\delta - \delta A - BA \\
&= \delta^2 + B\delta - A\delta + \delta(A) - BA \quad .
\end{aligned}$$

Let us require that

$$C D \nabla = C \Delta_1 \equiv C(\delta^2 + a_1 \delta + a_0)$$

$$= C\delta^2 + CB\delta - CA\delta + C\delta(A) - CBA \quad .$$

Hence

$$Ca_1 = CB - CA \qquad (4.12)$$

$$Ca_0 = C\delta(A) - CBA \quad . \qquad (4.13)$$

(4.12) implies that:

$$CBA = CA^2 + Ca_1 A \quad . \qquad (4.14)$$

Substitute (4.14) into (4.13):

$$Ca_0 = C\delta(A) - CA^2 - Ca_1 A$$

$$= \text{, using Cayley-Hamilton ,}$$

$$C\delta(A) - C\,\text{trace}(A)A + C\,\text{det}(A) - Ca_1 A \qquad (4.15)$$

Let us suppose that

$$C = (C_1, C_2)$$

$$A = \begin{pmatrix} A_{11} & A_{12} \\ A_{21} & A_{22} \end{pmatrix} \quad .$$

Then,

$$CA = (C_1 A_{11} + C_2 A_{21}, \ C_1 A_{12} + C_2 A_{22}) \quad .$$

Let us now introduce the "observability" concept.

<u>Definition</u>. The system (4.1) (in this simple case $n = 2$) is *observable* if, for all x, the 1×2 vectors

$$C, \nabla C$$

are *linearly independent*, with:

$$\nabla C = \frac{\partial}{\partial x}(C) - CA$$
$$= -CA(x) \qquad (4.16)$$

Remark. In [3], I have defined *controllability* of the system (4.1) in the following way: Set

$$\nabla B = \frac{d}{dx}(B) + AC$$

$$\nabla(\nabla B) = \frac{d}{dx}(\nabla B) + A B$$

etc.

The controllability condition means that, for each $x \in R$, the vectors

$$B(x), \nabla B(x), \nabla\nabla B(x), \ldots$$

span R^n. *Observability* is now naturally defined as *controllability* of the dual system.

With the observability condition satisfied, we can find (using linear algebra, i.e., satisfying linear equations) scalar functions $b_0(x)$, $b_1(x)$ such that

$$C \frac{\partial A}{\partial x} = b_0 C + b_1 CA \qquad (4.17)$$

With the aid of these functions, we can write a_0, a_1:

$$a_0 = b_0 + \det(A)$$
$$a_1 = b_1 - \operatorname{trace}(A) \qquad (4.18)$$

The steps are now reversible, i.e., if (4.17) is satisfied, then the operators D, Δ_1 and Δ_2 exist.

Exercise. Analyze what happens in the case observability is not satisfied:

a) If C, $CA(x)$ are linearly dependent for all x.
b) C, $CA(x)$ are linearly independent for all but a discrete set of x.

The discrete set of x's which occur in (b) correspond to "singular points" of the differential operator Δ_1. Find the conditions that Δ_1 be a Fuchsian differential operator, in the sense that all of its singular points are regular.

<u>Exercise</u>. Show that operators Δ_1, D satisfying (4.11) exist if the system (4.1) is *observable*.

5. ISOSPECTRAL DEFORMATION OF THE SYSTEMS AND INPUT-OUTPUT RELATIONS

Now, suppose the system (4.1) depends on a parameter τ. We see that the operator Δ_1, D determining the input-output relations also depend on τ. Suppose that the deformation is *isospectral*, in the sense that there is an $n \times n$ matrix function $E(x,t)$ such that:

$$\frac{\partial A}{\partial \tau} - \frac{\partial E}{\partial x} = [A, E] \quad . \tag{5.1}$$

Let

$$\nabla_\tau = \frac{\partial}{\partial x} - A \quad .$$

Thus, $\Delta_{1,\tau}$ and D_τ are defined by the relations (4.9), (4.10), i.e.,

$$C \Delta_{1,\tau} = C D_\tau \nabla_\tau$$

From now on, we leave the τ-subscript off, and just write

$$\nabla = \frac{\partial}{\partial x} - A \tag{5.2}$$

$$\Delta, C = CD\nabla \tag{5.3}$$

Thus,

$$\frac{\partial \nabla}{\partial \tau} = -\frac{\partial A}{\partial \tau}$$

$$= \quad , \text{ using } (5.1),$$

$$= -\frac{\partial E}{\partial x} + [E, A] \quad . \tag{5.4}$$

<u>Remark</u>. Keep in mind that this is an *operator* equation, considering each as operators on vector-valued functions of x.

The input-output relations of (4.1) can now be written as follows:

$$y = C\bar{V}^{-1} Bu \qquad (5.5)$$

$$\equiv Tu$$

with

$$T = C\bar{V}^{-1} B \qquad (5.6)$$

Thus,

$$\frac{\partial T}{\partial \tau} = -C\bar{V}^{-1} \frac{\partial \bar{V}}{\partial \tau} \bar{V}^{-1} B$$

$$= C\bar{V}^{-1}([E,A-\delta])\bar{V}^{-1} B$$

$$= -C\bar{V}^{-1}[E,\bar{V}]\bar{V}^{-1} B$$

$$= C[\bar{V}^{-1},E]B \quad . \qquad (5.7)$$

Is there a *scalar* operator S such that

$$\frac{\partial T}{\partial \tau} = [T,S] \quad ? \qquad (5.8)$$

Now,

$$[T,S] = C[\bar{V}^{-1},S]B \quad .$$

Hence, (5.7) would follow (at least formally) if the following conditions were satisfied:

$$C[\bar{V}^j,E]B = [C\bar{V}^j B, S] \quad . \qquad (5.9)$$

In turn, condition (5.9) will be related to the *isospectral deformation* of the input-output relation:

$$u \to Tu \quad .$$

6. EIGENVECTORS OF THE INPUT-OUTPUT RELATIONS

Continue with a scalar input-output linear system of the form:

$$\frac{dz}{dx} = A(x)z + Bu$$

$$y = Cz \quad ,$$

or, in the operator notation introduced in previous sections,

$$\nabla z = Bu$$

$$y = Cz \quad,$$

with

$$\nabla = \delta - A = \frac{d}{dx} - A \quad.$$

A function $x \to u(x)$ is *an eigenvector of the input-output relation with eigenvalue* λ, if there is a z such that

$$\begin{aligned} \nabla z &= Bu \\ \lambda u &= Cz \end{aligned} \quad. \tag{6.1}$$

Thus, also

$$\nabla z = \lambda^{-1} BCz \quad. \tag{6.2}$$

Then, the *values of* λ *such that*

$$(\nabla - \lambda^{-1} BC)$$

is not one-one are the *eigenvalues*.

Example. *Sturm-Liouville.*

$$n = 2$$

$$z = \begin{pmatrix} z_1 \\ z_2 \end{pmatrix}$$

$$B = \begin{pmatrix} b \\ 0 \end{pmatrix}; \quad C = (0,1)$$

$$A = \begin{pmatrix} 0, & q(x) \\ -1, & 0 \end{pmatrix}$$

$$BC\begin{pmatrix} z_1 \\ z_2 \end{pmatrix} = B(z_2) = \begin{pmatrix} z_2 \\ 0 \end{pmatrix}$$

$$\nabla z = \begin{pmatrix} \frac{dz_1}{dx} \\ \frac{dz_2}{dx} \end{pmatrix} - \begin{pmatrix} 0 & q \\ -1 & 0 \end{pmatrix}\begin{pmatrix} z_1 \\ z_2 \end{pmatrix} = \begin{pmatrix} \frac{dz_1}{dz} - qz_2 \\ \frac{dz_2}{dx} + z_1 \end{pmatrix}$$

SYSTEMS AND OPERATORS

Thus, (6.2) is satisfied if and only if:

$$\frac{dz_1}{dx} - qz_2 = \lambda^{-1} z_2$$

$$\frac{dz_2}{dx} + z_1 = 0$$

or

$$-\frac{d^2 z_2}{dx^2} - qz_2 = \lambda^{-1} z_2 \quad ,$$

which is just the Sturm-Liouville eigenvalue problem.

This "algebraization" of the Sturm-Liouville problem suggests a general systems-theoretic approach to "boundary value problems" and "spectra". This will be pursued in the next chapter.

BIBLIOGRAPHY

1. R. Brockett, <u>Finite Dimensional Linear Systems</u>, Wiley, 1970.

2. E.W. Kamen, "Representation and Realization of Operational Differential Equations with Time-Varying Coefficients", J. Franklin Institute <u>301</u> (1976), 559, 571.

3. R. Hermann, "On the Accessibility Problem of Control Theory", <u>Proc. Symp. on Differential Equations</u>, J. Lasalle (ed.), Academic Press, 1961.

Chapter 14

ISOSPECTRAL DEFORMATIONS OF BOUNDARY VALUE PROBLEMS
OF LINEAR SYSTEMS

1. INTRODUCTION

In "Interdisciplinary Mathematics", Volume 9, I have described how "boundary value problems" for linear differential equations may be set-up *geometrically*. I will now show how these concepts may be used to study linear "time-varying" systems. (However, we continue to denote "time" by "x", in order to keep with the *physics* notation.) We continue with systems of the following form:

$$\frac{dz}{dx} = A(x)z + Bu$$
$$y = Cz \quad .$$
(1.1)

Here z will be assumed to be a vector in a real finite dimensional vector space Z. u,y will be real valued functions of x.

$$C: Z \to R$$

$$B: R \to Z$$

are linear maps.

As with the last chapter, we want to consider eigenvectors of the input-output map, i.e., functions $x \to u(x)$ such that there exists a map $x \to z(x)$: $R \to Z$ and a real number λ such that

$$\lambda^{-1} u = Cz$$

$$\frac{dz}{dx} = Az + Bu \quad ,$$

or:

$$\frac{dz}{dx} = (A + \lambda BC)z \quad .$$
(1.2)

Thus, we must study properties of the solutions of the one-parameter family of *homogeneous* linear systems (1.2). This is, of course, a very classical topic; in this form it was most extensively studied in the work of G.D. Birkhoff; there are many relevant papers in Volume 1 of his *Collected Works*.

2. BOUNDARY CONDITIONS

The general concept of "boundary condition" was treated in Birkhoff's work. I gave a coordinate-free, abstract-vector space version in "Interdisciplinary Mathematics", Volume 9.

Consider a system of the form

$$\frac{dz}{dx} = A(x)z \tag{2.1}$$

$$a \leq x \leq b \quad .$$

For each integer r, let

$$G^r(Z)$$

be the Grassmann manifold of r-dimensional linear subspaces of Z.

Definition. A set of *boundary conditions* for the differential equation (2.1) is a pair

$$(\gamma_a, \gamma_b) \in G^r(Z) \times G^s(Z)$$

of linear subspaces of Z. A solution $x \to z(x)$ of (2.1) is said to *satisfy the boundary conditions* if:

$$\boxed{\begin{array}{l} z(a) \in \gamma_a \\ \\ z(b) \in \gamma_b \end{array}} \tag{2.2}$$

We can, of course, rephrase (2.2) in terms of the flow in $GL(Z)$ generated by $x \to A(x)$. Let $x \to g(x) \in GL(Z)$ be the curve such that

$$\frac{dg}{dx} = A(x)g(x) \tag{2.3}$$

$$g(a) = 1$$

Then,

$$z(x) = g(z)(z(a)) \quad ,$$

i.e., the solution $x \to z(x)$ to (2.1) is the *orbit* of the linear flow.

Thus, there is a non-zero solution of the boundary value problem if and only if

BOUNDARY VALUES   191

$$g(b)(\gamma_a) \cap \gamma_b \neq (0) .$$

Now, if

$$r + s > \dim Z \tag{2.4}$$

there will always be a solution by linear algebra. Hence,

$$r + s \leq \dim Z , \tag{2.5}$$

is a realistic situation. Most of the classical work is concerned with the case that:

$$r + s = \dim Z .$$

3. "LOCAL" ISOSPECTRAL DEFORMATION OF SYSTEMS

Consider again the system

$$\begin{aligned} \frac{dz}{dx} &= A(x)z + Bu \\ y &= Cx , \end{aligned} \tag{3.1}$$

together with the associated linear, one-parameter linear homogeneous system

$$\frac{dz}{dx} = (A + \lambda BC)z . \tag{3.2}$$

We will be concerned with families of such systems, depending on an additional parameter τ. For simplicity, assume that B and C are independent of both x and τ. Then, (3.2) takes the form:

$$\frac{dz}{dx} = (A(x,\tau) + \lambda BC)z . \tag{3.3}$$

Let us now postulate a τ-dependence of the following form:

$$\frac{dz}{d\tau} = D(x,\tau,\lambda)z . \tag{3.4}$$

Definition. The deformation is said to be *"locally" isospectral* if the composite system (3.3)-(3.4) is *completely integrable in the Mayer-Frobenius sense for each value of* λ. (We now add a "locality" to indicate that the "boundary conditions" have not yet been imposed.)

We can now work out the complete integrability conditions for (3.3)-(3.4) by cross-differentiating in the usual way:

$$\frac{\partial^2 z}{\partial x \partial \tau} = \frac{\partial A}{\partial \tau} z + (A + \lambda BC) \frac{\partial z}{\partial \tau}$$

$$= \frac{\partial A}{\partial \tau} z + ADz$$

$$= \frac{\partial D}{\partial x} z + D \frac{\partial z}{\partial x}$$

$$= \frac{\partial D}{\partial x} z + D(A + \lambda BC) z \quad .$$

Thus, it takes the familiar form:

$$\frac{\partial A}{\partial \tau} - \frac{\partial D}{\partial x} = [D, A + \lambda BC] \quad . \tag{3.5}$$

We have already investigated solutions to these equivalences in the 2×2 case where D is assumed to depend polynomially on λ.

4. ISOSPECTRAL DEFORMATION WITH BOUNDARY CONDITIONS

Continue with the notation of Section 3, *with* (3.5) *satisfied*. Let $(\gamma_a, \gamma_b) \in G^r(Z) \times G^s(Z)$ be a pair of boundary conditions. The *spectrum* is the set of all complex numbers λ such that there is a nonzero solution of (3.2) satisfying the boundary conditions. We ask:

> What are the further conditions on D assuring that the spectrum, as a subset of $\mathbb{C}$, is independent of τ?

In order to deal with this, suppose that, for λ fixed,

$$x \to z(x, \lambda)$$

is a curve in Z such that:

$$\frac{\partial z}{\partial x} = (A(x,0) + \lambda BC) z(x, \lambda) \tag{4.1}$$

$$z(a, \lambda) \in \gamma_a \tag{4.2}$$

$$z(b, \lambda) \in \gamma_b \quad . \tag{4.3}$$

Because of the complete integrability of the Equations (3.3) and (3.4), we can find a map such that:

$$z(x,0,\lambda) = z(x,\lambda) \tag{4.4}$$

$$\frac{\partial z}{\partial x} = (A(x,\tau,\lambda) + \lambda BC)z \tag{4.5}$$

$$\frac{\partial z}{\partial \tau} = D(x,\tau,\lambda)z \quad . \tag{4.6}$$

We know that z satisfies the following boundary conditions

$$z(a,0,\lambda) \in \gamma_a$$
$$z(b,0,\lambda) \in \gamma_b \tag{4.7}$$

We would like these boundary conditions to be satisfied for *all* τ, i.e.,

$$z(a,\tau,\lambda) \in \gamma_a$$
$$z(b,\tau,\lambda) \in \gamma_b \tag{4.8}$$

for *all* τ .

Using Equations (4.6), we see that the necessary and sufficient conditions for (4.8) are that the following conditions be satisfied:

$$\boxed{\begin{array}{c} D(a,\tau,\lambda)(\gamma_a) \subset \gamma_a \\ D(b,\tau,\lambda)(\gamma_b) \subset \gamma_b \end{array}} \tag{4.9}$$

Let us sum up as follows:

<u>Theorem 4.1</u>. Suppose conditions (4.9) are satisfied for all λ. Then, "local" isospectral deformation also preserves the spectrum of the system with boundary conditions.

So far, a and b have been finite real numbers. Obviously, the "singular" cases, i.e.,

$$a = 0, \quad b = \infty ,$$

should be considered also. The techniques discussed above will generalize in one form or another. (First, the *general* theory of "singular boundary value problems" à la Weyl, Titchmarsh and Kodaira must be generalized to this setting.) For example, the simplest case might be:

$$\gamma_a = Z$$

$$\gamma_\infty = (0) \ .$$

Obviously, the spectrum will be preserved under deformation by τ if

$$\lim_{x \to \infty} D(x,\tau,\lambda)$$

exists, in some appropriate sense.

For example, in the Sturm-Liouville problem, with the isospectral deformation that now classical one generated by the Korteweg-de Vries equation

$$D = \begin{pmatrix} -4\lambda^3 - 2\lambda q - \frac{\partial q}{\partial x}, & -\frac{\partial^2 q}{\partial x^2} - 2\lambda \frac{\partial q}{\partial x} - 4\lambda^2 q - 2q^2 \\ 4\lambda^2 + 2q, & 4\lambda^3 + 2\lambda q + \frac{\partial q}{\partial x} \end{pmatrix} \quad (4.10)$$

with

$$\frac{\partial q}{\partial \tau} + 6q \frac{\partial q}{\partial x} + \frac{\partial^3 q}{\partial x^3} = 0 \quad (4.11)$$

$$D(x,\tau,\lambda) \to \begin{pmatrix} -4\lambda^3, & 0 \\ 4\lambda^2, & 4\lambda^3 \end{pmatrix} \quad (4.12)$$

provided $q(x,\tau)$ and its space derivatives go to zero as $x \to \infty$. (The solution to the "Cauchy problem" obtained via the now classical Inverse-Scattering Technique-Gelfand-Levitan Equation will assure that this is satisfied for all τ if it is for one value.)

We see that the "singular boundary value theory" is another classical topic which has a very natural interpretation in the systems format, and whose general version does not yet exist in the literature. I plan to work on this in more detail later on.

Here is a final remark about the "regular" boundary value problem in the Sturm-Liouville case. The conditions that (4.9) be satisfied can be read off of (4.10). They require that the solutions of the Korteweg-de Vries equation (4.11) satisfy certain boundary conditions in x, at $x = a$ and $x = b$. Thus, existence of isospectral deformations for certain Sturm-Liouville boundary value problems can be deduced from this question about Korteweg-de Vries:

Question. Is the existence of such a solution of Korteweg-de Vries also necessary for isospectral deformation of Sturm-Liouville boundary value problems?

Chapter 15

THE KORTEWEG-DE VRIES-STURM-LIOUVILLE SYSTEM
FROM THE POINT OF VIEW OF ISOSPECTRAL
DEFORMATIONS OF SYSTEMS

1. INTRODUCTION

In previous chapters I have considered families of linear systems depending on a variable τ as a parameter (with x the independent variable) of the following form:

$$\frac{\partial z}{\partial x} = A(x,\tau)z + Bu$$
$$y = Cz \quad . \tag{1.1}$$

(A,B,C are constant matrices. u,z,y are the *input*, *state* and *output* vectors.) We say that this "deformation" of systems is *isospectral* if there is a trace zero matrix function $D(x,\tau)$ such that:

$$\frac{\partial A}{\partial \tau} - \frac{\partial D}{\partial x} = [A,D] \quad . \tag{1.2}$$

I will now specialize the system (1.1) to be the system which gives rise to the classical Sturm-Liouville equation. I will follow the material in the article by Wadati, Sanucki and Konno [1].

2. ISOSPECTRAL DEFOMRATIONS OF STURM-LIOUVILLE

We say the system (1.1) is of *Sturm-Liouville type* if A is a 2×2 matrix of the form:

$$A = \begin{pmatrix} \lambda & q(x,\tau) \\ -1 & -\lambda \end{pmatrix}$$

Note that, with this choice, the solutions of the zero-input equation are:

$$\frac{d}{dx}\begin{pmatrix} z_1 \\ z_2 \end{pmatrix} = \begin{pmatrix} \lambda & q \\ -1 & -\lambda \end{pmatrix}\begin{pmatrix} z_1 \\ z_2 \end{pmatrix} = \begin{pmatrix} \lambda z_1 + q z_2 \\ -z_1 - \lambda z_2 \end{pmatrix}$$

Thus

$$z_1 = -\frac{dz_2}{dx} - \lambda z_2 \quad,$$

$$-\frac{d^2 z_2}{dx^2} - \lambda \frac{dz_2}{dx} = \lambda \left(-\frac{dz_2}{dx} - \lambda z_2 \right) + q z_2 \quad,$$

or

$$\frac{d^2 z_2}{dx^2} = (\lambda^2 - q) z_2 \quad,$$

which is, of course, the classical Sturm-Liouville equation.

Write:

$$A = A_0 + \lambda A_1 \quad,$$

with

$$A_0 = \begin{pmatrix} 0 & q \\ -1 & 0 \end{pmatrix}$$

$$A_1 = \begin{pmatrix} 1 & 0 \\ 0 & -1 \end{pmatrix}$$

Let us look for the 2×2 matrix D satisfying (1.2) as a polynomial in λ

$$D = \sum_{j=0}^{n} D_j \lambda^j \quad.$$

Denote partial derivatives by subscripts. Thus

$$A_{0,\tau} + \lambda A_{1,\tau} - \sum_{j=0}^{n} D_{j,x} \lambda^j = \sum_{j=0}^{n} [A_0, D_j] \lambda^j + \sum_{j=0}^{n} [A_1, D_j] \lambda^{j+1}$$

$$= \sum_{j=0}^{n} [A_0, D_j] \lambda^j + \sum_{j=1}^{n+1} [A_1, D_{j-1}] \lambda^j$$

Now, let us impose condition (2.1) by equating powers of λ:

$$A_{0,\tau} - D_{0,x} = [A_0, D_0] \tag{2.2}$$

$$A_{1,\tau} - D_{1,x} = [A_0, D_1] + [A_1, D_0] \tag{2.3}$$

$$[A_1, D_n] = 0 \tag{2.4}$$

$$D_{j,x} + [A_0, D_j] + [A_1, D_{j-1}] = 0 \tag{2.5}$$

$$\text{for } 2 \leq j \leq n \tag{2.6}$$

Since A_1 is constant, we have:

$$-D_{1,x} = [A_0, D_1] + [A_1, D_0] \tag{2.7}$$

(2.4) requires that D_n be diagonal

$$D_n = a_n A_1 . \tag{2.8}$$

(2.5) and (2.8) combine to give:

$$0 = a_{n,x} A_1 + [A_0, a_n A] + [A_1, D_{n-1}]$$

$$= a_{n,x} A_1 + [A_1, D_{n-1} - a_n A_0] . \tag{2.9}$$

Here is a basic property of the Lie algebra of 2×2 matrices of trace zero

> A_1 generates a Cartan subalgebra. Hence, if α is a 2×2 trace zero matrix such that
> $$[\alpha, A_1] = cA_1 ,$$
> then $c = 0$, $\alpha = $ multiple of A_1.

Apply this to (2.9):

$$a_{n,x} = 0 \tag{2.10}$$

$$D_{n-1} - a_n A_0 = a_{n-1} A_1 \tag{2.11}$$

Now introduce the following canonical basis for the Lie algebra of 2×2 trace real matrices:

$$A_1 ; \quad A_+ = \begin{pmatrix} 0 & 1 \\ 0 & 0 \end{pmatrix} ; \quad A_- = \begin{pmatrix} 0 & 0 \\ 1 & 0 \end{pmatrix}$$

$$[A_1, A_+] = 2A_+$$

$$[A_1, A_-] = -2A_- \qquad (2.12)$$

$$[A_+, A_-] = A_1 \quad .$$

Also

$$A_0 = qA_+ - A_- \quad . \qquad (2.13)$$

Hence, (2.11) takes the form

$$D_{n-1} = a_{n-1}A_1 + a_n qA_+ - a_n A_- \quad . \qquad (2.14)$$

Differentiate (2.14) with respect to x:

$$D_{n-1,x} = a_{n-1,x}A_1 + a_n q_x A_+$$

$$= \text{, using (2.3),}$$

$$D - [aA_+ - A_-, D_{n-1}] + [A_1, D_{n-2}] \quad .$$

We begin to see how these identities can determine everything inductively on n. I will not proceed further with the general case, but will now examine the low values of n.

3. THE STURM-LIOUVILLE, $n = 1$ CASE

Specialize the formulas of Section 2 to this case, giving:

$$A_{1,x} = 0$$

$$D_1 = a_1 A_1$$

$$D_0 = a_0 A_1 + a_1 qA_+ - a_1 A_- \quad .$$

(2.2) now takes the following form:

$$q_\tau A_+ - a_{0,x}A_1 - a_1 q_x A_+ = [qA_+ - A_-, \; a_0 A_1 + a_1 qA_+ - a_1 A_-]$$

$$= -2qa_0 A_+ - qa_1 A_1 - 2a_0 A_- + a_1 qA_1$$

$$= -2qa_0 A_+ - 2a_0 A_-$$

whence:

KORTEWEG-DE VRIES 201

$$q_\tau - a_1 q_x = -2qa_0$$

$$a_0 = 0$$

$$\boxed{q_\tau = a_1 q_x}$$

is then the *equation of isospectral deformation*, in this case. Of course, from the Lax-Kruskal point of veiw, this is a trivial case. If

$$\Delta = \frac{d^2}{dx^2} + q$$

$$\Delta_1 = \frac{d}{dq} + C$$

then

$$[\Delta, \Delta_1] = \frac{dq}{dx}$$

The deformation of Δ_τ is then just basically that generated by *translation of the independent variable* x.

4. STURM-LIOUVILLE, n = 2

(2.10) and (2.11) give:

$$D_2 = a_2 A_1$$

$$a_{2,x} = 0$$

$$D_1 = a_2 A_0 + a_1 A_1 \quad .$$

Apply (2.7):

$$-D_{1,x} = [A_0, D_1] + [A_1, D_0]$$

$$= a_1 [A_0, A_1] + [A_1, D_0]$$

$$= -a_2 A_{0,x} + a_{1,x} A_1$$

$$= a_1 [qA_+ - A_-, A_1] + [A_1, D_0]$$

$$= -2a_1 qA_+ - 2a_1 A_- + [A_1, D_0]$$

$$= -a_2 q_x A_+ + a_{1,x} A_1$$

Again, the Cartan algebra property of A_1 forces:

$$a_{1,x} = 0$$

$$D_0 = a_0 A_1 + (-a_2 q_x + 2a_1 q) A_+ + 2a_1 A_-$$

Finally, use (2.2):

$$q_x A_+ = a_{0,x} A_1 - (-a_2 q_{xx} + a_1 q_x) A_+ = [qA_+ - A_-, a_0 A_1 + (-a_2 q_x + a_1 q) A_+ + a_1 A_-]$$

$$= -2a_0 qA_+ + a_1 qA_1 + 2a_0 A_-$$

$$+ (-a_2 q_x + a_1 q) A_1$$

This forces

$$a_0 = 0$$

$$q_x + a_2 q_{xx} - a_1 q_x = 0$$

$$2a_1 q - a_2 q_x = 0 \quad,$$

hence,

$$(1 - a_1) = -2a_1$$

$$a_1 = -1/3 \quad.$$

Again, this is a trivial deformation, corresponding to translation invariance of the independent variable.

5. STURM-LIOUVILLE, $n = 3$, AND KORTEWEG-DE VRIES

In this case, we hope to find a deformation equation for q which is *nonlinear*. The data of Section 2 specializes as follows

$$D_3 = a_3 A_1 \tag{5.1}$$

$$a_{3,x} = 0 \tag{5.2}$$

$$D_2 = a_3 A_0 + a_2 A_1 \quad . \tag{5.3}$$

Let us now use (2.5) for $n = 2$.

$$D_{2,x} = a_3 A_{0,x} + a_{2,x} A_1$$

$$= - [A_0, a_2 A_1] - [A_1, D_1] \quad .$$

Again, the Cartan algebra property of A_1 implies

$$a_{2,x} = 0 \quad ,$$

hence:

$$a_3 A_{0,x} = - [A_0, a_2 A_1] - [A_1, D_1] \tag{5.4}$$

Differentiate (5.4) again

$$A_3, A_{0,xx} = [a_2 A_1, A_{0,x}] + [D_{1,x}, A_1]$$

$$= \quad , \text{ using (5.4) and (2.3)},$$

$$[a_2 A_1, [a_2 A_1, A_0] + [D_1, A_1]] + [A_1, [A_0, D_1] + [A_1, D_0]] \tag{5.5}$$

Differentiate (5.5) with respect to x:

$$A_3 A_{0,xxx} = [a_2, A_1, A_{0,xx}] + [A_1, [A_{0,x}, D_1] + [A_0, D_{1,x}]]$$

$$+ [A_1, [A_1, D_{0,x}]]$$

$$= [a_2 A_1, A_{0,xx}] + [A_1, [A_{0,x}, D_1]]$$

$$+ [A_1, [A_0, [D_1, A_0] + [D_0, A_1]]] \tag{5.6}$$

$$+ (\text{Ad } A_1)^2 (A_{0,\tau} - [A_0, D_0])$$

To determine the deformation, one now uses Equations (5.4)-(5.6). I will not carry it out to the bitter end, since we can already see qualitatively what is involved. Since

$$A_{0,xxx} = q_{xxx} A \quad ,$$

(5.6) is (when (5.4)-(5.5) is used to determine D_0, D_1) a *third* order differential equation in x and a first order in τ. Of course, we know from the Wadati-Sanuki-Konno paper that this equation is Korteweg-de Vries. What is most important, the general pattern for solving Equations (2.2)-(2.8) should be clearer.

Bibliography

1. M. Wadati, H. Sanuki, and K. Konno, "Relationships among Inverse Method, Bäcklund Transformations and an Infinite Number of Conservation Laws", Prog. Theor. Phys. <u>53</u> (1975), 419-436.

Chapter 16

PLANE WAVE SOLUTIONS OF THE ISOSPECTRAL
DEFORMATION EQUATIONS

1. INTRODUCTION

If

$$D\psi(x,t) = 0$$

is a partial differential equation in two independent variables, a *plane wave solution* is one of the following form:

$$\psi(x,t) = \theta(x - vt)$$

with constant v.

For example, the one-soliton solutions of Korteweg-de Vries are of this form.

Let $\mathscr{G}$ be a real Lie algebra. The *isospectral deformation* equations are

$$A_t - B_x = [A,B] \quad , \tag{1.1}$$

to be solved for a pair of maps

$$(x,t) \to A(x,t), B(x,t)$$

$R^2 \to \mathscr{G}$. (Subscripts denote partial derivatives.)

2. THE PLANE WAVE SOLUTIONS

Consider Equation (1.1). Look for a solution of the form

$$A(x,t) = \alpha(x - vt)$$

$$B(x,t) = \beta(x - vt) \quad ,$$

where $\tau \to \alpha(\tau), \beta(\tau)$ are a pair of maps $R \to \mathscr{G}$. Then

$$A_t = -v\alpha_\tau(x - vt) \tag{2.1}$$

$$B_x = \beta_\tau(x - vt) \quad .$$

(1.1) reduces to the following *ordinary* differential equation:

$$\boxed{-v\alpha_\tau(\tau) - \beta_\tau(\tau) = [\alpha(\tau), \beta(\tau)]} \qquad (2.2)$$

3. GENERAL SOLUTION OF THE ORDINARY DIFFERENTIAL EQUATION FOR THE PLANE WAVE SOLUTIONS

Set

$$\gamma(\tau) = \alpha(\tau)v + \beta(\tau) \quad . \qquad (3.1)$$

We can then rewrite (2.2) as follows:

$$\gamma_\tau = [\beta, \alpha]$$
$$= [\beta + \alpha v, \alpha]$$

or

$$\boxed{\gamma_\tau = [\gamma, \alpha]} \qquad (3.2)$$

This equation is the *isospectral deformation equation for a matrix*, which is very familiar (e.g., as the "Lax equation") in soliton theory. In order to find its general solution, let us suppose that $\mathcal{G}$ is the Lie algebra of a group G of $n \times n$ real matrices: Write

$$\gamma(\tau) = g(\tau)\gamma(0)g(\tau)^{-1} \quad (\equiv \text{Ad } g(\gamma(\tau))) \qquad (3.3)$$

Then

$$\gamma_\tau = g_\tau g^{-1}\gamma - g\gamma(0)g^{-1}g_\tau g^{-1} \quad .$$

Thus, (3.2) will be satisfied if:

$$g_\tau g^{-1} = -\alpha \quad ,$$

or

$$\boxed{\begin{aligned} g_\tau &= -\alpha(\tau)g(\tau) \\ g(0) &= 1 \end{aligned}} \qquad (3.4)$$

Equation (3.4) is a very familiar *linear* equation. We can then write:

$$\alpha(\tau)v - \beta(\tau) = g(\tau)(\alpha(0)v - \beta(0)) \qquad (3.4)$$

Let us sum up as follows:

<u>Theorem 3.1</u>. Pick a curve $\tau \to \alpha(\tau)$ in $\mathcal{G}$. Solve the linear equation (3.3). Use (3.4) to define $\beta(\tau)$. Set

$$A(x,t) = \alpha(x-vt)$$

$$B(x,t) = \beta(x-vt) \quad .$$

This gives the general solution of the plane wave solution of the isospectral deformation problem.

GROUP IV

USEFUL CONCEPTS IN TIME-VARYING LINEAR SYSTEMS THEORY

INTRODUCTION

As we have seen, a central area for both systems theory and the theory of nonlinear waves is the theory of systems of linear, time-varying differential equations. This is a very classical topic, which was under intensive development from 1880-1920 (Klein, Poincaré, Halphen, Wilczynski, G.D. Birkhoff, Thomé, Frobenius, Riemann, Schlessinger, Picard, Vessiot, and everyone else!) An addition was made in the 1920's by Burchnall and Chaundy, which has turned out to be the essence of the Kruskal-Lax theory of Korteweg-de Vries. What must be done is to rework this material and extract and extend what we need. Here is a beginning.

Chapter 17

INFINITE DIMENSIONAL LIE GROUPS, HALPHEN-WILCZYNSKI
THEORY, AND LINEAR SYSTEMS

1. INTRODUCTION

One of the great accomplishments of 19th century Lie theory-differential geometry is the Halphen-Wilczynski theory of linear ordinary differential operators and "projective differential geometry" of curves in $P_n(R)$ or $P_n(\mathbb{C})$. This material is most readily accessible in the book *Projective Differential Geometry of Curves and Surfaces* by Wilczynski (reprinted by Chelsea). (This "accessibility" is only relative; it is very difficult for the modern mathematician to learn about the considerable work done in the 19th century. This is part of a historical phenomenon that has disturbed me for many years--most of the best work in 19th century differential geometry and geometric differential equation theory is now lost to us.) As Wilczynski points out in the first chapter, it may be thought of as a special case of Lie's general theory of *infinite Lie groups* (i.e., transformation groups defined by differential operators).

My aim in this chapter is to show that this material is closely related to the theory of linear, time-varying systems. To explain why, start with linear time-independent systems

$$\frac{dx}{dt} = Ax + Bu$$
$$y = Cx \, .$$
(1.1)

Certain transformations in (x,y,u,t)-space then take an *arbitrary* system of this type into another one. Thus, if Σ denotes the space of such systems, G the group of such transformations, the action

$$G \times \Sigma \to \Sigma$$

defines G as a *transformation group* on Σ. As I have explained already in this volume, and earlier ones (e.g., "Interdisciplinary Mathematics", Volumes 8, 9, 13) the important "systems-theoretic" questions are often also the important Lie-theoretic ones, e.g., the description of the orbits, invariants, etc.

In this case, Σ forms a finite dimensional manifold, and G is an "ordinary" Lie group. If one now generalizes to the case where the system is *time-varying*, everything immediately becomes *infinite dimensional*, which is much less well developed territory. Another feature of interest is that what the physicists call *gauge fields and groups* is closely related.

2. LINEAR TIME-VARYING (INPUT-OUTPUT) SYSTEMS OF CLASSICAL TYPE

Let X, U, Y be finite dimensional vector spaces, called *state*, *input*, *output* spaces. $L(X,U)$ denotes the vector space of linear maps $X \to U$.

An (input-output) *system* is of the form:

$$\frac{dx}{dt} = A(t)x + B(t)u$$
$$y = C(t)x \quad , \tag{2.1}$$

where

$$t \to A(t) \in L(X,X)$$
$$t \to B(t) \in L(U,X)$$
$$t \to C(t) \in L(X,U)$$

are curves in the space of linear maps. Let us denote such a system as

$$\underline{\sigma} = (\underline{A}, \underline{B}, \underline{C}) \tag{2.2}$$

(The underline should indicate "time-varying".)

"t" is a time-parameter varying over some fixed interval. Let $\underline{U}, \underline{Y}$ be the space of input and output curves, i.e., the space of curves

$$\underline{u}: t \to u(t)$$
$$\underline{y}: t \to y(t)$$

<u>Definition</u>. The system (2.2) is said to be of *classical type* if there are linear differential operators

$$\Delta_1: \underline{Y} \to \underline{Y}$$
$$\Delta_2: \underline{U} \to \underline{U}$$

such that the set of all $(\underline{u},\underline{y}) \in \underline{U} \times \underline{Y}$ for which there exists an $\underline{x} \in \underline{X}$ such that (2.1) is satisfied is describable as the set of solutions of the differential equations:

$$\Delta_1 y = \Delta_2 u \quad . \tag{2.3}$$

In words, the "classical systems" are those for which the input-output relations are given as the set of solutions of a linear differential equation.

For systems of classical type, we then have the mapping

$$(\underline{A},\underline{B},\underline{C}) \to (\Delta_1, \Delta_2)$$

This mapping is the basic mathematical object in our study.

3. LINEAR STATE-TIME TRANSFORMATIONS

Consider a system of type (1.1). It can be written as an exterior differential system $\mathscr{E}(\underline{A},\underline{B},\underline{C})$.

$$\begin{aligned} dx - (A(t)x + B(t)u)\,dt &= 0 \\ y - C(t)x &= 0 \end{aligned} \qquad (3.1)$$

As such, it lives on the manifold $X \times U \times Y \times T = M$ (T = time variable manifold). Let

$$\phi: M \to M$$

be a diffeomorphism. It is said to *transform system* $(\underline{A},\underline{B},\underline{C})$ *into system* $(\underline{A}',\underline{B}',\underline{C}')$ if

$$\phi^*(\mathscr{E}(\underline{A}',\underline{B}',\underline{C}')) = \mathscr{E}(\underline{A},\underline{B},\underline{C})$$

i.e., if ϕ defines an isomorphism of the exterior differential system defined by the two systems. The set of all such transformations (i.e., diffeomorphisms of M which *preserve* the *linear input-output systems*) forms a *group*. It is this group which is the main object of study. As we shall see, Wilczynski's book is a study of a piece of this group.

In this chapter, we are mainly interested in what I will call *linear state-time transformations*; namely, those of the form:

$$\phi(x,u,y,t) = (x',u',y',t') \quad ,$$

with

$$\begin{aligned} x' &= g(t)x \\ u' &= \alpha(t)u \\ y' &= \beta(t)y \\ t' &= h(t) \end{aligned} \qquad (3.2)$$

For each t, $g(t) \in GL(X)$, $\alpha(t) \in GL(U)$, $\beta(t) \in GL(Y)$. Notice that (3.2) allows for the possibility of change of time parameter t. We will denote the group of all such transformations as G.

Let us see how the transformation (3.2) affect input-output systems. Suppose $\mathcal{E}'$ is generated by

$$dx' - (A'(t')x' + B'(t')u') dt'$$

$$y' - C'(t')x' \quad .$$

Then

$$\phi*(dx' - (Ax' + B'u')dt')$$

$$= \alpha(g(t)x) - (A'(h(t))g(t)x + B'(h(t))\alpha u) \frac{dh}{dt} dt$$

$$g(t)^{-1}\phi*(dx' - (A'x' + B'u')dt')$$

$$= dx + g(t)^{-1} \frac{dg}{dt} x \, dt - g(t)^{-1} A'(h(t))g(t)x \frac{dh}{dt} dt$$

$$- g(t)^{-1} B'(h(t))\alpha(t) u \, dt$$

Similarly,

$$\phi*(y' - C'(t')x') = \beta(t)y - C'(h(t))g(t)x$$

Let us sum up as follows:

<u>Theorem 3.1.</u> Let $\phi : M \to M$ be the transformation given by formulas (3.2), i.e., ϕ is a *state-time transformation*. Then, ϕ sends the system

$$\underline{\sigma} = (\underline{A}, \underline{B}, \underline{C})$$

into

$$\underline{\sigma}' = (\underline{A}', \underline{B}', \underline{C}') \quad ,$$

where

$$A(t) = g(t)^{-1} A'(h(t))g(t) \frac{dh}{dt} - g(t)^{-1} \frac{dg}{dt} \tag{3.3}$$

$$B(t) = g(t)^{-1} B'(h(t))\alpha(t) \frac{dh}{dg} \tag{3.4}$$

$$C(t) = \beta(t)^{-1} C'(h(t))g(t) \tag{3.5}$$

One main goal of the theory I am developing here is to *describe the orbits and orbit spaces of the action of this group on the input-output systems.*

4. TIME-PRESERVING SYSTEM ISOMORPHISMS

Let $X \times U \times Y \times T = M$ be as before, and let G be the group of diffeomorphisms of M described above, which maps a linear time-varying system into another one. Let G_0 be the group of $g_0 \in G$ which *leave time invariant*, i.e., satisfy

$$g_0^*(t) = t = t' \quad .$$

(Physicists sometimes call this the group of "pure" gauge transformations.)

<u>Theorem 4.1.</u> G_0 is an invariant subgroup of G.

<u>Proof.</u> For $g \in G$, by definition, there is a function of one variable $t \to h(t)$ such that

$$g^*(t) = h(t) \quad . \tag{4.1}$$

Let $g_0 \in G_0$. Then

$$(gg_0 g^{-1})^*(t) = (g^{-1*} g_0^* g^*)(t)$$

$$= g^{-1*} g_0^*(h(t))$$

$$= g^{-1*} h(g_0^*(t))$$

$$= g^{-1*} h(t)$$

$$= \text{, using (4.1),}$$

$$t \quad .$$

This shows that $gg_0 g^{-1}$ belongs to G_0, which proves Theorem 4.1.

This "trivial" group theory plays the key role in Wilczynski's work. He first finds "canonical forms" of the action of G_0 on systems (which are called "semi-canonical forms"), then investigates how the quotient group G/G_0 (which is isomorphic to the group of diffeomorphisms of the time-interval) acts on the semi-canonical forms. I will go into the general features of this at a later point.

5. SOME ELEMENTARY PARTS OF THE WILCZYNSKI THEORY

I will now temporarily leave the general systems framework in order to present some useful material in Wilczynski's book.

Let "t" denote the independent variable. We will consider linear differential operators in t, acting in scalar functions

$$t \to y(t) \equiv y$$

$$d = \frac{d}{dt}$$

$$\Delta = a_n d^n + a_{n-1} d^{n-1} + \cdots + a_0 \tag{5.1}$$

The coefficients $a_0, \ldots, a_n$ are functions of t, in a certain interval which will remain fixed.

Let $\mathscr{F}$ denote the C^∞, scalar-valued functions $t \to f(t)$ on this interval. $\mathscr{F}^\#$ denotes the group (under multiplication) of those functions which are nonzero in this interval. Let the product group

$$\mathscr{F}^\# \times \mathscr{F}^\#$$

act on Δ as follows:

$$(f_1, f_2)(\Delta) = f_1 \Delta f_2^{-1} . \tag{5.2}$$

The group G of diffeomorphisms of the time interval also acts on these differential operators.

$$\phi(\Delta) = \phi^{-1*} \Delta \phi^* ,$$

$$\text{for } \phi \in G .$$

Thus, the product group

$$G \times \mathscr{F}^\# \times \mathscr{F}^\#$$

acts on the space of the linear differential operators. The aim of the Halphen-Wilczynski theory is to parameterize the orbits of this group and to find *canonical forms*.

There is also an important relation with *projective differential geometry*, the theory of differential invariants attached to curves in $P_n(R)$ or $P_n(\mathbb{C})$. However, I will not go into this at this point. (It is on the agenda for later discussion, since I believe it plays an important role in both system theory and the theory of nonlinear waves.)

Let us suppose (for simplicity) that in the time interval we consider the differential operator Δ has no singular points. Then, we can obviously arrange d, a gauge transformation (i.e., an action of $\mathscr{F}^\# \times \mathscr{F}^\#$) so that

$$a_n = 1 . \tag{5.3}$$

The gauge transformations which preserve this canonical form are those for which

$$f_1 = f_2 \quad , \tag{5.4}$$

i.e., the diagonal subgroup of $\mathscr{F}^{\#} \times \mathscr{F}^{\#}$.

Suppose that

$$\Delta = d^n + a_{n-1} d^{n-1} + \cdots \tag{5.5}$$

$$\Delta' = f^{-1} \Delta f$$

$$= d^n + a'_{n-1} d^{n-1} + \cdots \tag{5.6}$$

Now,

$$d^n(f\underline{y}) = \sum_{j=0}^{n} \binom{n}{j} d^{n-j}(f) \, d^j(\underline{y}) \quad ,$$

hence,

$$d^n f = \sum_{j=0}^{n} \binom{n}{j} d^{n-j}(f) d^j \tag{5.7}$$

$$\Delta' = f^{-1} \left(\sum_{j=0}^{n} \binom{n}{j} d^{n-j}(f) d^j + a_{n-1} \sum_{j=0}^{n-1} \binom{n}{j} d^{n-1-j}(f) d^j \right.$$

$$\left. + a_{n-2} \sum_{j=0}^{n-2} \binom{n-2}{j} d^{n-2-j}(f) d^j + \cdots \right)$$

$$= f^{-1} \left(fd^n + nd(f) d^{n-1} + \frac{n(n-1)}{2} d^2(f) d^{n-2} + \cdots \right.$$

$$\left. + a_{n-1}(fd^{n-1} + (n-1)d(f)d^{n-2} + \cdots + a_{n-2} fd^{n-2} + \cdots \right)$$

$$= d^n + (nf^{-1}d(f) + a_{n-1})d^{n-1} + a_{n-1}(n-1)f^{-1}d(f)d^{n-2} + a_{n-2}d^{n-2}$$

$$+ \frac{n(n-1)}{2} d^2(f) d^{n-2} + \cdots$$

Thus, we can choose f so that Δ' takes the next *canonical form*:

$$\Delta' = d^n + a'_{n-2} d^{n-2} + \cdots \tag{5.8}$$

Let us look for gauge transformations (5.2) which preserve the canonical form

$$\Delta'' = f_1^{-1} \Delta' f_1$$

$$= d^n + n f_1^{-1} d(f_1) d^{n-1} + (n-1) f_1^{-1} d(f_1) d^{n-2}$$

$$+ a_{n-2} d^{n-2} + \frac{n(n-1)}{2} d^2(f) d^{n-2} + \cdots \qquad (5.9)$$

Let us require that Δ' also have this canonical form, i.e., *the coefficients of* d^{n-1} *vanishes*. This requires that

$$f_1 = constant \; .$$

All the rest of the coefficients are *invariants* of the action of $\mathscr{F}^\# \times \mathscr{F}^\#$

The next step in the process is to consider the action of the *group of changes of variable in* t. Rather than go into details about this, in the classical way, I will now transform the problem into input-output form, i.e., put System Theory into the picture.

6. THE INPUT-OUTPUT FORM OF SCALAR DIFFERENTIAL OPERATORS

$$\Delta = \frac{d^n}{dt^n} + a_{n-1} \frac{d^{n-1}}{dt^n} + \cdots + a_0 \qquad (6.1)$$

consider the input-output relation:

$$\Delta y = u \; . \qquad (6.2)$$

We know at least one way to realize them as *input-output relations*

$$\frac{dx}{dt} = Ax + Bu$$
$$y = Cx \qquad (6.3)$$

with

$$x = \begin{pmatrix} x_1 \\ \vdots \\ x_n \end{pmatrix} \in R^n \; .$$

Namely, in the relation (6.2) set

$$x_1 = y$$

$$x_2 = \frac{dy}{dt} = \frac{dx_1}{dt}$$

$$\vdots$$

$$x_n = \frac{dx_{n-1}}{dt} = \frac{d^{n-1}y}{dt^{n-1}}$$

Then

$$\frac{dx_n}{dt} = \frac{d^n y}{dt^n} = -a_{n-1}x_{n-1} - \cdots - a_0 x_1 + u$$

We can then write these in form (6.3):

$$\frac{dx_1}{dt} = 0 \cdot x_1 + x_2$$

$$\frac{dx_2}{dt} = 0 \cdot x_1 + 0 \cdot x_2 + x_3$$

$$\vdots$$

$$\frac{dx_{n-1}}{dt} = 0 \cdot x_1 + \cdots + x_n$$

$$\frac{dx_n}{dt} = -a_0 x_1 - a_1 x_2 - \cdots - a_{n-1} x_{n-1} + u$$

Set:

$$x = \begin{pmatrix} x_1 \\ \vdots \\ x_n \end{pmatrix}$$

$$\frac{dx}{dt} = \begin{pmatrix} 0 & 1 & 0 & \cdots & 0 \\ 0 & 0 & 1 & 0 & \cdots & 0 \\ \vdots & & & & \\ 0 & \cdots & & & 1 \\ -a_0 & -a_1 & \cdots & & -a_{n-1} \end{pmatrix} \begin{pmatrix} x_1 \\ \vdots \\ x_n \end{pmatrix} + \begin{pmatrix} 0 \\ \vdots \\ 1 \end{pmatrix} u$$

$$y = C \begin{pmatrix} x_1 \\ \vdots \\ x_n \end{pmatrix} = (1 \cdots 0) \begin{pmatrix} x_1 \\ \vdots \\ x_n \end{pmatrix}$$

Hence:

$$A = \begin{pmatrix} 0 & 1 & 0 & \cdots & 0 \\ 0 & 0 & 1 & \cdots & 0 \\ \vdots & & & & \\ -a_0 & -a_1 & & \cdots & -a_{n-1} \end{pmatrix}$$

$$B = \begin{pmatrix} 0 \\ \vdots \\ 1 \end{pmatrix}$$

$$C = (1 \;\cdots\; 0)$$

Notice that

B and C are *time-independent* .

Now, the gauge transformations

$$y \to f_1 y$$

$$u \to f_2 u$$

obviously introduce linear, time-dependent transformations on the state vectors x. I will now consider this in a somewhat less explicit form.

7. LINEAR TRANSFORMATIONS OF THE INPUT-OUTPUT SYSTEM WHICH ARE GAUGE TRANSFORMATIONS OF THE INPUT-OUTPUT RELATIONS

Consider a scalar input-output system of the form

$$\frac{dx}{dt} = A(t)x + Bu$$
$$y = Cx \; . \tag{7.1}$$

Suppose the input-output relations are

$$\Delta y = u \;, \tag{7.2}$$

where Δ is an n-th order linear (time-varying) differential operator; $x \in R^n$; $u, y \in R$.

Consider a transformation to another such system of the form:

$$f_2 y = y' \;,$$

$$x' = g(t)x'$$

$$u = f_1(t)u' \quad .$$

Then,

$$\frac{dx'}{dt} = \frac{dg}{dt}g^{-1}x' + gAg^{-1}x' + Bf_1u'$$

$$y' = f_2Cx$$

$$= (f_2Cg^{-1})x'$$

Thus,

$$\frac{dx'}{dt} = A'(t)x' + B'u'$$

$$y' = C'x' \quad , \tag{7.3}$$

with:

$$A' = \frac{dg}{dt}g^{-1} + fAg^{-1} \tag{7.4}$$

$$B' = gBf_1 \tag{7.5}$$

$$C' = f_2Cg^{-1} \quad . \tag{7.6}$$

Let us now enquire about the conditions (A,B,C) must satisfy in order that (7.2) be satisfied, i.e., *the input-output relations do not involve derivatives of the inputs*.

$$\frac{d^2x}{dt^2} = \frac{dA}{dt}x + A\frac{dx}{dt} + B\frac{du}{dt}$$

$$= \frac{dA}{dt}x + A(Ax + Bu) + B\frac{du}{dt}$$

$$\frac{dy}{dt} = C\frac{dx}{dt}$$

$$= CAx + CBu$$

$$= CAx \tag{7.7}$$

since

$$CB = (1 \cdots 0) \begin{pmatrix} 0 \\ \vdots \\ \vdots \\ 1 \end{pmatrix}$$

$$= 0 \tag{7.8}$$

$$\frac{d^2 y}{dt^2} = C \frac{dA}{dt} x + CA(Ax + Bu)$$

$$= C \left(\frac{dA}{dt} + A^2 \right) x + CABu$$

$$\frac{d^3 y}{dt^3} = C \left(\frac{d^2 A}{dt^2} + A \frac{dA}{dt} + \frac{dA}{dt} A \right) x + C \left(\frac{dA}{dt} + A^2 \right) (Bx + Cu)$$

$$+ C \frac{dA}{dt} Bu + CAB \frac{du}{dt}$$

Continuing in this way, we see that the condition for the input-output relations to take the form (7.2) (i.e., with no derivatives of u) is that:

$$\boxed{\begin{aligned} CA^j B &= 0 \\ \text{for } 0 &\leq j \leq n-2 \end{aligned}} \tag{7.9}$$

Suppose A' is given by (7.3). Let us find the condition that $t \to g(t)$ must satisfy (in addition to conditions (7.4)-(7.6)) which guarantee the following relations be satisfied:

$$C(A')^j B = 0 \tag{7.10}$$

for $0 \leq j \leq n-2$

$$CA'B = C \frac{dg}{dt} g^{-1} B + CgAg^{-1} F$$

$$= \text{, using relation (4.6),}$$

$$C \frac{dg}{dt} g^{-1} (B + f_2 CAB f_1)$$

$$= C \frac{dg}{dt} g^{-1} B \quad . \quad .$$

Thus, the first condition

$$\boxed{C\frac{dg}{dt}g^{-1}B = 0} \qquad (7.11)$$

must be satisfied.

$$C(A')^2 B = C\left(\frac{dg}{dt}g^{-1} + gAg^{-1}\right)^2 B$$

$$= C\left(\frac{dg}{dt}g^{-1} + gAg^{-1}\right)\left(\frac{dg}{dt}g^{-1} + gAg^{-1}\right)B$$

$$= C\left(\frac{dg}{dt}g^{-1} + gAg^{-1}\right)\left(\frac{dg}{dt} + gA\right)Bf_1$$

$$= C\left(\frac{dg}{dt}g^{-1}\frac{dg}{dt} + gAg^{-1}\frac{dg}{dt} + \frac{dg}{dt}A + gA^2\right)Bf_1$$

$$= \left(C\frac{dg}{dt}g^{-1}\frac{dg}{dt} + f_2 CAg^{-1}\frac{dg}{dt} + C\frac{dg}{dt}A + f_2 CA^2\right)Bf_1$$

Now,

$$C\frac{dg}{dt} = \frac{d}{dt}(Cg)$$

$$= \frac{d}{dt}(f_2 C)$$

$$= \frac{df_2}{dt}C \quad .$$

Hence, finally

$$C(A')^2 B = \left(\frac{df_2}{dt}Cg^{-1}\frac{dg}{dt} + f_2 CAg^{-1}\frac{dg}{dt} + \frac{df_2}{dt}CA\right)Bf_1$$

$$= f_2 CAg^{-1}\frac{dg}{dt}B \quad .$$

Continuing in this way, we see that

$$t \to g(t)$$

must satisfy the following conditions

$$\boxed{\begin{array}{c} CA^j g^{-1}\frac{dg}{dt}B = 0 \\ \text{for } 0 \le j \le n-2 \end{array}} \qquad (7.12)$$

Theorem 7.1. With (7.9) satisfied, i.e., the input-output relation of the system (7.7) given by relations (7.2), the coefficient of

$$\frac{d^{n-1}}{dt^{n-1}}$$

in Δ vanishes if and only if the following condition is satisfied:

$$\text{trace}(A) = 0 \ . \tag{7.13}$$

Exercise. Prove Theorem 7.1.

8. WILCZYNSKI EQUIVALENT TIME-VARYING LINEAR SYSTEMS

I will now consider one general version of certain material in Wilczynski's book. Consider two *scalar* input-output (for simplicity) systems of the form

$$\frac{dx}{dt} = A(t)x + Bu$$
$$y = Cx \tag{8.1}$$

$$\frac{dx'}{dt} = A'(t)x' + Bu'$$
$$y' = Cx' \ . \tag{8.2}$$

Notice that they have the *same* B and C matrices, which are constant. Suppose that (8.1) and (8.2) are related via a diffeomorphism of $X \times U \times T$ of the following form:

$$\phi : \begin{pmatrix} x \\ u \\ y \\ t \end{pmatrix} \to \begin{pmatrix} x' \\ u' \\ y' \\ t' \end{pmatrix}$$

with

$$\boxed{\begin{aligned} x' &= g(t)x \\ u' &= f_1(t)u \\ y' &= f_2(t)y \\ t' &= h(t) \end{aligned}} \tag{8.3}$$

We will then say that systems (8.1) and (8.2) are W-*equivalent*.

Notice that the set of diffeomorphisms of $M = X \times U \times T$ of the form (8.3) is a *group*. (Of course there are domain problems; technically, it only forms a *pseudogroup*. For the moment, I will not worry about this.) It is a typical "infinite Lie group". The condition that B and C be the *same* on both (8.1) and (8.2) further constrains ϕ. It is not necessarily obvious a priori that these constraints form a *subgroup*. In fact, we will now show that this is so.

Write (8.1)-(8.2) as exterior differential systems

$$dx - (Ax + Bu)dt = 0$$
$$y - Cx = 0 \qquad (8.4)$$

$$dx - (A'x + B'u)dt = 0$$
$$y - Cx = 0 \qquad (8.5)$$

Let $\phi: M \to M$ be the diffeomorphisms determined by formulas (8.3), i.e.,

$$\phi^*(x) = g(t)x$$
$$\phi^*(g) = f_2 y$$
$$\phi^*(u) = f_1 u \qquad (8.6)$$
$$\phi^*(t) = h(t)$$

The condition that ϕ maps (8.1) into (8.2) means that ϕ^* pulls back the ideal of forms generated by those in the left hand side of (8.5) into those of (8.4). Now, using (8.6) (subscripts denote partial derivatives):

$$\phi^*(dx - (A'x + Bu)dt) = d(gx) - (A'(h(t))gx + B(h(t))f_1 u)h_t dt$$
$$= g_t dtx + gdx - (A'(h(t))gx + B(h(t))f_1 u)h_t dt$$

$$g^{-1}\phi^*(dx - (A'x + Bu)dt) = g^{-1}g_t xdt + dx - (g^{-1}A'gx + g^{-1}Bf_1 u)h_t dt \quad .$$

Thus,

$$\boxed{-g(t)^{-1}g_t(t) + g(t)^{-1}A'(h(t))g(t)h_t(t) = A(t)} \qquad (8.7)$$

$$g(t)^{-1}Bf_1(t)h_t(t) = B \quad ,$$

or

$$\boxed{g(t)B = f_1(t)Bh_t(t)} \qquad (8.8)$$

Also,

$$\phi^*(y - Cx) = f_2 y - Cgx \quad ,$$

$$f_2^{-1}\phi^*(y - Cx) = y - f_2^{-1}Cgx$$

$$= y - Cx \quad ,$$

or

$$\boxed{Cg(t) = f_2(t)C} \qquad (8.9)$$

(8.7)-(8.9) are the relations which must be satisfied if ϕ is to be a W-transformation between systems. Notice that (8.8) and (8.9) are *conditions* on ϕ, while (8.7) serves to define A in terms of A'. (8.8) says that B is an *eigenvector* of $g(t)$, while (8.9) says that C is an eigenvector of $g(t)$ dual. These conditions obviously define the W-transformations as a *group*. Notice that each element of the group is parameterized by functions of t:

$$(h, g, f_1, f_2) \quad .$$

These functions are linked by relations (8.7)-(8.9).

9. THE W-GROUP AS A TRANSFORMATION GROUP ON THE SPACE OF SYSTEMS AND ITS DIFFERENTIAL INVARIANTS

Continue with the notation of Section 8. Now, change the emphasis; let $\Sigma_{B,C}$ denote the space of *all* systems of type (8.1) with fixed constant matrices B,C. (Continue with scalar input-output, for simplicity.) Let G be the space of all W-transformations. By its definition, G acts on $\Sigma_{B,C}$ as a transformation group. The transformation group map

$$G \times \Sigma_{B,C} \to \Sigma_{B,C}$$

is that defined by formulas (8.7). The ultimate aim is to calculate the *orbits* of this group acting on this space. One method of doing this (basically due to Lie) is to find *differential invariants*, i.e., another space Σ', another transformation group G' on Σ', a pair of maps

$$\alpha: G \to G'$$

$$\pi: \Sigma \to \Sigma'$$

such that:

    a) α is a group homomorphism

    b) π intertwines the transformation group action

    c) Σ', G' are parameterized by functions of t, so that α, π are given by *differential operations*.

Example. *The trace.*

If $\sigma = (A(t), B, C)$ is an element of Σ, set:

$$\pi(\sigma) = \text{trace}(A) \quad .$$

Σ' is now the space of real-valued functions $t \to a(t)$. Let us apply the trace to both sides of (8.7).

$$\text{trace}(A(t)) = h_t(t)\, \text{trace}(A'(h(t))) - \text{trace}(g(t)^{-1} g_t(t)) \qquad (9.2)$$

Let

$$\Sigma' = \text{space of all real valued functions of } t.$$

$$G = G' \quad .$$

Let

$$g = (A, h, f_1, f_2) \in G$$

act on Σ' as follows:

$$g \cdot f = h_t f(h(t)) - \text{trace}(g^{-1} g_t) \quad . \qquad (9.3)$$

It is readily seen that this defines the required transformation group action.

Theorem 9.1. Given $\sigma \in \Sigma$, there is a $g \in G$ such that

$$\pi(g\sigma) = 0$$

$$g = (h=t, g, f_1, f_2)$$

Proof. This follows from (9.2). $\text{trace}(A) = 0$ and $h = t$ require that

$$\text{trace}(g(t)^{-1} g_t(t)) = \text{trace}(A) \quad .$$

Such a g can obviously be found. (Of course, it is not unique!)

Remark. As we have seen, "trace $A = 0$" is related to the differential operator determining the input-output relations having $(n-1)$-st order terms equal to zero.

Theorem 9.2. Let $g = \sigma = \sigma'$, with $\sigma, \sigma' \in \Sigma_{A,B}$, $g \in G$ elements such that:

$$\pi(\sigma) = \pi(\sigma') = 0 \quad .$$

Then,

$$\det g(t) = \text{constant} \quad . \tag{9.4}$$

Proof. Left as an exercise.

Remark. These results can be described in alternate form as follows. Let $\Sigma^0_{B,C}$ be the space of all systems $(A(t), B, C)$ with

$$\text{trace}(A) = 0 \quad . \tag{9.5}$$

Then,

a) Each orbit of G on $\Sigma_{B,C}$ touches $\Sigma^0_{B,C}$

b) The intersection of each orbit of G with $\Sigma^0_{B,C}$ consists of an orbit of the subgroup of G determined by condition (9.4).

The trace is the easiest invariant. One can see from the special cases treated in Wilczynski's book that the others are more complicated nonlinear combinations of A and its derivatives. As an aid to understanding their character, we shall now introduce an auxilliary concept.

10. THE PROLONGATIONS OF A LINEAR, TIME-VARYING SYSTEM

Consider a linear system of the usual form:

$$\frac{dx}{dt} = A(t)x + Bu \qquad (10.1)$$

$$y = Cx \qquad (10.2)$$

For simplicity, continue with the assumption that inputs and outputs are scalars, and that B, C are constant. To facilitate computation, write:

$$\delta = \frac{d}{dt} \quad.$$

Then, (10.1) is rewritten as

$$\delta x = Ax + Bu \quad.$$

Differentiate further:

$$\delta^2 x = \delta(A)x + A\delta x + B\delta u$$

$$= \delta Ax + A(Ax + Bu) + B\delta u$$

$$= (\delta A + A^2)x + ABu + B\delta u$$

$$\delta^3 x = \delta(\delta A + A^2)x + (\delta A + A^2)(Ax + Bu) + (\delta A)Bu + AB\delta u + B\delta^2 u$$

and so on.

We see that we have relations of the following type:

for $j \geq 1$:

$$\delta^j x = A_j x + B\delta^j u + B_{j,0} u + \cdots + B_{j,j-1} \delta^{(j-1)} u \qquad (10.3)$$

The $A_j, B_{j,0}, \ldots, B_{j,j-1}$ are time-varying matrices which are constructed from A, its $(j-1)$-st derivatives, and B.

We can now add relations (10.2):

$$\delta^j y = CA_j x + CB\delta^j u + CB_{j,0} u + \cdots + CB_{j,j-1} \delta^{j-1} u \qquad (10.4)$$

We can now change variables in the Wilczynski manner:

$$\boxed{\begin{aligned} x'(t) &= g(t)x(h(t)) \\ f_1(t)u'(t) &= u(h(t)) \\ f_2(t)y'(t) &= y(h(t)) \end{aligned}} \qquad (10.5)$$

$$\begin{aligned} \delta x' &= \delta g g^{-1} x' + g(A(h(t))x(h(t)) + h_t + Bu(h(t)))h_t \\ &= (\delta g g^{-1})x' + g(t)A(h(t))g(t)^{-1}x'(t)h_t + g(t)Bf_1(t)u'(t)h_t \end{aligned}$$

$$\boxed{= A'(t)x'(t) + Bu'(t)} \qquad (10.6)$$

with:

$$\boxed{\begin{aligned} A'(t) &= (\delta g g^{-1})(t) + g(t)A(h(t))g(t)^{-1}h_t \\ h_t g(t)Bf_1(t) &= B \end{aligned}} \qquad (10.7)$$

(10.6) gives the transformation of the system (10.1) under the W-transformation (10.5). We can, of course, also calculate the prolonged system:

$$\boxed{\delta^j x' = A'_j x' + B\delta^j u' + B'_{j,0} u' + \cdots} \qquad (10.8)$$

It is the transform under the W-transformation of the prolonged system (10.3).

We see that the W-group acts on the coefficients of the prolonged system. It is the covariants of this action which we must compute. I will now turn to the 2×2 case.

11. THE (2 × 2)-CASE

Consider

$$\delta x = A(t)x + Bu$$
$$y = Cx$$

$A(t)$ is a 2×2 matrix of *trace zero*. (We have seen that we can reduce to this case.) Thus,

$$\delta^2 x = \delta Ax + A(Ax + Bu) + B\delta u$$
$$= (\delta A + A^2)x + ABu + B\delta u$$

$$\delta y = CAx + CBu$$

$$\delta^2 y = C(\delta A + A^2)x + CABu + CB\delta u \quad .$$

Let $t \to a(t)$ be a scalar function of t. Then,

$$\delta^2 y + a(t)y = (C(\delta A + A^2) + aC)x + CABu + CB\delta u \qquad (11.1)$$

Using Cayley-Hamilton

$$A^2 - (\text{trace } A)A + \det A = 0 \quad .$$

Since we have assumed

$$\text{trace } A = 0 \quad ,$$

we have:

$$A^2 = -\det A \quad . \qquad (11.2)$$

Combine (11.1) with (11.2):

$$\delta^2 y + ay = (C(\delta A - \det A) + aC)x + CABu + CB\delta u \qquad (11.3)$$

We can now *choose* a so that the first term in the right hand side vanishes. The input-output relations then take the following "canonical" form:

$$\boxed{\delta^2 y + ay = CABu + CB\delta u} \qquad (11.4)$$

a must then satisfy the following relation:

$$C\delta A = ((\det A) - a)C \quad . \tag{11.5}$$

Thus, a is an *eigenvalue* of $(\det A \cdot 1 - \delta A)$.

Now, make a W-transformation via formula (10.7). The relation

$$y'(t) = Cx'(t)$$

goes into

$$y(h(t)) = f_2(t)Cg(t)x(h(t))$$
$$= Cx(h(t)) \quad ,$$

whence:

$$\boxed{f_2(t)Cg(t) = C} \tag{11.6}$$

whence:

$$\delta f_2 Cg + f_2 C \delta g = 0$$

or

$$\boxed{f_2^{-1} \delta f_2 C = -C \delta g g^{-1}} \tag{11.7}$$

Apply δ to both sides of (10.7):

$$\delta A' = \delta^2 gg^{-1} - (\delta gg^{-1})^2 + \delta gA(h(t))g^{-1}h_t + g\delta A(h(t))g^{-1}h_t^2$$

$$- gAg^{-1}\delta gg^{-1}h_t + gAg^{-1}h_{tt}$$

$$(A')^2 = (\delta gg^{-1})^2 + h_t^2 g(t)g(t)A(h(t))^2 g(t)^{-1} + gAg^{-1}\delta gg^{-1}h_t$$

$$+ \delta gAg^{-1}h_t$$

Hence,

$$\boxed{\begin{aligned}\delta A' + (A')^2 &= (\delta^2 g)g^{-1} + h_t^2 g(t)(\delta A(h(t)) + A^2)g(t)^{-1} \\ &+ 2\delta gA(h(t))g^{-1}h_t + gA(h(y))g^{-1}h_{tt}\end{aligned}} \tag{11.8}$$

Remark. (11.8) says that $\delta A + A^2$ has a more *nearly* "tensorial" transformation law under W-transformation than δA or A^2 alone. $\delta A + A^2$ (and analogous expressions for higher derivatives) seems to play a key role in the Wilczynski theory. It is, in fact, very reminiscent of the "curvature two-form" of a connection. Notice also that (11.8) is independent of the dimensionality of the state space.

Now we shall combine (11.8) with the relations (11.6)-(11.7).

$$C((\delta A') + (A')^2) = C\delta^2 gg^{-1} + h_t^2 Cg(\delta A + A^2)g^{-1} + 2C\delta g A g^{-1} h_t + CgAg^{-1} h_{tt}$$

$$= \delta^2(f_2^{-1})f_2 C + h_t^2 f_2^{-1} C(\delta A + A^2)g^{-1} + 2C\delta(f_2^{-1})Ag^{-1} h_t$$

$$+ f_2^{-1} h_{tt} CA \quad .$$

We can now *choose* the change of variable $t \to h(t)$ so that the last two terms on the right hand side of (11.9) (which "break" the tensorial properties) *vanish*, i.e.,

$$\frac{h_{tt}}{h_t} = -2f_2 \delta(f_2^{-1})$$

$$= 2 \frac{\delta f_2}{f_2}$$

or

$$\delta \log \delta(h) = \delta 2 \log f_2$$

$$\boxed{\delta h = c f_2^2} \qquad (11.10)$$

If this restriction on the W-transformation is made, then:

$$C(\delta A' + (A')^2) = -a'C$$

$$C(\delta A + A^2) = -aC$$

$$\boxed{-a' = \delta^2(f_2^{-1})f_2 - h_t^2 a} \qquad (11.11)$$

This indicates that the mapping

$$(A,B,C) \to a$$

is a *differential invariant* with respect to the action of the subgroup of the W-group defined by conditions (11.10).

As an application of this result, notice that we can choose f_2 (and hence also h) so that

$$a' = 0 \quad.$$

This requires solving a third order nonlinear ordinary differential equation for h. (It involves the classical "Schwarzian" expression.) This enables us to state one concrete result.

Theorem 11.1. Let

$$\frac{dx}{dt} = A(t)x + Bu$$

$$y = Cx$$

be a system with scalar input-output and two-dimensional state space. Suppose in addition that the input-output relations are of the form

$$\Delta_1 y = \Delta_2 u \quad,$$

with Δ_1 a second order linear differential operator, Δ_2 a first order one. One can then change the system by a Wilczynski transformation so that

$$\Delta_1 = \delta^2 \equiv \frac{d^2}{dt^2} \quad. \tag{11.12}$$

Thus, (11.12) is a sort of "canonical form" under the Wilczynski group. We can also investigate when which W-transformations *preserve* the canonical form (11.12). This requires that

$$a = a' = 0$$

in (11.11), i.e.,

$$\delta^2(f_2^{-1}) = 0 \quad,$$

or that

$$f_2 = \frac{1}{c_1 t + c_2}$$

$$h_t = \left(\frac{c}{c_1 t + c_2}\right)^2$$

$$h = \int \left(\frac{c}{c_1 t + c_2}\right)^2 dt$$

$\equiv$ linear fractional transformation in t.

In fact, the appearance of the "projective group" is no accident. a is the "projective curvature" in the sense explained by Cartan in his book *Les espaces a connexion projective*.

Chapter 18

TIME-VARYING INPUT-OUTPUT SYSTEMS GOVERNED BY
LAPLACE INTEGRAL OPERATORS WITH
ALGEBRAIC "SYMBOLS"

1. INTRODUCTION

The job of "systems theorists" is to study *input-output relations*. Most of the serious effort has gone into those governed by linear, time-invariant differential equations of the form:

$$\frac{dx}{dt} = Ax + Bu$$
$$y = Cx \quad . \tag{1.1}$$

In this case, the mapping (input-output) (with zero initial data) is of the form

$$D\left(\frac{d}{dt}\right) y = N\left(\frac{d}{dt}\right) u \tag{1.2}$$

where $D(\ )$, $N(\ )$ are linear, constant coefficient differential operators.

In this chapter I will *begin* a study of systems of the form (1.1) with time-varying coefficient that satisfy a second order linear ordinary differential equation. For example, the *Bessel equation*

$$\frac{d^2y}{dt^2} + t^{-1}\frac{dy}{dt} + \left(1 - \frac{\lambda^2}{t^2}\right)y = u \tag{1.3}$$

is a good one to keep in mind, since its solution can be written down explicitly in terms of integrals of exponentials of *algebraic* functions. These expressions are known to workers in partial differential operators as *Fourier integral operators*. They also have some features which are of great interest from the point of view of *algebraic geometry*, using the techniques developed by Clyde Martin and myself.

2. INPUT-OUTPUT RELATIONS FOR SECOND ORDER, LINEAR DIFFERENTIAL EQUATIONS

Let

$$\Delta_t y = a_2 y'' + a_1 y' + a_0 y \tag{2.1}$$

be a second order, linear differential operator. (y denotes a function of the variable t. A prime denotes derivatives with respect to t.) Let y_1, y_2

be two linearly independent solutions of the equation

$$\Delta(y_1) = 0 = \Delta(y_2) \tag{2.2}$$

We will use the traditional "variation of parameters" formulas to solve

$$\Delta(y) = u \tag{2.3}$$

for arbitrary input u.

Look for a solution of (2.2) of the following form:

$$y = Ay_1 + By_2 , \tag{2.4}$$

where A,B are functions of t, such that

$$A'y_1 + B'y_2 = 0 . \tag{2.5}$$

$$\Delta y = a_2(A''y_1 + 2A'y_1' + Ay_1'' + B''y_2 + 2B'y_2' + By_2'')$$

$$+ a_1(A'y_1 + Ay_1' + B'y_2 + By_2')$$

$$+ a_0(Ay_1 + By_2)$$

$$= , \text{ using (2.2) and (2.5),}$$

$$a_2(A''y_1 + 2A'y_1' + B''y_2 + 2B'y_2') . \tag{2.6}$$

Differentiate (2.5):

$$A''y_1 + A'y_1' + B''y_2 + B'y_2' = 0 . \tag{2.7}$$

Substitute (2.7) into (2.5), and use (2.3):

$$a_2(-A'y_1' - B'y_2' + 2A'y_1' + 2B'y_2') = u ,$$

hence

$$\boxed{\begin{aligned} A'y_1' + B'y_2' &= u/a_2^{-1} \\ A'y_1 + B'y_2 &= 0 \end{aligned}} \tag{2.8}$$

LAPLACE INTEGRAL OPERATORS   239

Now, solve (2.8) with Cramer's rule:

$$A' = \frac{\begin{vmatrix} u & y_2' \\ 0 & y_2 \end{vmatrix}}{\begin{vmatrix} y_1' & y_2' \\ y_1 & y_2 \end{vmatrix}} = \frac{uy_2/a_2^{-1}}{\begin{vmatrix} y_1' & y_2' \\ y_1 & y_2 \end{vmatrix}}$$

$$B' = \frac{\begin{vmatrix} y_1' & u \\ y_1 & 0 \end{vmatrix}}{\begin{vmatrix} y_1' & y_2' \\ y_1 & y_2 \end{vmatrix}} = \frac{-uy_1/a_2^{-1}}{\begin{vmatrix} y_1' & y_2' \\ y_1 & y_2 \end{vmatrix}}$$

or

$$A(t) = \int_0^t \frac{u(t_1)y_2(t_1)a_2(t_1)^{-1} \, dt_1}{W(t_1)} + A(0)$$

$$B(t) = \int_0^t \frac{-u(t_1)y_1(t_1)a_2(t_1)^{-1} \, dt_1}{W(t_1)} + B(0)$$

$$W(t) = y_1'y_2 - y_2'y_1 \equiv \text{the Wronskian}$$

Now,

$$W' = y_1''y_2 - y_2''y_1$$

$$= \frac{-(a_1y_1' + a_0y_1)y_2 - (a_1y_2' + a_0y_2)y_1}{a_2}$$

$$= -\frac{a_1}{a_2} W$$

$$\frac{dW}{W} = -\frac{a_1}{a_2} dt \quad ,$$

$$W = e^{-\int \frac{a_1}{a_2} dt}$$

or

$$W(t) = W(0) e^{-\int_0^t \frac{a_1(t_1)}{a_2(t_2)} dt_1}$$

Hence,

$$A(t) - A(0) = W(0) \int_0^t u(t_1) y_2(t_1) a_2(t_1)^{-1} \exp\left(\int_0^{t_1} \frac{a_1(t_2)}{a_2(t_2)} dt_2\right) dt_1$$

$$B(t) - B(0) = W(0) \int_0^t - u(t_1) y_1(t_1) a_2(t_1)^{-1} \exp\left(\int_0^{t_1} \frac{a_1(t_2)}{a_2(t_2)} dt_2\right) dt_1$$

Let us look for the solution $y(t)$ of (2.3) with

$$y(0) = 0 = y'(0) \quad . \tag{2.11}$$

(2.11) requires that:

$$A(0) y_1(0) + B(0) y_2(0) = 0$$

$$A(0) y_1'(0) + B(0) y_2'(0) = 0$$

forcing

$$A(0) = 0 = B(0) \quad . \tag{2.12}$$

Hence,

$$W(0)^{-1} y(t) = \int_0^t (y_1(t) y_2(t_1) - y_2(t) y_1(t_1)) a_2(t_1)^{-1} u(t_1) \exp\left(\int_0^{t_1} \frac{a_1(t_2)}{a_2(t_2)} dt_2\right) dt_1 \tag{2.13}$$

Relation (2.13) is the definitive input-output relation.

3. THE INPUT-OUTPUT RELATIONS IN LAPLACE INTEGRAL FORM

Continue with the notation of Section 2. Further suppose that y_1, y_2 can be written in the following form:

$$y_1(t) = \int_{\gamma_1} e^{st} f_1(s)\, ds$$

$$y_2(t) = \int_{\gamma_2} e^{st} f_2(s)\, ds \qquad (3.1)$$

γ_1, γ_2 are curves in the complex s-plane. The integrals are line-integrals. $s \to f_1(s)$, $s_2 \to f_2(s)$ are defined and *holomorphic* in some neighborhood of the curves γ_1, γ_2.

Remark. The possibility of such a representation of the solutions of linear ordinary differential equations is a very standard subject in the *19th century* theory of differential equations, which is now forgotten. Ince's treatise is the best place to learn this material.

$$y_1(t) y_2(t_1) = \iint e^{s_1 t + s t_1} f_1(s) f_2(s_1)\, ds\, ds_1$$

$$y_2(t) y_1(t_1) = \iint e^{s_1 t + s_1 t_1} f_2(s) f_1(s_1)\, ds\, ds_1$$

$$y_1(t) y_2(t_1) - y_2(t) y_1(t_1) = \iint e^{st + s_1 t_1} (f_1(s) f_2(s_1) - f_2(s) f_1(s_1))\, ds\, ds_1 \qquad (3.2)$$

Set:

$$F(s, s_1) = f_1(s) f_2(s_1) - f_2(s) f_1(s_1) \qquad (3.3)$$

Then

$$y(t) = W(0) \int\!\!\int_{\gamma_1 \times \gamma_2} F(s, s_1)\, e^{st}\, \hat{u}(s_1, t)\, ds\, ds_1 \qquad (3.4)$$

with:

$$\hat{u}(s_1,t) = \int_0^t e^{st_1} a_2(t_1)^{-1} \exp\left(\int_0^{t_1} \frac{a_1(t_2)}{a_2(t_2)} dt_2\right) u(t_1) \, dt_1 \quad (3.5)$$

Now, suppose that

$$u(t) = \frac{1}{2\pi i} \int_\gamma e^{s_2 t} \mathscr{L}(u)(s_2) \, ds_2 \quad ,$$

i.e., $\mathscr{L}(u)$ is the Laplace transform of u.

$$\hat{u}(s,t) = \frac{1}{2\pi i} \int_0^t \int_\gamma e^{(s+s_2)t_1} a_2(t_1)^{-1}$$

$$\exp\left(\int_0^{t_1} \frac{a_1(t_2)}{a_2(t_2)} dt\right) \mathscr{L}(u)(s_2) \, dt_1 \, ds_2 \quad (3.6)$$

<u>Definition</u>. The function $S(s,s_1,t)$ such that

$$y(t) = \frac{1}{2\pi i} \int_{\gamma \times \gamma_1} e^{st} S(s,s_1,t) \mathscr{L}(u)(s_1) \, ds_1 \, ds \quad (3.7)$$

where γ, γ_1 are some curves in the complex plane, is called the *symbol of the input-output relation* (or the differential equation). The operator (3.7) is called a *Laplace integral operator*.

<u>Remark</u>. This is, of course, modelled after the notion of "Fourier integral operator" in analysis [2]. Formulas (3.5)-(3.6) exhibit the input-output relation of a second-order, linear differential operator whose solution can be found by the classical Laplace integral method on this "Laplace integral" form.

<u>Example</u>. Suppose

$$y(t) = \int_0^t k(t-t_1) u(t_1) \, dt_1 \quad . \quad (3.8)$$

Then,

$$y(t) = \frac{1}{2\pi i} \int e^{st} (\mathscr{L}y)(s) \, ds$$

By the well-known property of Laplace transform,

$$\mathscr{L}y(s) = (\mathscr{L}k)(s) \; (y)(s)$$

Hence,

$$y(t) = \frac{1}{2\pi i} \int e^{st} \mathscr{L}k(s) \mathscr{L}(w) \, ds$$

The symbol is essentially then given as a distribution

$$S(s,s_1) = \mathscr{L}k(s)\delta(s-s_1) \quad . \tag{3.7}$$

Remark. This is about the only *general* case where the symbol is at all calculable in some effective way. In a later chapter, we will consider some classical examples where *some* calculation is feasible. However, my main aim is to attempt to discover some algebro-geometric properties of the symbol which might be useful in a study of linear, time-varying systems.

4. THE BESSEL EQUATION

Consider

$$(y) = y'' + \frac{1}{t} y'' + \left(1 - \frac{2}{t^2}\right) y \tag{4.1}$$

the *Bessel operator*. The two classical Laplace integral solutions of the homogeneous equations are

$$y_1(t) = \frac{1}{2\pi i} \int e^{st} \frac{\left(\sqrt{s^2+1} - s\right)^{\lambda}}{\sqrt{s^2+1}} \, ds$$

$$y_2(t) = \frac{1}{2\pi i} \int e^{st} \frac{\left(\sqrt{s^2+1} - s\right)^{-\lambda}}{\sqrt{s^2+1}} \, ds \tag{4.2}$$

Now, $t = 0$ is the singularity of these conditions, hence in order to make sense of formula (3.5), the origin $t = 0$ must be moved elsewhere, say to t_0. Now,

$$f_1'(s) = \frac{1}{2\pi i} \frac{\left(\sqrt{s^2+1} - s\right)^\lambda}{\sqrt{s^2+1}}$$

$$f_2(s) = \frac{1}{2\pi i} \frac{\left(\sqrt{s^2+1} - s\right)^{-\lambda}}{\sqrt{s^2+1}}$$

$$F(s,s_1) = f_1(s)f_2(s_1) - f_2(s)f_1(s_1)$$

$$= \frac{1}{4\pi^2} \frac{1}{\sqrt{(s^2+1)(s_1^2+1)}} \left(\left(\frac{\sqrt{s^2+1}-s}{\sqrt{s_1^2+1}-s_1}\right)^\lambda - \left(\frac{\sqrt{s_1^2+1}-s_1}{\sqrt{s^2+1}-s}\right)^\lambda \right)$$

$$= -\frac{1}{4\pi^2} \frac{1}{\sqrt{(s^2+1)(s_1^2+1)}} \left(\frac{\left(\sqrt{s^2+1}-s\right)^{2\lambda} - \left(\sqrt{s_1^2+1}-s_1\right)^{2\lambda}}{\left(\sqrt{s_1^2+1}-s_1\right)^\lambda \left(\sqrt{s^2+1}-s\right)^\lambda} \right)$$

Also,

$$a_2 = 1$$

$$a_1 = \frac{1}{t}$$

Hence,

$$\exp\left(\int_{t_0}^{t_1} \frac{a_1(t_2)}{a_2(t_2)} dt_2\right) = \exp\left(\int_{t_0}^{t_1} \frac{1}{t_2} dt_2\right)$$

$$= \exp\left(\frac{1}{2}\left(t_0^{-2} - t_1^{-2}\right)\right)$$

Hence, following the notation of formulas (3.4), (3.5),

$$\hat{u}(s,t) = \exp\frac{1}{2}t_0^{-2} \int_{t_0}^{t} \exp(st_1 - t_1^{-2}) u(t_1) \, dt_1 \tag{4.4}$$

Suppose

$$u(t) = \frac{1}{2\pi i} \int_\gamma e^{s_2 t} \mathscr{L}(u)(s_2) \, ds_2 \quad .$$

Then,

$$\hat{u}(s,t) = \frac{\exp \frac{1}{2} t_0^{-2}}{2\pi i} \int_\gamma \left(\int_{t_0}^{t} \exp(st_1 - t_1^{-2} + s_2 t_1) \, dt_1 \right) \mathscr{L}(u)(s_2) \, ds_2$$

Hence,

$$y(t) = \frac{W(t_0)}{2\pi i} \exp\left(\frac{1}{2} t_0^{-2}\right) \int_{\gamma_1 \times \gamma_2 \times \gamma} F(s,s_1) e^{st} \left(\int_{t_0}^{t} \exp(s_1 t_1 - t_1^{-2} + s_2 t_1) dt_1 \right)$$

$$\mathscr{L}(u)(s_2) \, ds_2 \, ds \, ds_1$$

Up to factors, the symbol is given by the following formula:

$$\boxed{S(s,s_2) = \int_{\gamma_1} \int_{t_0}^{t} F(s,s_1) \exp(s_1 t_1 - t_1^{-2} + s_2 t_1) \, dt_1 \, ds_1}$$

Chapter 19

THE ASYMPTOTIC BEHAVIOR OF A LINEAR, TIME-VARYING SYSTEM
AND TRANSFER FUNCTION

1. INTRODUCTION

Consider a linear system of the form

$$\frac{dx}{dt} = A(t)x + B(t)u$$

$$y = C(t)x \quad .$$

(1.1)

Our aim is to develop notions generalizing the "transfer function" of a time-invariant system, and to apply concepts of vector bundle theory to it.

In case (1.1) has constant coefficients, the input-output relations are governed by the "transfer function"

$$T(s) = C(s-A)^{-1}B \quad .$$

(1.2)

In the time-varying cases we have tried to generalize this using "Laplace integral operators". An alternate method is to study the asymptotic behavior as $t \to \infty$ of solutions of (1.1). We will consider systems of form (1.1) which have polynomial coefficients. Then, asymptotic behavior as $t \to \infty$ can be converted into behavior as $s \to 0$ of the Laplace transformed system [1,2]. (This is known as "Watson's lemma".) This will provide us with methods of studying systems-theoretic perspectives of certain time-varying systems. First let us deal with the constant coefficient case from this point of view.

2. ASYMPTOTIC BEHAVIOR OF TIME-INDEPENDENT SYSTEMS

Consider

$$\frac{dx}{dt} = Ax + Bu$$

$$y = Cx$$

(2.1)

with constant matrices A,B,C. Suppose that:

> A is *stable* in the sense that all its eigenvalues have negative real parts. (2.2)

Thus, if $t \to x(t)$, $x_1(t)$ are two solutions with the same input $t \to u(t)$, their difference satisfies the homogeneous equation, hence goes to zero as $t \to 0$. Hence, to study asymptotic behavior we can choose the solution such that

$$x(0) = 0 . \quad (2.3)$$

Set

$$\hat{x}(s) = \int_0^\infty e^{-st} x(t) \, dt$$

$$\equiv \text{Laplace transform} .$$

Then,

$$\hat{y}(s) = T(s)\hat{u}(s) , \quad (2.4)$$

with

$$T(s) = C(s-A)^{-1} B$$

$$\equiv \textit{transfer function} \quad (2.5)$$

Now, large t behavior of $y(t)$ is equivalent to small s behavior of $s\hat{y}(s)$. (See [1,2].) Using (2.1)

$$T(s) = -\frac{CA^{-1}}{1 - A^{-1}s} B$$

$$= -CA^{-1}(1 + A^{-1}s + A^{-2}s^2 + \cdots) B$$

$$= -CA^{-1}B - CA^{-2}Bs - \cdots$$

> $$\hat{y}(s) \sim -CA^{-1}B\hat{u}(s) - CA^{-2}Bs\hat{u}(s)$$
> $$- CA^{-3}Bs^2\hat{u}(s) - \cdots$$

ASYMPTOTIC BEHAVIOR

We see that the "transfer function" of the system (1.1) is contained in the asymptotic relations (2.6), which characterize the asymptotic behavior of (2.4) as $t \to 0$.

The coefficients $CA^{-1}B$, $CA^{-2}B$,... play the same role that the "Hankel matrix" plays in realization theory. (See "Interdisciplinary Mathematics, Volume 8.) In particular, the familiar results of linear systems theory enable one to "realize" any set of input-output relations of form (2.6). *Our goal is to generalize to time-varying systems.*

We will now examine certain classical, linear, variable-coefficient differential equations from the point of view of their asymptotic behavior.

3. INPUT-OUTPUT BEHAVIOR OF CERTAIN EQUATIONS WHOSE LAPLACE TRANSFORMS ARE FIRST ORDER

Let us consider scalar input-scalar output systems where the relations between input $t \to u(t)$ output $t \to y(t)$ are given by an equation of the form

$$\left(af\left(-\frac{d}{dt}\right) + bg\left(-\frac{d}{dt}\right)\right)(y) = u , \qquad (3.1)$$

where $s \to f(s)$, $s \to g(s)$ are polynomials in the variables s, a and b are constants.

Look for solutions of (3.1) be taking the Laplace transform of both sides, assuming $t \to y(t)$ is a smooth function on $0 \leq t \leq \infty$, of at most exponential growth as $t \to \infty$. Let

$$\hat{y}(s) = \int_0^\infty e^{-st} y(t)\, dt$$

$$\hat{u}(s) = \int_0^\infty e^{-st} u(t)\, dt .$$

Let us look for solutions g of (3.1) with *zero Cauchy data*, in the sense that

$$\int_0^\infty f\left(-\frac{d}{dt}\right) e^{-st} y(t)\, dt = f(s)\hat{y}(s)$$

$$\int_0^\infty g\left(-\frac{d}{dt}\right) e^{-st} y(t)\, dt = g(s)\hat{y}(s) .$$

Then,

$$\frac{d}{ds}(f(s)\hat{y}(s)) = \int_0^\infty -tf\left(\frac{d}{dt}\right)e^{-st}y(t)\,dt \quad .$$

Hence, (3.1) takes the form:

$$a\frac{d}{ds}(f(s)\hat{y}(s)) + bg(s)\hat{y}(s) = u(s) \quad . \tag{3.2}$$

We can, of course, solve this equation by elementary methods:

Set

$$z = f\hat{y}$$

$$az' + b\frac{g}{f}z = \hat{u}, \qquad (z' = dz/ds)$$

$$z = \alpha e^{-\int \frac{bg}{fa}\,ds} \tag{3.3}$$

where α is a function of s to be determined.

$$z' = \alpha'\alpha^{-1}z - z\frac{bg}{fa}$$

$$\alpha' = e^{\int \frac{bg}{fa}\,ds}\,\hat{u},$$

$$\alpha = \int e^{\int \frac{bg}{fa}\,dx}\,\hat{u}\,ds$$

$$\hat{y} = f^{-1}\int \left(e^{\int \frac{bg}{fa}\,ds}\,\hat{u}\,ds\right)e^{-\int \frac{bg}{fa}\,ds}$$

$$\hat{y}(s) = f(s)^{-1}\int_a^s a^{F(s')-F(s)}\,\hat{u}(s')\,ds' \tag{3.4}$$

with

$$F(s) = \int_b^s \frac{bg(s')}{f(s')^a}\,ds' \quad . \tag{3.5}$$

ASYMPTOTIC BEHAVIOR   251

Asymptotic expansions of these integrals as $s \to 0$ can be obtained by "steepest descent" methods [1,2]. However, rather than proceed with this generality, I will now turn to some special examples suggested by the classical Laplace transform theory.

4. INPUT-OUTPUT RELATIONS GOVERNED BY LAPLACE EQUATIONS

Consider scalar input-output relations governed by the following differential equation:

$$t \frac{d^2 y}{dt^2} + (at+b) \frac{dy}{dt} + cy = u(t) \tag{4.1}$$

$$y(0) = y \; ; \quad \frac{dy}{dt}(0) = 0 \; ;$$

a,b,c are constants. Take the Laplace transform of both sides:

$$\hat{u}(s) = \int_0^\infty e^{-st} u(t) \, dt$$

$$\int_0^\infty e^{-st} t \frac{d^2 y}{dt^2} \, dt = -\frac{d}{ds} \int_0^\infty e^{-st} \frac{d^2 y}{dt^2} \, dt$$

$$= -\frac{d}{ds} (s^2 \hat{y}(s)) \quad .$$

Hence, (4.1) takes the form:

$$-\frac{d}{ds}(s^2 \hat{y}) + a \frac{d}{ds}(s\hat{y}) + bs\hat{y} + c\hat{y} = \hat{u}$$

$$(as - s^2) \frac{d\hat{y}}{ds} + (a - 2s + bs + c)\hat{y} = \hat{u} \quad . \tag{4.2}$$

Following the classical method of "variation of parameters", let us first solve (4.2) with $\hat{u} = 0$:

$$(as - s^2) \frac{d\hat{y}_0}{ds} + (a - 2s + bs + c)\hat{y}_0 = 0$$

$$\frac{d\hat{y}_0}{\hat{y}} = -\frac{(a - 2c + bs + c)}{(as - s^2)} \, ds$$

$$\hat{y}_0 = e^{-\int \left(\frac{a-2c+bs+c}{as-s^2}\right)ds} \tag{4.3}$$

Now, look for a solution of (4.2) of the form

$$\hat{y} = \alpha \hat{y}_0,$$

$$(as-s^2)\frac{d\hat{y}}{ds} + a + s(b-2) + c)\hat{y} = (as-s^2)\frac{d\alpha}{ds}\hat{y}_0$$

$$= \hat{u},$$

$$\frac{d\alpha}{ds} = \frac{\hat{u}}{\hat{y}_0(as-s^2)} \tag{4.4}$$

Now, the integrand in (4.3) can be expanded in partial fractions and the integral evaluated. The result is that y_0 is of the following form:

$$\hat{y}_0 = s^{-\lambda_1}(a-s)^{-\lambda_2}, \tag{4.5}$$

with complex numbers λ_1, λ_2. Hence,

$$\frac{d\alpha}{ds} = \hat{u} s^{\lambda_1 - 1}(a-s)^{\lambda_2 - 1}$$

$$\alpha = \int_{s_0}^{s} \hat{u}(\tau)\tau^{\lambda_1 - 1}(a-\tau)^{\lambda_2 - 1} d\tau$$

$$\hat{y}(s) = \int_{s_0}^{s} \hat{u}(\tau) \frac{(\tau/s)^{\lambda_1}((a-\tau)/(a-s))^{\lambda_2}}{\tau(a-\tau)} d\tau \tag{4.6}$$

For example, an exponential input $\hat{u}(t) = e^{\lambda t}$ would give a $\hat{u}(\tau)$ of the form

$$\hat{u}(\tau) = \frac{1}{\tau - \lambda},$$

and (4.6) would be a familiar integral in classical analysis.

Since we are particularly interested in asymptotic behavior as $s \to 0$, it is much more direct to use *power series* methods. If $a \neq 0$, Equation (4.2) has a *regular singular point*, with expansions of the form:

$$\hat{y}(s) = s^k(\hat{z}(s)), \tag{4.7}$$

ASYMPTOTIC BEHAVIOR   253

with $\hat{z}(s)$ analytic in the neighborhood of $s = 0$. A classical result of
Thomé (see Ince [3], p. 423) gives the general form:

$$\hat{y} = e^{Q(s)} s^k \hat{z}(s) \quad , \qquad (4.8)$$

with Q an analytic function of s, k a complex number, $\hat{z}(s)$ analytic at
$s = 0$.

The coefficients of the power series solution then give us a "steady state" asymptotic expansion of the system (4.1) with coefficients which are *analytic functions* of the coefficients a, b, c. The "singularity structure" of these functions is what we should study in order to generalize the "transfer func-function" analysis of the usual system theory. I will now go over more of this material from this "Thomé point of view".

5. LINEAR SYSTEMS FROM THE POINT OF VIEW OF THOMÉ

Return to the following general situation. A scalar input-output linear system has the following relation between input-output:

$$\left(\tau f\left(-\frac{d}{dt} \right) + g\left(-\frac{d}{dt} \right) \right) y = u \quad .$$

Laplace transformed, this becomes:

$$\frac{d}{ds}(f(s)\hat{y}) + g(s)\hat{y} = \hat{u} \quad . \qquad (5.1)$$

Let $\hat{y}_0$ satisfy:

$$\frac{d}{ds}(f\hat{y}_0) + g\hat{y}_0 = 0 \qquad (5.2)$$

$$\hat{y} = z\hat{y}_0$$

$$\frac{d}{ds}(fz\hat{y}_0) + gz\hat{y}_0 = \hat{u}$$

$$\frac{dz}{ds} f\hat{y}_0 = \hat{u}$$

$$z = \int \hat{u} \, f^{-1} \hat{y}_0^{-1} \, ds \qquad (5.3)$$

Now, (5.2) takes the form:

$$f \frac{d\hat{y}_0}{ds} + \left(\frac{df}{ds} + g \right) \hat{y}_0 = 0$$

$$s \frac{d\hat{y}_0}{ds} + \left(\frac{df}{ds} + g\right)\left(\frac{s}{f}\right)\hat{y}_0 = 0 \quad,$$

or

$$s \frac{d\hat{y}_0}{ds} + p(s)\hat{y}_0 = 0$$

with

$$p(s) = \frac{ma_1}{s^m} + \frac{(m-1)a_2}{s^{m-1}} + \cdots + \frac{a_m}{s} - \rho - \phi(s) \tag{5.4}$$

(We use the notation in Ince [3], p. 423.)

$$\phi(0) = 0 \quad.$$

Then, *Thomé's formula* is:

$$\hat{y}_0(s) = s^\rho e^{(Q(s)+\Phi(s))} \tag{5.5}$$

with

$$Q(s) = \frac{a_1}{s^m} + \frac{a_2}{s^{m-1}} + \cdots + \frac{a_m}{s} \tag{5.6}$$

$$\Phi(s) = \int \frac{\phi(s)}{s} ds \quad. \tag{5.7}$$

Thus, the input-output relations are of the following form:

$$z(s) = \int \hat{u}(s) f(s)^{-1} s^{-\rho} e^{-(Q(s)+\Phi(s))} ds$$

$$\hat{y}(s) = \int_{s_0}^{s} \hat{u}(\tau) f(\tau)^{-1} \left(\frac{s}{\tau}\right)^\rho e^{(Q(\tau)-Q(s)+\Phi(\tau)-\Phi(s))} d\tau \tag{5.8}$$

Notice the general pattern to this:

The input-output relations are given in "Laplace integral" form

$$\hat{y}(s) = \int_{\text{(path in Riemann surface above variable is } \tau)} \omega_s \hat{u}(\tau) d\tau \tag{5.9}$$

They are "Abelian integrals", in the sense of Riemann surface theory; for fixed s, ω_s is a differential form in τ.

Another virtue of the Thomé form of the solution is that it is the starting point for generalization to higher order equations. See the treatises by Ince [3] and Hille [4]. An immense amount of material, which has never been considered before (as far as I can see) in the systems theory context, is in a book by Lappo-Danilevsky [5].

Bibliography

1. A Deutsch, *Handbuch der Laplace Transformationen*, Vols. I-III, Vorlag Birkhausen, Bosel, 1950-55.
2. B. Davies, *Integral Transforms and Their Applications*, Springer-Verlag, 1978.
3. E. L. Ince, *Ordinary Differential Equations*, Dover, 1956.
4. E. Hille, *Ordinary Differential Equations in the Complex Domain*, Wiley, 1976.
5. J.A. Lappo-Danilevsky, *Systèmes des Equations Différentielles Linéaires*, Chelsea, New York, 1953.

Chapter 20

CONSERVATION LAWS AND ASYMPTOTIC FORMAL POWER SERIES
SOLUTIONS OF DIFFERENTIAL EQUATIONS

1. INTRODUCTION

There are at least *four* remarkable properties of the Korteweg-de Vries equation:

a) The Inverse Scattering relation between Korteweg-de Vries and the Sturm-Liouville equation
b) Korteweg-de Vries admits a Bäcklund transformation
c) Korteweg-de Vries admits an infinite number of conservation laws. Their densities are polynomials in the unknown function and its space derivatives. Their Poisson brackets are zero with respect to the natural symplectic structure defined by the variational principle.
d) Korteweg-de Vries admits an Estabrook-Wahlquist prolongation with structure group $SL(2,R)$.

The extent to which these properties are interrelated *in general* is not known; investigating these links is an important topic of research. In this chapter I want to develop the relation between (a) and (d). I want to do this with an eye on general differential-geometric principles which might extend to other situations.

A new differential geometric feature of this material is that it requires extension of connection ideas to certain infinite dimensional manifolds. They are not "modelled on Banach spaces", as usual in "Global Analysis", but are *projective limits* of ordinary manifolds. The "infinite order jet spaces" of Ehresmann are about the right level of generality, and we shall concentrate on them. This extension on differential-geometric ideas is, I believe, widely applicable to the theory of "asymptotic solutions" of differential equations. In fact, we shall see that the relation between property (c) and (d) of Korteweg-de Vries involves just such ideas.

2. ORDER OF CONTACT OF MAPS

Let X,Y be manifolds in the usual sense. (Say, of differentiability class C^∞.) $T(X)$ denotes its *tangent vector bundle*. Let $\mathcal{M}(X,Y)$ be the space of (C^∞) maps

$$\phi: X \to Y \quad .$$

For each $\phi \in \mathcal{M}(X,Y)$,

257

$$\phi_*: T(X) \to T(Y)$$

denotes the extension of ϕ to the tangent vectors. We will now define the basic concept of the theory of Ehresmann jets, that of *maps agreeing to a certain order*, or that they have a *certain order of contact at a point*.

<u>Definition</u>. Let $\phi, \phi_1 \in \mathcal{M}(X,Y)$, $x \in X$.

a) ϕ and ϕ_1 *agree to the 0th order at* x if $\phi(x) = \phi_1(x)$

b) ϕ and ϕ_1 *agree to the first-order at* x if

$$\phi_*(v) = \phi_{1*}(v)$$

for all $v \in X_x$.

c) Continue by induction of y to define where ϕ, ϕ_1 agree to the j-th order. Suppose we have defined what it means for ϕ, ϕ_1 to agree to the (j-1)-st order, $j \geq 2$. Consider $\phi_*: T(X) \to T(Y)$. Let us say that ϕ, ϕ_1 *agree to the j-th order* at x if ϕ_*, ϕ_{1*} agree to the (j-1)-st order at all points $v \in X_x$.

d) Let us now say that ϕ, ϕ_1 agree to *an infinite order at* $x \in X$ if ϕ, ϕ_1 agree to the j-th order for all integers $j \geq 0$.

We can put these conditions in a more familiar form in case X and Y are finite dimensional vector spaces. Suppose also that the point x is the zero point of the vector space. $T(X)$ can then be identified with $X \times X$. If $\phi: X \to Y$ is a map

$$\phi_*: X \times X \to Y \times Y$$

is defined by the following formula:

$$\phi_*(x,v) = \left(\phi(x), \frac{d}{dt} \phi(x+tv) \Big|_{t=0} \right) \tag{2.1}$$

$$(\phi(x), \phi_x(x) \cdot v) \tag{2.2}$$

for $x \in X$, $v \in X$.

(ϕ_x denotes the vector of partial derivatives.) We see that the maps ϕ, ϕ_1 agree to the *first* order at $x = 0$ if and only if

$$\boxed{\begin{aligned} \phi(0) &= \phi_1(0) \\[6pt] \phi_x(0) &= (\phi_1)_x(0) \end{aligned}} \tag{2.3}$$

Let us now regard ϕ_x as a map between vector spaces $X \times X \to Y \times Y$ and regard $X \times X$ as the *direct sum* of two copies of the vector space X.

$$(\phi_*)_{(x,v)}(x',v') = \frac{d}{dt}\phi_*((x,v) + t(x',v'))\Big|_{t=0}$$

$$= \frac{d}{dt}(\phi_*(x+tx', v+tv'))\Big|_{t=0}$$

$$= \text{, using (2.2),}$$

$$\frac{d}{dt}(\phi(x+tx'), \phi_x(x+tv')(v+tv'))$$

$$= (\phi_x x', \phi_x(x)(v)) + (\phi(x), \phi_{xx}(x)(v') + \phi_x(x), v'))$$

(2.4)

Let us suppose that ϕ, ϕ_1 agree to the first order at $x = 0$, i.e., conditions (2.3) are satisfied. We see from (2.4) that, for all $v \in X$,

$$(\phi_*)_{(0,s)} = (\phi_{1*})_{(0,v)}$$

if and only if, in addition to (2.3),

$$\boxed{\phi_{xx}(0) = (\phi_1)_{xx}(0)}$$

Continuing in this way (left as an exercise for the reader) we see that we can restate the condition for j-th order contact in the following more traditional (and simpler) form:

Theorem 2.1. Let X, Y be finite dimensional, real vector spaces and let $\phi, \phi_1: X \to Y$ be C^∞ maps. (It suffices, in fact, to suppose that ϕ, ϕ_1 are only defined in some neighborhood of 0.) Then ϕ, ϕ_1 *agree to the j-th order* at $x = 0$ if and only if

$$\boxed{\begin{aligned} \phi(0) &= \phi_1(0) \\ \phi_x(0) &= \phi_{1,x}(0) \\ &\vdots \\ \phi_{x_1\cdots x_j}(0) &= \phi_{1,x_1\cdots x_j}(0) \end{aligned}}, $$

> for all $x_0, x_1, \ldots, x_j \in X$

i.e., if all partial derivatives of ϕ and ϕ_1 agree to order $\leq j$ at $x = 0$. Another way to put this is to say that:

> $\phi - \phi_1$ is a map $X \to Y$ whose Taylor series about the point $x = 0$ is of the form:
>
> $$(\phi - \phi_1)(x) = f_{j+1}(x) + f_{j+2}(x) + \cdots$$
>
> where f_k, $k = j+1, \ldots, \infty$ polynomials of degree k.

3. THE JET SPACES AND THEIR DIFFERENTIAL GEOMETRY

Let X and Y continue as manifolds with $\mathcal{M}(X,Y)$ the space of maps $\phi: X \to Y$. For each integer r,

$$0 \leq r \leq \infty \quad ,$$

introduce an equivalence relation in $X \times \mathcal{M}(X,Y)$

$$(x, \phi) \sim (x_1, \phi_1)$$

if and only if

> $$x = x_1$$
> $$\phi(x) = \phi_1(x_1)$$
> ϕ and ϕ_1 agree to order j at x

(3.1)

It is readily seen that this *is* an equivalence relation. The quotient space is denoted by

$$J^r(X,Y)$$

and is called the *space of r-jets of maps*: $X \to Y$.

For $r < s$, notice that equivalence classes are contained in the r equivalence classes, i.e., there is a map

CONSERVATION LAWS 261

$$\pi_{s,r}: J^s(X,Y) \to J^r(X,Y)$$

This maps $J^s(X,Y)$ as fiber space over $J^r(X,Y)$.

Remark. In previous work, I have sometimes used the notation

$$\mathcal{M}^r(X,Y)$$

and the terminology *"space of r-th order mapping elements"*. In fact, this is basically Lie's terminology. However, in further work I will use Ehresmann's notation and terminology, which is that used in the current literature.

Now it is readily seen that $J^r(X,Y)$ for $0 \leq r \leq \infty$, is finite dimensional and a manifold in the usual sense. The system of maps

$$\pi_{\infty,r}: J^\infty(X,Y) \to J^r(X,Y)$$

exhibits $J^\infty(X,Y)$ as a *projective limit* of manifolds. We will define a "differential geometry" on $J^\infty(X,Y)$ in accordance with the general framework of Volume 16 of "Interdisciplinary Mathematics". First define a commutative associative algebra $\mathscr{F}(J^\infty(X,Y))$ of real-valued functions on $J^\infty(X,Y)$ which is to be called the *space of C^∞ functions*, then define *algebraically* such basic differential-geometric notions as *vector fields*, *differential forms*, *connection*, etc.

Definition. $\mathscr{F}(J^\infty(X,Y))$ consists of the real-valued functions f on $J^\infty(X,Y)$ of the form

$$f = \pi_{\infty,r}^*(f'),$$

for some integer r, $f \in \mathscr{F}(J^r(X,Y))$.

$\mathscr{F}(J^\infty(X,Y))$ is a commutative associative algebra, and the "differential geometry" (e.g., differential forms, vector fields,...) are built up using it. Let us say that an f of this form is of degree r, and denote by $\mathscr{F}^r(J^\infty(X,Y))$ the set of these elements. Thus, we have

$$\mathscr{F}^0(J^\infty(X,Y)) \subset \mathscr{F}^1(J^\infty(X,Y)) \subset \cdots \qquad (3.2)$$

i.e.,

> $\mathscr{F}(J^\infty(X,Y))$ has an ascending filtration.

Remark. Ultimately, it will be necessary (e.g., to deal with the notion of "asymptotic solutions of differential equations") to "complete" $\mathscr{F}(J^\infty(X,Y))$, putting a topology on $\mathscr{F}(J^\infty(X,Y))$ using the filtration (3.1), and then completing $\mathscr{F}(J^\infty(X,Y))$ is a topological vector space. This "completion" of $\mathscr{F}(J^\infty(X,Y))$ plays a role in differential equation theory analogous to the role that the "formal power series" (e.g., "Puiseux expansions") play in *algebraic* geometry. Often one passes to this completion, operates there for a while, then returns back to the uncompleted object for the definitive result.

4. THE CONTACT STRUCTURE ON $J^\infty(X,Y)$

Having defined "functions" on $J^\infty(X,Y)$, we can, of course, define "differential forms", "vector fields" etc. by the usual algebraic algorithms of differential geometry. (The vector fields are the derivations, the differential forms are the dual modules to the vector fields, etc.) I will now describe the "contact structure", which involves *differential forms*.

For notational symplicity, I will deal with

$$X = Y = R \quad .$$

To deal with general manifolds merely means (since jet concepts are "local") allowing X and Y to be finite dimensional manifolds, and generalizing the formulas given below to *vectorial* notation. Points of X are denoted as x, those of Y as y.

An element $\mathscr{M}(X,Y)$ is then a mapping which we denote as

$$\underline{y}: x \to \underline{y}(x) \quad .$$

Define real-valued functions

$$x, y, y', y'', \ldots \text{ on } X \times \mathscr{M}(X,Y)$$

as follows:

$$\begin{aligned} x(x,\underline{y}) &= x \\ y(x,\underline{y}) &= \underline{y}(x) \\ y'(x,\underline{y}) &= \frac{d\underline{y}}{dx}(x) \end{aligned}$$

CONSERVATION LAWS

$$y''(x,\underline{y}) = \frac{d^2y}{dx^2}(x)$$
$$\vdots$$
(4.1)

These formulas define $x, y, y', y'', \ldots$ as real-valued functions on $X \times \mathcal{M}(X,Y)$.

Now, notice that:

The functions $x, y, y', \ldots, y^{(r)}$ remain constant on the equivalence relations on $X \times \mathcal{M}(X,Y)$ whose quotient is $J^r(X,Y)$. In fact, $(x,\underline{y})$ and $(x_1,\underline{y}_1)$ are equivalent (i.e., $x = x_1$, and $\underline{y}, \underline{y}_1$ have r-th order contact at x) if and only if

$$x(x,\underline{y}) = x(x_1,\underline{y}_1)$$

$$y(x,\underline{y}) = y(x_1,\underline{y}_1)$$

$$y'(x,\underline{y}) = y'(x_1,\underline{y}_1)$$

$$\vdots$$

$$y^{(r)}(x,\underline{y}) = y^{(r)}(x_1,\underline{y}_1) \quad .$$

Thus, the functions $(x, y, y', \ldots, y^{(r)})$ define coordinate systems for $J^r(X,Y)$.

We can now pass to $r = \infty$. *All* the functions $x, y, y', y'', \ldots$ are constant on the equivalence classes, hence pass to the quotient to define functions on $J^\infty(X,Y)$

$\mathscr{F}(J^\infty(X,Y))$ is the algebra generated by the function $x, y, y', y'', \ldots$

Definition. The *contact structure* is the ideal $\mathscr{C}$ of differential forms on $J^\infty(X,Y)$ generated by the following 1-forms:

$$\theta'_1 = dy - y' \, dx$$
$$\theta'' = dy' - y'' \, dx$$
$$\vdots$$
(4.2)

Here is the main geometric property of these forms. If

$$\phi: X \to Y$$

is a map, its *jet* is a map

$$j^\infty(\phi): X \to J^\infty(X,Y)$$

is defined as follows. For $x \in X$, consider the element

$$(x,\phi) \in X \times \mathcal{M}(X,Y) \quad .$$

Then

$$j^\infty(\phi)(x) \equiv \text{quotient of } (x,\phi) \text{ under the map } X \times \mathcal{M}(X,Y) \to J^\infty(X,Y)$$

Then, we have:

$$\boxed{j^\infty(\phi)^*(\mathscr{C}) = 0 \quad .} \qquad (4.3)$$

5. PROLONGATION OF VECTOR FIELDS

There is a "dual" geometric structure on vector fields, which plays a key role in Lie's work. $\mathscr{V}(X \times Y)$ denotes the vector fields on $X \times Y \equiv J^0(X,Y)$. For $V \in \mathscr{V}(X \times Y)$ there is a vector field $V^\infty \in \mathscr{V}(J^\infty(X,Y))$, which is characterized by the following properties:

$$\boxed{\begin{array}{l} \mathscr{L}_V(\mathscr{C}) \subset \mathscr{C} \\ \\ \pi_{\infty,0*}(V^\infty) = V \end{array}} \qquad (4.4)$$

In fact, we can readily calculate V^∞ using (5.1). Suppose

$$V = a(x,y) \frac{\partial}{\partial x} + b(x,y) \frac{\partial}{\partial y} \quad .$$

Then,

$$V^\infty(x) = a, \qquad V^\infty(y) = b$$

$$0 = \mathscr{L}_{V^\infty}(\theta') = \mathscr{L}_{V^\infty}(dy - y'dx)$$

$$= d(V^\infty(y)) - V^\infty(y')dx - y'dV^\infty(x)$$

$$= da - V^\infty(y')dx - y'da$$

$$= \frac{\partial b}{\partial x} dx + \frac{\partial b}{\partial y} dy - V^\infty(y')dx - y'\left(\frac{\partial a}{\partial x} dx + \frac{\partial a}{\partial y} dy\right)$$

$$= \frac{\partial b}{\partial x} dx + \frac{\partial b}{\partial y}(d\theta') + \frac{\partial b}{\partial y} dx\, y'dx - V^\infty(y')dx$$

$$- y'\frac{\partial a}{\partial x} dx - y'\frac{\partial a}{\partial y} y'dx + y'\frac{\partial a}{\partial y} \theta'$$

Thus, in order that

$$\mathscr{L}_{V^\infty}(\theta') \in \mathscr{C}$$

we must have:

$$\boxed{V^\infty(y') = \frac{\partial b}{\partial x} + \frac{\partial b}{\partial y} y' - y'\frac{\partial a}{\partial x} - (y')^2 \frac{\partial a}{\partial y}} \qquad (5.2)$$

This formula determines the action V^∞ on $\mathscr{F}^1(J^\infty(X,Y))$. Continue in this way; we see that it may be defined on all of $\mathscr{F}(J^\infty(X,Y))$.

Remark. We have the additional property

$$\mathscr{L}_{V^\infty}(\mathscr{F}^r(J^\infty(X,Y)) \subset \mathscr{F}^r(J^\infty(X,Y)) \quad ,$$

which means geometrically that the prolonged vector field acts on $J^r(X,Y)$.

Now, let us descend from these generalities to a very concrete situation, but one which is of the greatest importance for soliton-nonlinear wave theory.

6. ASYMPTOTIC SOLUTIONS OF THE RICCATI EQUATION. (LIOUVILLE-GREEN-POINCARÉ-LANGER-JEFFRIES-WENTZEL-KRAMERS-BRILLOIN)

Consider a Sturm-Liouville differential equation:

$$\psi''(x) + y(x)\psi(x) + \lambda^2 \psi(x) = 0 \qquad (6.1)$$

$x \to y(x)$ is a *given* real valued function. $x \to \psi(x)$ is the unknown function.

Convert this *second order, linear,* differential equation into a *first order, nonlinear* one by the usual trick:

$$z = \frac{\psi'}{\psi} \qquad . \qquad (6.2)$$

Then,

$$z' = \frac{(-y\psi - \lambda^2\psi)}{\psi} - \frac{(\psi')^2}{\psi^2} ,$$

or

$$\boxed{z' + z^2 + y + \lambda^2 = 0} \qquad (6.3)$$

(6.3) is a *Riccati equation*.

λ is a real parameter. We are interested in large λ-behavior. Let us look for a *formal power series* solution in the form:

$$z = \lambda \left(\sum_{j=0}^{\infty} z_j \lambda^{-j} \right) \qquad (6.4)$$

The z_j are *functions of* x.

$$z' = \lambda \sum_{j=0}^{\infty} z_j' \lambda^{-j}$$

$$z^2 = \lambda^2 \sum_{j,k=0}^{\infty} z_j z_k \lambda^{-(j+k)}$$

$$= \lambda^2 \sum_{j=0}^{\infty} \sum_{k=0}^{j} (z_k z_{j-k}) \lambda^{-j}$$

$$z' + z^2 + y + \lambda^2 = \sum_{j=0}^{\infty} \left(\lambda z_j' + \lambda^2 \sum_{k=0}^{j} z_k z_{j-k} \right) \lambda^{-j} + y + \lambda^2 \qquad (6.5)$$

Let us look for the conditions that z satisfy (6.3) *term-by-term in* λ, i.e., as a "formal power series". The first step is to equate to zero the terms which involve λ:

$$0 = z_0' + \sum_{k=0}^{1} z_k z_{1-k} ,$$

or

CONSERVATION LAWS 267

$$z_0' + 2z_0 z_1 = 0 \qquad (6.6)$$

Next, the coefficients of λ^0 and λ^2:

$$z_1' + \sum_{k=0}^{2} z_k z_{j-k} = -y \, ,$$

or

$$z_1' + z_0 z_2 + z_1 z_1 + z_2 z_0 = -y$$

or

$$z_1' + 2z_0 z_2 + z_1^2 = -y \qquad (6.7)$$

$$z_0^2 = -1 \qquad (6.8)$$

(6.8) and (6.6) force

$$z_1 = 0 \qquad (6.9)$$

(6.7) and (6.9) force

$$2z_0 z_2 = -y$$

Thus, the three lowest order conditions are equivalent to the following conditions as the first three coefficients of the formal power series:

$$\begin{aligned} z_0 &= \pm i \\ z_1 &= 0 \\ z_2 &= \frac{\mp iy}{2} \end{aligned} \qquad (6.10)$$

Beyond the coefficients of $\lambda^0, \lambda, \lambda^2$, we are in the "stable" range for the equations for the coefficients of the power series

$$z_2' + \sum_{k=0}^{3} z_k z_{j-k} = 0 \quad,$$

or

$$z_2' + 2z_0 z_3 + 2z_1 z_2 = 0 \quad,$$

or, using (6.10),

$$z_3 = -\frac{1}{2} z_0^{-1} z_2$$

$$= -\frac{1}{2} (\mp 1) \left(\frac{\mp iy}{2}\right)^1$$

or

$$\boxed{z_3 = \frac{1}{4} y'} \tag{6.11}$$

$$z_3' + \sum_{k=0}^{4} z_k z_{4-k} = 0$$

$$= z_3' + z_0 z_4 + z_1 z_3 + z_2 z_2 + z_3 z_1 + z_0 z_4$$

$$= z_3' + 2z_0 z_4 + 2z_1 z_3 + z_2^2$$

$$= z_3' \pm 2iz_4 - \frac{i}{4} y^2$$

or

$$\boxed{z_4 = \pm \frac{i}{2} \left(\frac{1}{4} y'' - \frac{i}{4} y^2\right)} \tag{6.12}$$

$$z_4' + \sum_{k=0}^{5} z_k z_{5-k} = 0 \quad,$$

$$z_4' + z_0 z_5 + z_1 z_4 + z_2 z_3 + z_3 z_2 + z_4 z_1 + z_5 z_0 = 0 \quad,$$

CONSERVATION LAWS

$$z_4' + 2z_0 z_5 + 2z_1 z_4 + 2z_2 z_3 = 0$$

$$-2z_0 z_5 = z_4' + \frac{1}{2} y' \left(\frac{\mp iy}{2}\right)$$

$$\boxed{-z_0 z_5 = \frac{z_4'}{2} \mp \frac{i}{8} yy'} \qquad (6.13)$$

$$z_5' + \sum_{k=0}^{6} z_k z_{6-k} = 0$$

$$z_5' + 2(z_0 z_6 + z_1 z_5 + z_2 z_4) + z_3^2 = 0$$

$$\boxed{2z_0 z_6 + z_5' + (\mp iy)\left(\frac{\pm i}{2}\right)\left(\frac{1}{4} y'' - \frac{i}{4} y^2\right) = 0} \qquad (6.14)$$

Continuing in this way, we see that the relations give a *recursive relation*, determining each z_j in terms of $z_0, \ldots, z_{j-1}$ and *their derivatives*. It is the involvement of the derivatives which destroys *convergence* of the series— one knows that a bound on the function gives no bound on the *derivatives* of the function. However, the series are perfectly good formal *power* series.

Such expansions were first systematically considered in the 19th century. I understand from Nayfeh's book, *Perturbation Methods*, Liouville and Green have priority. These formulas keep reappearing, with the names listed in the title of this section among those that I have noticed. Physicists call it "WKB" or "JWKB"; often with the variant where the perturbation parameter λ would be that involved in "Planck's Constant", rather than the "eigenvalue". One of the best and clearest papers written on this is that by G.D. Birkhoff and R.E. Langer, "The Boundary Problems and Developments Associated with a System of Ordinary Linear Differential Equations of the First Order", Proc. Amer. Acad. Arts and Sci. 58 (1923), ;;. 49-128. It is to be most conveniently found in Birkhoff's *Collected Works*, Volume 1, page 345.

These formulas have turned out to be basic to the theory of the Korteweg-de Vries equation for reasons I will explain later. There, we are interested in the integrals:

$$\int_a^b z_j(x) \, dx \quad , \qquad (6.15)$$

between fixed end-points. They turn out--miraculously!--to be the "conserved quantities" of the Korteweg-de Vries equation. This extremization determines the "solitons". Now, derivatives do not contribute to these integrals. Hence compute

$$\int z_2 : \int y$$

$$\int z_4 : \int y^2$$

$$\int z_6 : \int (yy'' + (\text{constant}) y^3)$$

$$\vdots$$

These are just the usual "conserved quantities" of the Korteweg-de Vries equation.

7. JWBK (AND LIOUVILLE-GREEN-BIRKHOFF-LANGER) TYPE SOLUTIONS TO THE STURM-LIOUVILLE EQUATION

Given the close relation between the scalar Riccati equation and the Sturm-Liouville (6.1), we would expect that there would be formal power series solutions of the latter. It is instructive to work it out.

Let z be a solution of (6.3) of the form (6.4). Write it as:

$$z = z_0 \lambda + w\left(x, \frac{1}{\lambda}\right) \tag{7.1}$$

(Of course, w is not a "real" function, but only one defined by a "formal power series".) Let ψ be a function such that

$$\frac{\psi_x}{\psi} = z \quad ,$$

or

$$\log \psi = \int z$$

$$\psi = e^{\int z \, dx} \tag{7.2}$$

$$= e^{\lambda z_0 x} e^{\int w(x, (1/\lambda)) \, dx} \tag{7.3}$$

CONSERVATION LAWS   271

As we have seen, z_0 may have two values $\pm i$, hence, there are two solutions of form (7.3)

$$\psi^{\pm} = e^{\pm ix} e^{\int w(x,(1/\lambda))\,dx} \qquad (7.4)$$

Thus, (7.4) represents a formal power series for the two linearly independent solutions of (6.1), i.e., the eigenfunctions of Sturm-Liouville, in a maximally convenient form exhibiting the perturbation effects of the function y.

8. THE LGBLJWKB-EXPANSION INTERPRETED GEOMETRICALLY AS A "VIRTUAL MAPPING" DEFINED ON THE INFINITE ORDER JET BUNDLE

Consider X and Y as one-dimensional manifolds with coordinates x and y. $J^{\infty}(X,Y)$ is then a space with functions $x,y,y',y'',\ldots$ defined on it. It is readily seen that these functions define a *coordinate system*, i.e., they label points of $J^{\infty}(X,Y)$.

The LGBLJWKB-series is that defined by (6.4). We have determined the coefficients $z_0, z_1, z_2, \ldots$ involved in the series as functions of $y, y', \ldots$. Each z_j only involves a finite number of the variables $y, y', y'', \ldots$. However, the series itself does not converge, hence there is no *mapping*

$$\phi_\lambda : J^{\infty}(X,Y) \to R$$

defined by these formulas. However, the pull-back mapping

$$\phi_\lambda^* = \sum_j \phi_{\lambda,j}^* \lambda^j$$

is well-defined as *formal power series*. We will call such a gadget a *virtual or asymptotic mapping*. Perhaps it is worth our while to develop some general concepts related to this notion.

9. THE FORMAL LAURENT SERIES ALGEBRA OF AN ALGEBRA AND ITS DIFFERENTIAL FORM ALGEBRA

In order to encompass material like the LGBLJWKB-expansion in a systematic differential geometric way, I will pause to develop some algebraic ideas within the general framework discussed in "Interdisciplinary Mathematics", Volume 16. Let $\mathcal{A}$ be a commutative associative algebra over the reals as fields of scalars. (As in IM, Vol. 16, the "commutative" aspect can be relaxed. However, this simplifying assumption is adequate for my immediate purposes.) Denote the algebraic product as

$$(A_1, A_2) \to A_1 A_2 \quad .$$

A *differential form algebra* based on $\mathcal{A}$ is a *graded associative algebra*

$$\mathcal{D}(A) = \mathcal{D}^0(A) \oplus \mathcal{D}^1(A) \oplus \cdots \quad ,$$

with the algebra product denoted by $\wedge$, and a linear mapping d such that:

$$d: \mathcal{D}(A) \to \mathcal{D}(A)$$

$$d(\mathcal{D}^r(A)) \subset \mathcal{D}^{r+1}(A)$$

$$d^2 \equiv 0$$

$\mathcal{D}^0(\mathcal{A}) = \mathcal{A}$, with the algebra product $\wedge$ and $\mathcal{D}^0(A)$ agreeing with that given in $\mathcal{A}$.

$\mathcal{D}(A)$ is generated by A, under the operation $d, \wedge$.

$$d(A_1 A_2) = (dA_1) A_2 + A_1 dA_2$$

for $A_1, A_2 \in \mathcal{A} = \mathcal{D}^0(A)$

$$\mathcal{D}^r(\mathcal{A}) \wedge \mathcal{D}^s(\mathcal{A}) \supset \mathcal{D}^{r+s}(\mathcal{A})$$

$$\omega_1 \wedge \omega_2 = (-1)^{r+s} \omega_2 \wedge \omega_1$$

for $\omega_1 \in \mathcal{D}^r(\mathcal{A})$, $\omega_2 \in \mathcal{D}^s(\mathcal{A})$.

Now, let $\mathcal{A}$ be a commutative associative algebra and let λ be a real variable. (Algebraically, it is what algebraicists call an "indeterminant".) A *formal Laurent series on* λ, *with coefficients in* $\mathcal{A}$, is an object of the form

$$A = \sum_{n=-\infty}^{\infty} A_n \lambda^n \quad , \tag{9.1}$$

$A_n \in \mathcal{A}$ and $A_n = 0$, for n sufficiently large, considered abstractly, independent of its possible convergence properties. Let

$$\underline{A} = \sum A_n \lambda^n$$

$$\underline{B} = \sum B_n \lambda^n$$

be two such series. They are *equal* if and only if

$$A_n = B_n, \quad \text{for all } n.$$

They are *added* as

$$\underline{A} + \underline{B} = \sum (A_n + B_n)\lambda^n,$$

and *multiplied* as follows

$$\underline{AB} = \sum_{n=-\infty}^{\infty} \left(\sum_{m=-\infty}^{\infty} (A_m B_{n-m}) \right) \lambda^n$$

Denote the commutative associative algebra which results in this as

$$\underline{\mathscr{A}}$$

(The notation $\mathscr{A}((\lambda))$ is more commonly used in the literature of algebra. $\mathscr{A}[[\lambda]]$ then denotes those formal power series where the indices only have non-negative values.)

If $\mathscr{D}(\mathscr{A})$ is a differential form algebra for $\mathscr{A}$, then we can construct a differential form algebra for $\mathscr{D}(\underline{\mathscr{A}})$ for $\underline{\mathscr{A}}$ as follows.

$\mathscr{D}^r(\underline{\mathscr{A}})$ is the formal power series of the form

$$\underline{\omega} = \sum_n \omega_n \lambda^n,$$

$$\omega_n \in \mathscr{D}^r(\mathscr{A}).$$

$$d\underline{\omega} = \sum_n d\omega_n \lambda^n$$

$$\underline{\omega} \wedge \underline{\omega}' = \sum_n \left(\sum_{m=-\infty}^{\infty} \omega_m \wedge \omega'_{n-m} \right) \lambda^n$$

We see that $\mathcal{D}(\mathcal{A})$ is the appropriate algebraic gadget to carry out the LGPBLJWKB-type of expansions. Notice that defining "differential forms" enables us to think of these operations in a more "geometric", coordinate-free way. In fact, such a "calculus" is rather implicit in the classic work of Poincaré and Birkhoff, and is quite useful for a systematic treatment (which I hope to present some day) of scattered ideas presented in Nayfeh's recent book, *Perturbation Methods*. (This book is, in fact, a sort of explicit, "low-brow" elaboration of these ideas as developed originally by Poincaré and G.D. Birkhoff.)

Remark. Numerous elaborations of this construction are possible and possibly also very useful. Here are two linear ones.

a) Power series in more variables $\lambda_1, \lambda_2, \ldots$

$$\underline{\omega} = \sum_{n_1, n_2, \ldots} \omega_{n_1 \ldots} \lambda_1^{n_1} \lambda_2^{n_2} \ldots$$

b) λ is another *variable*, instead of playing the fole of a "parameter". Then

$$d\underline{\omega} = \sum d\omega_n \lambda^n + \omega_n \wedge n\lambda^{n-1} d\lambda \quad .$$

However, the simpler situations are presented here as adequate for our immediate purposes.

10. THE LGPBLJWKB-SOLUTION TO THE RICCATI EQUATION WITHIN THE CONTEXT OF DIFFERENTIAL FORM ALGEBRAS

Return to the Riccati equation (6.3). However, consider it now as a differential form "equation"

$$dz + (z^2 + y + \lambda^2) dx = 0 \tag{10.1}$$

We have seen that it then is a formal power series "solution" of the

$$z = \sum_{n=1}^{-\infty} z_n \lambda^n \quad , \tag{10.2}$$

with the z_n functions of the variables $y, y', \ldots, y^{(n)}$. There are many ways of integrating these formulas. Here is one which is most closely linked to the Estabrook-Wahlquist ideas.

Let X, Y be one-dimensional manifolds with variables x, y. Let

$$M = J^\infty(X,Y) \equiv \text{the infinite order jet space}.$$

Let $\mathcal{D}(M)$ be the differential form algebra of the space (based on $\mathcal{A} = \mathcal{F}(J^\infty(X,Y))$). It is then generated by:

functions of $x, y, y', y'', \ldots$

$dx, dy, dy', dy'', dy''', \ldots$

Let

$\theta', \theta'', \theta''', \ldots$

be the contact forms:

$\theta' = dy - y' \, dx$

$\theta'' = dy - y'' \, dx$

$\vdots$

Now, let E be the space of variables $(z, x, y, \ldots)$, i.e., $\mathbb{R} \times J^\infty(X,Y)$. The Cartesian projection map

$\mathbb{R} \times J^\infty(X,Y) \equiv E$

$\downarrow$

$J^\infty(X,Y)$

defining it is a *fiber space* over $J^\infty(X,Y)$. Consider it as an $SL(2,\mathbb{R})$-fiber bundle ($SL(2,\mathbb{R})$ acts on $\mathbb{R}$ via linear fractional transformations).

Let $\mathcal{F}(E)$ be the algebra of C^∞, real-valued functions on E. So far, λ has been an extraneous parameter. Introduce it as the "Laurent series" variable described in Section 8. Thus, we have the differential form "Laurent series" algebra $\underline{\mathcal{D}}$,

$$\underline{\omega} = \sum_n \omega_n \lambda^n ,$$

with ω_n differential forms on the variables $(z, x, y, y', y'', \ldots)$. Let

$\underline{\mathcal{D}}(J^\infty(X,Y))$

be the power series differential forms without the variable z.

Consider

$$\omega = dz + (z^2 + y + \lambda^2)\, dx \in \underline{\mathcal{D}}$$

Define an algebra homomorphism

$$\alpha: \underline{\mathcal{D}} \to \mathcal{D}(J^\infty(X,Y))$$

as follows

$$\alpha(x) = x\,; \quad \alpha(y) = y\,; \quad \alpha(y') = y'\,; \quad \text{etc.}$$

$$\alpha(z) = \sum_{n=1}^{-\infty} z_n \lambda^n,$$

$$z_n \in \mathcal{F}(J^\infty(X,Y))$$

Notice now that the LGPBLJWKB-formulas mean that the z_n are to be chosen so that:

$$\alpha(\omega) = \text{linear combinations of the contact forms } \theta', \theta'', \ldots$$

Thus, in a sense, α is a "virtual" integral manifold of a certain exterior differential system.

11. ASYMPTOTIC MAPS DEFINED BY FORMAL POWER SERIES IN THE MANNER OF POINCARE AND BIRKHOFF

Now I will put this material into a general differential geometric context. (This is a generalization of classical material primarily due to Poincaré and G.D. Birkhoff.) Let X and Y be manifolds, λ a real parameter. Suppose for the moment that, for each value of λ, ϕ_λ is a map

$$\phi_\lambda: X \to Y \quad .$$

Let $\mathcal{D}(Y)$, $\mathcal{D}(X)$ be the differential forms on Y and X. Then, for each $\omega \in \mathcal{D}(Y)$,

$$\lambda \to \phi_\lambda^*(\omega)$$

is an element of $\mathcal{D}(X)$ which depends on λ as a parameter. *Suppose it is expandable as a Laurent series in* λ:

$$\phi_\lambda^*(\omega) = \sum_{n=-\infty}^{\infty} \theta_n \lambda^n \quad . \tag{9.1}$$

Let $\underline{\mathcal{D}(X)}$ be the formal power series based on the differential form algebra $\mathcal{D}(X)$. Thus,

$$\delta \to \phi_\lambda^*(\omega)$$

is a *homomorphism* from $\underline{\mathcal{D}(Y)}$ to $\underline{\mathcal{D}(X)}$.

We are now prepared to generalize. Suppose given a map

$$\mathcal{D}(Y) \to \underline{\mathcal{D}(X)} \quad ,$$

$$\omega \to \sum_{n=-\infty}^{\infty} \theta_n \lambda^n \quad ,$$

which is a *homomorphism of exterior algebra* and commutes with d. It will be called an *asymptotic map* (or, strictly, a "virtual deformation" of maps) in the sense of Poincaré and Birkhoff.) If one needs an abbreviation, perhaps

$$\boxed{\text{P-B asymptotic maps}}$$

would be appropriate.

Of course, one could throw away the "geometric" intuition completely, and regard a P-B asymptotic map as a discrete family of linear maps

$$\phi_n : \mathcal{D}(Y) \to \mathcal{D}(X)$$

such that:

$$\sum_{m=-\infty}^{\infty} \phi_m(\omega_1) \wedge \phi_{n-m}(\omega_2) = \phi_n(\omega_1 \wedge \omega_2)$$

$$d\phi_n(\omega) = \phi_n(d\omega)$$

12. THE LGPBLJWKB-FORMULAS AND THE CONSERVATION LAWS FOR THE KORTEWEG-DE VRIES EQUATION

So far, we have been considering a single Sturm-Liouville equation

$$\psi_{xx} + (y + \lambda^2)\psi = 0 \quad . \tag{12.1}$$

The independent variable is x. However, the essence of the "Inverse Scattering Technique" is to let y also depend on another variable t, and to set up a partial differential equation

$$y_t = K(y) \, , \qquad (12.2)$$

where $y \to K(y)$ is a nonlinear differential operator on x alone. Thus, if

$$z = \frac{\psi_x}{\psi} \, ,$$

and if one adopts the formal power series solution

$$z = \sum_{n=1}^{-\infty} z_n \lambda^n \, ,$$

the z_n are determined by the $y, y_x, y_{xx}, \ldots$. The main point is that the integrals

$$\int_{-\infty}^{\infty} z_n(y(x,t), y_x(x,t), \ldots) \, dt$$

are *independent of* t if y satisfies (6.2), i.e., they are "conservation laws".

Chapter 21

THE POISSON-MOYAL BRACKET AND THE BURCHNALL-CHAUNDY
THEORY OF LINEAR ORDINARY
DIFFERENTIAL OPERATORS

1. INTRODUCTION

It is now recognized that a good deal of the material developed in the last ten years concerning nonlinear waves-inverse scattering is closely related to work in the 1920's by Burchnall and Chaundy [1] in "commutative" and "semi-commutative" linear, ordinary differential operators. In "Interdisciplinary Mathematics", Volume 14, I have developed the point that a certain amount of this material takes its most natural form in terms of the "Poisson-Moyal" bracket of quantum mechanics. (See "Interdisciplinary Mathematics", Volume 16, for development of the general differential-geometric principles underlying this material.) In this chapter I want to develop some of the Burchnall-Chaundy material in terms of the Poisson-Moyal bracket.

Remark. Boris Kuperschmidt informs me that this material has been extensively treated in the Russian literature by Krichever and Manin in papers that are not yet translated. As far as I can understand this work from the short announcements which have been translated, the *methodology* I present here is quite different from theirs. Of course, this method goes back to work of mine [2] of several years ago.

2. THE TRANSVECTION DIFFERENTIAL OPERATOR

For simplicity, deal with only time independent variables which we label "t" and "y". $f(t,p)$ is scalar valued function. $\mathscr{F}$ is the vector space of all such functions (say, C^∞).

$$\partial_t, \partial_p: \mathscr{F} \to \mathscr{F}$$

are partial derivatives

$$\partial_t^n, \partial_p^n, \partial_t^m, \partial_p^m, \ldots$$

are higher order partial derivatives. Another notation is:

$$a = (n,m) \in Z_+ \times Z_+$$

$$\partial_a(f) = \partial_p^n \partial_t^m(f)$$

(Z_+ denotes non-negative integers.) For $f_1, f_2 \in \mathcal{F}$, each $j \in Z_+$, set:

$$T^j(f_1, f_2)^j = \sum_{k=0}^{j} (-1)^j \binom{j}{k} \partial_p^{j-k} \partial_t^k(f_1) \partial_t^{j-k} \partial_p^k(f_2) \qquad (2.1)$$

T^j is called the j-*the transvection operator*. This terminology derives from classical invariant theory. The German word is "Überschiebung". It is an $SL(2,R)$-invariant differential operator. If

$$\alpha = \begin{pmatrix} \alpha_{11} & \alpha_{12} \\ \alpha_{21} & \alpha_{22} \end{pmatrix}$$

is a 2×2 real matrix with $\det \alpha = 1$, if

$$\alpha(f)\binom{x}{p} = f\left(\alpha^{-1}\binom{x}{p}\right) \qquad (2.2)$$

then

$$\alpha(T^j(f_1, f_2)) = T^j(\alpha f_1, \alpha f_2) \qquad (2.3)$$

For special values of j, these are familiar operators:

$$T^0(f_1, f_2) = f_1 f_2 \qquad (2.4)$$

$$T^1(f_1, f_2) = \partial_p(f_1)\partial_t(f_2) - \partial_t(f_1)\partial_p(f_2) \qquad (2.5)$$

$$T^2(f, f) = \partial_p^2(f)\partial_t^2(f) - 2(\partial_{pt}(f))^2 + \partial_t^2(f)\partial_p^2(f)$$

$$= 2(\partial_{pp}(f)\partial_{tt}(f) - \partial_{pt}(f)^2) \qquad (2.6)$$

T^1 is the *Poisson bracket*. In this particular case, it is also the *Jacobian* of the mapping

$$(p,t) \to (f_1(p,t), f_2(p,t))$$

of $R^2 \to R^2$. T^2 is the *Hessian*.

Here is a neat way to construct the transvection, which is akin to the "symbolic method" of classical invariant theory. Consider $\mathcal{F}$ as a real vector space and construct its real-scalar tensor product with itself:

$$\mathcal{F} \otimes \mathcal{F} \quad .$$

$$\partial_p \otimes \partial_t : \mathcal{F} \otimes \mathcal{F} \to \mathcal{F} \otimes \mathcal{F}$$

$$\partial_t \otimes \partial_p : \mathcal{F} \otimes \mathcal{F} \to \mathcal{F} \otimes \mathcal{F}$$

are the R-linear maps defined as follows:

$$\partial_p \otimes \partial_t (f_1 \otimes f_2) = \partial_p(f_1) \otimes \partial_t(f_2)$$

$$\partial_t \otimes \partial_p (f_1 \otimes f_2) = \partial_t(f_1) \otimes \partial_p(f_2)$$

Let

$$M : \mathcal{F} \otimes \mathcal{F} \to \mathcal{F}$$

be the R-linear map defined by multiplication

$$M(f_1 \otimes f_2) = f_1 f_2 \quad .$$

Then,

$$T^0(f_1, f_2) = M(f_1 \otimes f_2) \tag{2.7}$$

$$T^1(f_1, f_2) = M((\partial_p \otimes \partial_t - \partial_t \otimes \partial_p) f_1 \otimes f_2) \tag{2.8}$$

$$(\partial_p \otimes \partial_t - \partial_t \otimes \partial_p)^2 \equiv (\partial_p \otimes \partial_t - \partial_t \otimes \partial_p)(\partial_p \otimes \partial_t - \partial_t \otimes \partial_p)$$

$$= \partial_p^2 \otimes \partial_t^2 - \partial_t \partial_p \otimes \partial_p \partial_t - \partial_p \partial_t \otimes \partial_t \partial_p + \partial_t^2 \otimes \partial_p^2$$

$$= \partial_p^2 \otimes \partial_t^2 - 2\partial_p \partial_t \otimes \partial_p \partial_t + \partial_t^2 \otimes \partial_p^2$$

Thus, we see that:

$$T^2(f_1, f_2) = M(\partial_p \otimes \partial_t - \partial_t \otimes \partial_p)^2 (f_1 \otimes f_2) \tag{2.9}$$

Continuing in this way we see that:

$$\boxed{T^j(f_1, f_2) = M(\partial_p \otimes \partial_t - \partial_t \otimes \partial_p)^j (f_1 \otimes f_2) \\ \text{for all } j \in Z_+ .} \tag{2.10}$$

Proof of (2.10). Notice that:

$$(\partial_p \otimes \partial_t)(\partial_t \otimes \partial_p) = \partial_{pt} \otimes \partial_{tp}$$

$$= (\partial_t \otimes \partial_p)(\partial_p \otimes \partial_t) \ .$$

In words,

> The operators
> $$\partial_p \otimes \partial_t, \ \partial_t \otimes \partial_p : \mathscr{F} \times \mathscr{F} \to \mathscr{F} \otimes \mathscr{F}$$ (2.11)
> commute .

If A,B are *commuting* linear operators, the

$$(A - B)^j = \sum_{k=0}^{j} (-1)^k \binom{j}{k} A^{j-k} B^k \quad (2.12)$$

Hence,

$$(\partial_p \otimes \partial_t - \partial_t \otimes \partial_p)^j = \sum_k (-1)^k \binom{j}{k} (\partial_p \otimes \partial_t)^{j-k} (\partial_t \otimes \partial_p)^k$$

$$= \sum_k (-1)^k \binom{j}{k} (\partial_p^{j-k} \otimes \partial_t^{j-k})(\partial_t^k \otimes \partial_p^k)$$

$$= \sum_k (-1)^j \binom{j}{k} \partial_p^{j-k} \partial_t^k \otimes \partial_t^{j-k} \partial_p^k \quad (2.13)$$

This formula proves (2.10).

3. THE TRANSVECTION APPLIED TO POLYNOMIALS

$f \in \mathscr{F}$ is an m-*th degree polynomial* (in p) if and only if

$$\partial_p^{m+1}(f) = 0 \ . \quad (3.1)$$

Let

$$\mathscr{P}_m$$

denote the set of f's satisfying (3.1). Then,

$$(\partial_p \otimes \partial_t)(\mathcal{P}_m \otimes \mathcal{P}_n) \subset \mathcal{P}_{m-1} \otimes \mathcal{P}_n$$

$$(\partial_p \otimes \partial_t - \partial_t \otimes \partial_p)(\mathcal{P}_m \otimes \mathcal{P}_n) \subset \mathcal{P}_{m-1} \otimes \mathcal{P}_n + \mathcal{P}_m \otimes \mathcal{P}_{n-1} \qquad (3.2)$$

Iterate this relation:

$$(\partial_p \otimes \partial_t - \partial_t \otimes \partial_p)^2 (\mathcal{P}_m \otimes \mathcal{P}_n)$$

$$\subset (\partial_p \otimes \partial_t - \partial_t \otimes \partial_p)(\mathcal{P}_{m-1} \otimes \mathcal{P}_n + \mathcal{P}_m \otimes \mathcal{P}_{n-1})$$

$$\subset \mathcal{P}_{m-2} \otimes \mathcal{P}_n + \mathcal{P}_{m-1} \otimes \mathcal{P}_{n-1} \otimes \mathcal{P}_m \otimes \mathcal{P}_{n-2}$$

Finally:

$$(\partial_p \otimes \partial_t - \partial_t \otimes \partial_p)^j (\mathcal{P}_m \otimes \mathcal{P}_n) \subset \mathcal{P}_{m-j} \otimes \mathcal{P}_n + \mathcal{P}_{m-j+1} \otimes \mathcal{P}_{n-1} + \cdots$$

(3.3)

<u>Theorem 3.1</u>. Suppose that:

$$j > m + n \qquad (3.4)$$

Then,

$$T^j(\mathcal{P}_n, \mathcal{P}_m) = 0 \qquad (3.5)$$

In words, the j-th transvection applied to a pair of polynomials will vanish if j is sufficiently large.

<u>Proof</u>. (3.5) will follows from (3.3) and (3.4) if:

$$\begin{array}{l} m - j + k < 0 \\ \text{or} \\ n - k < 0 \end{array} \qquad (3.6)$$

for *all* k, such that $0 \leq k \leq j$. Thus we must show that (3.4) *implies* (3.6).

Suppose (3.6) is *not* satisfied, i.e., for some k such that $0 \leq k \leq j$,

$$n \geq k$$

and

$$k \geq j - m \quad .$$

These two combine to give

$$m \geq j - m \quad ,$$

which contradicts (3.4).

4. THE LIE ALGEBRA OF LINEAR DIFFERENTIAL OPERATORS AND THE POISSON-MOYAL LIE ALGEBRA

The transvections are *nonlinear* differential operators in *two* variables. They are closely related to the *scalar-coefficient linear* differential operators in *one* variable, i.e., those of the form:

$$D = a_n(t) \frac{d}{dt} + \cdots + a_0 \quad . \tag{4.1}$$

Let $\mathscr{D}$ denote the vector space of such operators, with n arbitrary. $\mathscr{D}$ is a *Lie algebra* under commutation.

$$\langle D_1, D_2 \rangle = D_1 D_2 - D_2 D_1$$

The Burchnall-Chaundy (and Lax) theory is basically a beginning toward a theory of *structure* for the algebraic properties of $\mathscr{D}$.

Let $\mathscr{P}$ be the vector space of functions $f(p,t)$ of *two* variables p and t, which are polynomials in p:

$$f = b_n(t) p^n + \cdots + b_0(t) \quad .$$

It can be made to a *Lie algebra* via the *Poisson-Moyal bracket*:

$$[f_1, f_2] = \sum_{j=1}^{\infty} \frac{1}{(2j-1)!} T_{2j-1}(f_1, f_2) \tag{4.2}$$

where

$$T_j : \mathscr{P} \times \mathscr{P} \to \mathscr{P}$$

is the j-th transvection operation, as defined in previous sections.

There is now a Lie algebra *isomorphism*

$$\rho: \mathscr{P} \to \mathscr{D}$$

defined as follows: First, represent the functions "p" and "t" as

$$\rho(p) = \frac{d}{dt} \quad (4.3)$$

$$\rho(t) = \text{multiplication by } t. \quad (4.4)$$

$$\rho(e^{ap}) = e^{a\frac{d}{dt}} \quad (4.5)$$

$$\rho(e^{at}) = \text{multiplication by } e^{at}. \quad (4.6)$$

If

$$f(t,p) = \iint \hat{f}(s,q)\, e^{st+qp}\, ds\, dq \quad (4.7)$$

then

$$(f) = \iint \hat{f}(s,q)\, \exp\!\left(q\frac{d}{dt} + st\right) ds\, dq \quad (4.8)$$

This is just the "Weyl formula" for *quantization of a classical observable* $f(t,p)$.

5. COMMUTATIVITY AND SEMI-COMMUTATIVITY IN THE POISSON-MOYAL LIE ALGEBRA

Let $\mathscr{P}_n$ be the linear subspace of elements of $\mathscr{P}$ which are of degree at most n in p. Thus,

$$[\mathscr{P}_n, \mathscr{P}_m] \subset \mathscr{P}_{n+m-1} \quad (5.1)$$

In particular,

$$[\mathscr{P}_1, \mathscr{P}_1] \subset \mathscr{P}_1, \quad (5.2)$$

i.e., $\mathscr{P}_1$ forms a Lie subalgebra of $\mathscr{P}$.

Let $f \in \mathscr{P}_n$. The condition that $f_1 \in \mathscr{P}_m$ commute with f, i.e.,

$$[f, f_1] = 0 \quad (5.3)$$

is a linear differential equation for the coefficient of f_1.

$$[f,f_1] = \sum_{j=0}^{\infty} \frac{1}{(2j+1)!} T_j(f,f_1)$$

$$= \sum_{j=0}^{n+m} \frac{1}{(2j+1)!} T_j(f,f_1)$$

$$= \sum_{j=0}^{n+m} \frac{1}{(2j+1)!} \sum_{k=0}^{2j+1} (-1)^k \partial_p^{2j+1-k} \partial_t^k(f) \partial_t^{2j+1-k} \partial_p^k(f_1)$$

(5.4)

Setting $[f,f_1] = 0$ gives $(n+m)$ equations for $(m+1)$ unknowns, the coefficients of f_1. It is then *over determined*.

Instead of commutativity, let us impose the condition

$$[f,f_1] \in \mathscr{P}_k \ .$$

(5.5)

This determines

$$\boxed{n+m-1-k}$$

equations for the

$$\boxed{m+1}$$

coefficients of f_1. The system will then (by this crude "counting") be "determined" when:

$$m+1 = n+m-1-k \ ,$$

or

$$\boxed{k = n-2}$$

(5.6)

<u>Definition</u>. f and f_1 are said to be *semi-commutative* when (5.5) and (5.6) are satisfied.

For example, for

$$n = 2,$$

notice that (5.6) gives the condition

$$k = 0,$$

i.e., f and f_1 are semi-commutative if

$$[f,f_1] \in \mathscr{P}_0 \ .$$

This is the basic condition of the Lax paper on inverse scattering. Notice how naturally it comes out of the Burchnall-Chaundy paper!

6. SEMI-COMMUTATIVITY FOR THE CASE OF SECOND ORDER POLYNOMIALS

Suppose

$$f = p^2 + a(t) \in \mathscr{P}_2 \ . \qquad (6.1)$$

Let us look for

$$g = a_3 p^3 + a_2 p^2 + a_1 p + a_0 \in \mathscr{P}_3, \qquad (6.2)$$

which is semi-commutative with f, i.e., such that

$$[g,f] \in \mathscr{P}_0 \ . \qquad (6.3)$$

We will now compute the transvections applied to f and g.

$$\begin{aligned}
T_1(g,f) &= \text{Poisson bracket} \\
&= g_p f_t - g_t f_p \\
&= (3a_3 p^2 + 2a_2 p + a_1)(a') - (a_3' p^3 + a_2' p^2 + a_1' p + a_0')(2p) \\
&= -2 p a_3' p^4 - 2 a_2' p^3 + p^2(3a_3 a' - a_1') + (2a_2 a' - 2a_0')p + a_1 a' \ .
\end{aligned}$$

(' means differentiation with respect to t.)

$$\begin{aligned}
T_3(g,f) &= \partial_p^3 g \partial_t^3 f - 3 \partial_p^2 \partial_t g \partial_t^2 \partial_p f + 3 \partial_p \partial_t^2 g \partial_t \partial_p^2 f - \partial_t^3 g \partial_p^3 f \\
&= 6 a_3 a'''
\end{aligned}$$

$$T_5(g,f) = 0 \quad \ldots$$

$$[g,f] = T_1(g,f) + \frac{1}{6} T_3(g,f) \quad .$$

Now, $T_3(g,f) \in \mathscr{P}_0$. Hence, condition (6.3) is equivalent to

$$T_1(g,f) \in \mathscr{P}_0 \quad . \tag{6.4}$$

Thus, "semi-commutivity" is a "Poisson bracket" condition. Explicitly, (6.4) means that

$$a_3' = 0 = a_2'$$

$$3a_3 a' = a_1'$$

$$a_2 a' = a_0' \quad ,$$

or

$$a_3 = \text{constant} \equiv c_3$$

$$a_2 = \text{constant} \equiv c_2$$

$$a_1 = 3c_3 a + c_1$$

$$a_0 = c_2 a + c_0$$

($c_0 \cdots c_3$ are constants).

Theorem 6.1. Let

$$D = \frac{d^2}{dt^2} + a(t)$$

be a Sturm-Liouville differential operator. Then, a third operator D' is semi-commutative with respect to D, i.e., $[D,D']$ is a zero-th order operator if and only if D' is the image under the Weyl representation of the function

$$g = c_c p^3 + c_2 p^2 + (3c_3 a + c_1)p + (c_2 a + c_0) \tag{6.5}$$

with constants c_0, c_1, c_2, c_3. Thus,

$$D' = c_3 \frac{d^3}{dt^3} + c_2 \frac{d^2}{dp^2} + 3c_3\left(a\frac{d}{dt} + \frac{1}{2}a'\right) + c_1 \frac{d}{dt} + c_2 a + c_0$$

(6.6)

7. THE COMMUTATIVE CONDITION IN THE STURM-LIOUVILLE CASE

Given a differential operator D, the idea of Burchnall-Chaundy is that there will be operators D' which are semi-commutative relative to D. The condition that such a D' actually *commute* with D is then a *further* condition of D. Let us work it out in the context of the calculations of Section 6: f and g are given by (6.1) and (6.4), then the condition

$$[f,g] = 0$$

is *equivalent* to the condition that:

$$\frac{1}{6} T_3(g,p) + (3c_3 a + a)a' = 0$$

or that:

$$c_3 a''' + (3c_3 a + c_1)a' = 0 ,$$

or

$$\boxed{c_3 a'' + \frac{3c_3 a^2}{2} + c_1 a = c_4} \qquad (6.6)$$

This is a differential equation for a. (It is essentially just that for the time-independent solutions of the Korteweg-de Vries equation.) Let us sum up the qualitative information we derive from this argument.

<u>Theorem 6.1.</u> Let

$$D = \frac{d^2}{dt^2} + a(t)$$

be a Sturm-Liouville differential operator. Then, there is a third order differential operator which commutes with it if and only if there are constants c_1, c_2, c_3, c_4 such that a satisfies the second order nonlinear differential equation (6.6). (Note that the set of all such D's then depends on *six* parameters.)

BIBLIOGRAPHY

1. J.L. Burchnall and T.W. Chaundy, Commutative ordinary differential equations, Proc. London Math. Soc. $\underline{21}$ (1923), 420-440.

2. R. Hermann, The inverse scattering technique of soliton theory, Lie algebras, the quantum-mechanical Poisson-Moyal bracket, and the rotating rigid body, Phys. Rev. Lett. $\underline{37}$ (1976), p. 1591.

Chapter 22

THE BIRKHOFF-LANGER EXPANSION OF LINEAR SYSTEMS

1. INTRODUCTION

Although there is a vast mathematical literature on the "asymptotic solution" to differential equations, it has not been deeply explored in the systems theory literature. One of the most interesting papers in the classical mathematical literature is "The Boundary Problems and Developments Associated with a System of Ordinary Differential Equations of the First Order", by G.D. Birkhoff and R.L. Langer, and reprinted on page 345 of Volume 1 of Birkhoff's *Collected Works*. I will now develop certain topics from this paper which seem especially interesting for systems-theoretic purposes. This material also has profound relations with "nonlinear physics".

2. THE FORMAL ASYMPTOTIC SERIES

Consider a time-varying system of linear, homogeneous, ordinary differential equations depending on a real parameter λ of the following form

$$x' = (\alpha(t)\lambda + \beta(x))x \qquad (2.1)$$

$t \to \alpha(t)$, $\beta(t)$ are $n \times n$ matrix valued functions of t, $t \to x(t)$ is an $n \times 1$-vector function of t. Primes denote partial derivatives with respect to t. Let us, following Birkhoff and Langer, look for a solution in the following formal power series form:

$$x(t) = e^{\lambda \gamma(t)} \sum_{n=0}^{\infty} x_n(t) \lambda^{-n} \qquad (2.2)$$

($t \to \gamma(t)$ is a scalar function of t.) Let us substitute (2.2) and derivatives into (2.1) and equate powers of λ:

$$x' = \lambda \gamma' x + e^{\lambda \gamma} \sum_n x_n'(t) \lambda^{-n}$$

$$= (\alpha \lambda + \beta) \left(e^{\lambda \gamma} \sum_n x_n \lambda^{-n} \right)$$

$$= \alpha e^{\lambda \gamma} \sum x_n \lambda^{-n+1} + \beta e^{\lambda \gamma} \sum x_n \lambda^{-n}$$

$$= \alpha e^{\lambda \gamma} \sum_{n=-1}^{\infty} x_{n+1} \lambda^{-n} + \beta e^{\lambda \gamma} \sum_{n=0}^{\infty} x_n \lambda^{-n}$$

$$= \gamma' e \sum_{n=-1}^{\infty} x_{n+1} \lambda^{-n} + e^{\lambda \gamma} \sum_{n=0}^{\infty} x'_n \lambda^{-n}$$

First calculate the $n = -1$ term:

$$(\alpha(t) - \gamma'(t))(x_0(t)) = 0 \tag{2.3}$$

Now, equate the rest of the terms:

$$(\alpha(t) - \gamma'(t)) x_{n+1}(t) = x'_n(t) - \beta(t) x_n(t) \tag{2.4}$$

This is a recursion relation determining (in the "generic" case) all $x_n(t)$. Here are the first two terms:

$$x_1(t) = (\alpha - \gamma')^{-1}(x'_0 - \beta x_0) \tag{2.5}$$

$$x'_1 = -(\alpha - \gamma')^{-1}(\alpha' - \gamma'')(\alpha - \gamma')^{-1}(x'_0 - \beta x_0)$$
$$+ (\alpha - \gamma')^{-1}(x''_0 - \beta' x_0 - \beta x'_0) \tag{2.6}$$

$$x_2 = (\alpha - \gamma')^{-1}(x'_1 - \beta x_0) \quad . \tag{2.7}$$

The general formulas are extremely complicated, but the pattern is clear: x_{n+1} is expressed in terms of the first $(n+1)$ derivatives of x_0 and γ, and the first n derivatives of α and β. In the paper cited above, Birkhoff and Langer discuss conditions under which these series represent *asymptotic series* for the true solutions. I will now turn to a systems theoretic setting.

3. THE VOLTERRA-BIRKHOFF-LANGER SERIES FOR BILINEAR SYSTEMS

For this section, I will use the standard systems theoretic notation. Suppose given

$$\frac{dx}{dt} = Ax + \lambda u(t)bx \quad . \tag{3.1}$$

$t \to x(t)$ is the *state*, A and B are square matrices which are independent of t. $t \to u(t)$ is the *input*, which is supposed, for simplicity, to be scalar. One can look for solutions of (3.1) of the form:

$$x(t) = \sum_{n=0}^{\infty} x_n(t) \lambda^n \tag{3.2}$$

They are called *Volterra series* in the systems theoretic literature. However, we can now look for the Birkhoff-Langer series in powers of λ^{-n}. In terms of the formulas of Section 2, set:

$$\boxed{\begin{aligned} \alpha &= u(t)B \\ \beta &= A \end{aligned}} \tag{3.3}$$

Equations (2.3) and (2.4) take the following form:

$$\boxed{\begin{aligned} (u(t)B - \gamma'(t))x_0(t) &= 0 \\ (u(t)B - \gamma'(t))x_{n+1} &= x_n'(t) - Ax_n(t) \end{aligned}} \tag{3.4} \tag{3.5}$$

(3.4) shows us how $x_0(t)$ is determined by the equations and eigenvalues of B, in the "generic" case that B is non-singular and has distinct eigenvalues.

Let us work out the first few terms of the recursion relation (3.5), and think about its "systems-theoretic" meaning.

$$x_0' = -(uB - \gamma')^{-1}(u'B - \gamma'')x_0$$

$$x_1 = (uB - \gamma')^{-1}(x_0' - Ax_0)$$

$$= (uB - \gamma')^{-1}(uB - \gamma')^{-1}(u'B - \gamma'')x_0 - (uB - \gamma')^{-1}Ax_0 \quad .$$

The general pattern should be clear:

> $x_n(t)$ is expressed in terms of the first n derivatives of the input u. (3.6)

This is to be compared with the usual Volterra series, where $x_n(t)$ involves n *integrals* of u.

4. BIRKHOFF-LANGER EXPANSIONS FOR LINEAR INPUT SYSTEMS

The Birkhoff-Langer formulas can be immediately generalized to the "non-homogeneous" situation,

$$x' = \frac{dx}{dt} = (\alpha\lambda + \beta)x + y \quad, \qquad (4.1)$$

where y are given functions of t,

$$x = e^{\gamma\lambda} \sum_{n=0}^{\infty} x_n \lambda^{-n} \quad. \qquad (4.2)$$

The recursion relations (2.3)-(2.4) for the zero-input case now generalize as follows:

$$(\alpha(t) - \gamma'(t))(x_0(t)) = 0 \qquad (4.3)$$

$$\gamma' x_1 + x_0' = \alpha x_1 + \beta x_0 + y \qquad (4.4)$$

$$\gamma' x_{n+1} + x_n' = \alpha x_{n+1} + \beta x_n \qquad (4.5)$$

for $n \geq 1$

Again, we see that the series is analogous to the usual formulas of Systems Theory, with the difference that the *derivatives* of the input occur instead of the *integrals*. Since the operation of derivation is *not* "smoothing", these series will not converge and are in general only "asymptotic".

Chapter 23

THE BIRKHOFF-LANGER EXPANSION AND THE CONSERVATION LAWS OF NON-LINEAR WAVE EQUATIONS

1. INTRODUCTION

The paper "Relationships Among Inverse Method, Bäcklund Transformation and Conservation Laws", by M. Wadati, H. Sanuki and K. Konno (*Prog. Theor. Phys*. 53 (1925), 419-436) presents a very convenient unification of the methods that have been utilized to solve non-linear partial differential equations in two variables. In this chapter, I will show that their technique is closely related to the Birkhoff-Langer asymptotic expansion considered in previous chapters.

Now we must deal with vector and matrix functions of two real variables. In non-linear wave theory, these variables are usually "space" and "time", denoted as "x" and "t". However, for our purposes it is more convenient to denote them as "s" and "t". Partial derivatives with respect to s and t are denoted by subscripts. In addition, there will be a parameter λ to define formal power series.

2. COMPLETELY INTEGRABLE (IN THE SENSE OF MAYER-FROBENIUS) LINEAR SYSTEMS OF TWO INDEPENDENT VARIABLES

x denotes an n-vector; A,B denote $n \times n$ matrices. Consider the system:

$$\boxed{\begin{aligned} x_s &= Ax \\ x_t &= Bx \end{aligned}} \qquad (2.1)$$

The "integrability conditions" are obtained by equating cross-derivatives

$$(x_s)_t = (x_t)_s \quad .$$

They are:

$$A_t - B_s = [A,B] \quad . \qquad (2.2)$$

Let us suppose now that A,B are polynomial functions of the parameter λ.

$$A = A_0 + \lambda A_1 \tag{2.3}$$

$$B = \sum_{j=0}^{N} B_j \lambda^j \, .$$

We can now substitute (2.3) into (2.2) and equate powers of λ:

$$A_{0,t} + \lambda A_{1,t} - \sum_{j=0}^{N} B_{j,s} \lambda^j = \sum_{j=0}^{N} [B_j, A_0] \lambda^j + [B_j, A_1] \lambda^{j+1}$$

$$= \sum_{j=0}^{N} [B_j, A_0] \lambda^j + \sum_{j=1}^{N+1} [B_{j-1}, A_1] \lambda^j$$

Equate the coefficients of powers of λ:

$$A_{0,t} - B_{0,s} = [B_0, A_0] \tag{2.4}$$

$$[B_N, A_1] = 0 \tag{2.5}$$

$$A_{1,t} - B_{1,s} = [B_1, A_0] + [B_0, A_1] \tag{2.6}$$

$$-B_{j,s} = [B_j, A_0] + [B_{j-1}, A_1] \tag{2.7}$$

$$2 \le j \le N$$

For example, for $N = 3$ (which is the number for the Korteweg-de Vries equation) in the Wadati-Sanuki-Konno formalism, these equations are:

$$A_{0,t} - B_{0,s} = [B_0, A_0] \tag{2.8}$$

$$[B_s, A_0] = 0 \tag{2.9}$$

$$A_{1,t} - B_{1,s} = [B_1, A_0] + [B_0, A_1] \tag{2.10}$$

$$-B_{2,s} = [B_2, A_0] + [B_1, A_1] \tag{2.11}$$

$$-B_{3,s} = [B_3, A_0] + [B_2, A_1] \tag{2.12}$$

3. THE BIRKHOFF-LANGER EXPANSION

Let us now look for an asymptotic expansion of (2.1) of the Birkhoff-Langer type:

$$x = e^{\lambda \gamma} \sum_{n=0}^{\infty} x_n \lambda^{-n}$$

$$x_t = \lambda \gamma_t e^{\lambda \gamma} \sum_{n=0}^{\infty} x_n \lambda^{-n} + e^{\lambda \gamma} \sum_{n=0}^{\infty} x_{n,t} \lambda^{-n}$$

$$= e^{\lambda \gamma} \sum_{n=-1}^{\infty} \gamma_t x_{n+1} \lambda^{-n} + e^{\lambda \gamma} \sum_{n=0}^{\infty} x_{n,t} \lambda^{-n}$$

or

$$\boxed{x_t = e^{\lambda \gamma} \sum_{n=0}^{\infty} (\gamma_t x_{n+1} + x_{n,t}) \lambda^{-n} + e^{\lambda \gamma} \gamma_t x_0 \lambda} \qquad (3.1)$$

$$e^{-\lambda \gamma}(A_0 + \lambda A_1) x = \sum_{n=0}^{\infty} A_0 x_n \lambda^{-n} + \sum_{n=0}^{\infty} A_1 x_n \lambda^{-n+1}$$

$$= \sum_{n=0}^{\infty} A_0 x_n \lambda^{-n} + \sum_{n=-1}^{\infty} A_1 x_{n+1} \lambda^{-n},$$

or

$$\boxed{(A_0 + \lambda A_1) x = e^{\lambda \gamma} \sum_{n=0}^{\infty} (A_0 x_n + A_1 x_{n+1}) \lambda^{-n} + \lambda A_1 x_0)}$$

$$\boxed{x_s = e^{\lambda \gamma} \sum_{n=0}^{\infty} (\gamma_s x_{n+1} + x_{n,s}) \lambda^{-n} + e^{\lambda \gamma} \gamma_s x_0 \lambda} \qquad (3.2)$$

$$\sum_{j=0}^{N} (B_j \lambda^j) x = e^{\lambda \gamma} \sum_{j=0}^{N} \sum_{n=0}^{\infty} B_j x_n \lambda^{-n+j}$$

$$B(\lambda)x = e^{\lambda\gamma} \sum_{n=-N}^{\infty} \sum_{j=0}^{N} B_j x_{j+n} \lambda^{-n} \qquad (3.3)$$

Thus, the equation

$$x_s = B(\lambda)x$$

requires that:

$$\sum_{n=0}^{\infty} (\gamma_s x_{n+1} + x_{n,s}) \lambda^{-n} + x_0 \gamma_s = \sum_{n=-N}^{\infty} \sum_{j=0}^{N} B_j x_{j+n} \lambda^{-n} ,$$

or

$$\gamma_s x_{n+1} + x_{n,s} = \sum_{j=0}^{N} B_j x_{j+n} , \qquad (3.4)$$
$$n \geq 0$$

$$x_0 \gamma_s = \sum_{j=0}^{N} B_j x_{j-1} \qquad (3.5)$$

$$\sum_{j=n}^{N} B_j x_{j-n} = 0 , \qquad \text{for } 0 < n \leq N \qquad (3.6)$$

or

$$B_N x_0 = 0$$
$$B_N x_1 + B_{N-1} x_0 = 0 \qquad (3.7)$$
etc.

For example, for $N = 3$ (which is the Korteweg-de Vries situation)

CONSERVATION LAWS

$$
\begin{aligned}
B_3 x_0 &= 0 \\
B_3 x_1 + B_2 x_0 &= 0 \\
B_3 x_2 + B_2 x_1 + B_1 x_0 &= 0
\end{aligned}
\tag{3.8}
$$

Similarly, the conditions that

$$x_t = Ax$$

are:

$$\gamma_t x_{n+1} + x_{n,s} = A_0 x_n + A_1 x_{n+1} \quad,$$

or

$$x_{n+1} = (\gamma_t - A_1)^{-1}(x_{n,t} - A_0 x_n) \tag{3.9}$$

$$\text{for } n \geq 0$$

and

$$(A_1 - \gamma_t)(x_0) = 0 \tag{3.10}$$

Combine (3.9) and (3.4):

$$x_{n,s} + \gamma_s (\gamma_t - A_1)^{-1}(x_{n,t} - A_0 x_n) = \sum_{j=0}^{N} B_j x_{j+n} \tag{3.11}$$

To construct "conservation laws" we now find functions

$$F(x_1, \ldots, x_n)$$

which have the property that

$$\frac{d}{ds} \int_{-\infty}^{\infty} F(x_1(s,t), \ldots, x_n(s,t)) \, dt = 0 \tag{3.12}$$

for solutions $(s,t) \to x(s,t)$ which vanish at infinity at t. Equation (3.11) shows that these should *exist* as *rational* functions, but how to effectively construct them is not so obvious. Of course, in the examples of the Wadati-Sanuki-Konno paper cited above, x is a 2-vector, and these formulas are explicitly given. I would like to pose as a *conjecture* that they can be extended to higher dimensions.

GROUP V

SOME WORK IN GEOMETRIC CONTROL THEORY

INTRODUCTION

I see part of my job as an "applied differential geometer" to be that of pushing ahead with the development of tools and methodology in control theory. Here is work of this sort.

I especially point out the material in Chapter 27. I briefly describe a *modification* of standard system-theoretic ideas to handle certain interesting *nonlinear* systems from the input-output point of view. I believe they may be useful in modeling biological systems (e.g., "biological clocks") with "periodic inputs" and "close to periodic" outputs. The idea presented is in a sense *new*, although it is just a formalization of the classical method used by Van der Pol.

Chapter 24

BOLTZMANN-HAMEL EQUATIONS
OF ANALYTICAL MECHANICS

1. INTRODUCTION

Since the introduction of "state space" methods in the 1950's, control and mathematical systems theory has developed into a powerful analytical tool for a variety of engineering disciplines. However, it must be admitted that the sort of problem that can be routinely treated with the machinery developed and available today is relatively restricted in the context of the whole spectrum of engineering problems that are potentially amenable. In particular, the continual astounding growth of computer technology offers ever-more tantalizing possibilities. (I still recall the exciting atmosphere in the late 1950's when the "state space methods" were quickly accepted and learned because they were so obviously better suited to implementation on computers.)

Thus, there is now a considerable emphasis in control theory journals on "large scale systems", "decentralized control", etc., i.e., areas that attempt to adapt the standard techniques to more complicated and sophisticated areas of application. An underlying theme in this development is the notion of "computational complexity", taken perhaps not so much in the technical sense in which that term is used by computer scientists, but more intuitively and informally as the feasibility of implementing the mathematical schemes in terms of computer technology.

The aim of this paper is to begin the systematic study of what I call *geometric complexity*. I will not attempt a formal definition at this stage--it is meant to capture the "geometric" features of systems like helicopters, bicycles, gyroscopes, space craft, etc., which involve complicated and interesting geometric configurations. Perhaps the notion of "geometrical complexity" can only be described relatively, i.e., as a "partial ordering" on systems. Thus, a rotating rigid body is geometrically more complex than a particle moving on a line, a helicopter is geometrically more complex than an airplane, etc. Perhaps ultimately this could be made precise in terms of Lie group theory and/or algebraic topology. Thus, the rotating rigid body has a configuration space which is $SO(3,R)$, a semisimple Lie group, while a particle moving on the line has configuration space an abelian group. From the viewpoint of topology, the configuration space of the rotating rigid body is also more complex than is that of the particle on the line, since the former is a compact manifold with various non-trivial homology and homotopy groups, while the latter is topologically one of the simplest of spaces. These ideas should ultimately be reflected at the computer level also, although at present this is very implicit and primitive (as are most possibilities of developing sophisticated geometric ideas on the computer).

My thesis is that the "modern" differential geometry (emphasizing differential manifolds, differential forms, etc.) is the appropriate framework for much of this. Such a possibility was certainly glimpsed by Gabriel Kron and the RAAG memoirists, but unfortunately their work and ideas were not adequately appreciated by their contemporaries, for reasons I do not completely understand and in any case will not go into here. Certainly one problem was the mathematical formalism they were forced to use--tensor analysis--is very complicated and forbidding. Post World War II differential geometry has emphasized a more systematic and algebraic formalism, much closer to the mainstream of the rest of mathematics, which I believe is much better adapted to these ends *when it is more widely understood*. (One major difficulty is the lack of adequate means of instruction in this formalism outside of a small group. Even many mathematicians do not know it.)

Now, the hallmark of this new approach is that (so far as is feasible) *only operations are used that are independent of coordinates*, the so called "natural" differential operations of Lie derivation and exterior derivative. Although I will not be able to go into it in this paper, I believe that these operations are ideally suited to implementation in terms of the symbolic and algebraic manipulation techniques being developed in today's computers, e.g., the MACSYMA system at MIT. The rapid development of computer graphics also offers many possibilities of realizing the geometric side of the formalism in exciting ways. The mathematical goal of the formalism is to "think geometrically", and the enhancement of our geometric intuition provided by computer graphics is yet to be widely explored.

In this paper, I will mainly restrict attention to traditional material of "analytical mechanics" (AKA "classical mechanics", "analytical dynamics") in order to illustrate the ideas. A certain amount of this material has already been worked into the language of modern differential geometry but, as Roger Brockett remarks, the integration is still incomplete because treatises do not deal with such important questions as forces, constraints, quasi-coordinates, Boltzmann-Hamel equations, etc. Another applied topic of some importance which is well suited to treatment via differential forms is nonlinear and singular perturbation theory. Thus, I would foresee the lengthy calculations necessary to carry out traditional perturbation theory "mechanized" via differential forms, then carried out on a computer via the symbolic and algebraic manipulation systems.

A technical feature of this paper is that I shall emphasize the role of *differential forms*, and de-emphasize the role of *vector fields*. There is no strong logical reason for doing this--partly it is because I do not want to hinder the reader from learning the basic ideas by introducing too much jargon. The theory of differential forms is a readily digestible subset of the whole business, that is in fact strongly emphasized and preferred in Elie Cartan's work, and that is the foundation for much of "modern" differential geometry. Also, "differential forms" are natural objects in terms of mechanics; for example, "forces" are really "differential forms", as are "quasi-coordinates". In Cartan's way of doing mechanics-- which is now called the *symplectic manifold approach*--certain second degree differential forms play the basic role.

2. DIFFERENTIAL FORMS WITH SPECIAL ATTENTION TO THOSE OF FIRST AND SECOND DEGREE

The objects of interest in "modern" differential geometry are; first, *differentiable manifolds*, and second, certain types of *tensor-fields* which "live" on them. The manifolds are the basic geometric objects, and the tensor fields enable us to study geometric properties using the tools of differential and integral calculus. The historical prototype of much of this is Descartes' "analytical geometry"; the study of geometric figures ("curves and surfaces") by translating "geometry" into "algebra" via the *equations* which define the geometric figures. I will not define "differential manifold" here; there are many references for this materials that do it far more brilliantly than I could ever manage. The intuitive idea is that a "manifold" *locally* looks like Euclidean space, but that several "coordinate patches" may be needed to describe the whole space. Think of the surface of the sphere

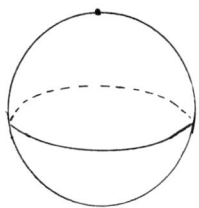

Either "spherical coordinates" θ, ϕ are used, or the "north" and "south" hemispheres. Thus, the basic idea is much the same as Descartes'--"parametrizing" points of space with real numbers, with the added proviso that fairly arbitrary choices may be made, reminiscent of what Einstein called "General Covariance". Of course, Euclidean space R^n is also a manifold.

In modern differential geometry, certain special types of tensor fields play a special role: For a manifold X:

$\mathscr{F}(X)$ = space of real-valued infinitely differentiable functions on X (Elements of $\mathscr{F}(X)$ are called *scalar fields* in tensor analysis.)

$\mathscr{D}^r(X)$ = space of r-*th order differential forms*, $r \geq 1$
≡ space of r-th degree *covariant*, skew-symmetric tensor fields

$\mathscr{V}(X)$ = space of *vector fields* on X
≡ space of *one-contravariant* tensor fields.

In this paper we shall mainly be concerned with $\mathscr{F}(X)$ (which often is identified with $\mathscr{D}^0(X)$), $\mathscr{D}^1(X)$ and $\mathscr{D}^2(X)$.

For the moment I will work with a manifold X with a single, everywhere defined coordinate system

$$(x^1, \ldots, x^n)$$

Thus, if p is a point of X, $x^1(p), \ldots, x^n(p)$ are the real numbers one assigns to the point p. For example, X could be R^3, "our" space, with a fixed Euclidean coordinate system,

x^1 = x - component of the (x,y,z)

$x^2 = y$

$x^3 = z$.

Alternately, x^1, x^2, x^3 could be identified with the familiar physical coordinates (r, θ, ϕ) of a point.

The set of real-valued functions on X which are *infinitely differentiable* (C^∞) in these variables will be denoted by

$$\mathscr{F}(X) \ .$$

("C^∞" is mainly chosen for its technical superiority and convenience. The formalism can be readily adapted to other possibilities.) Since two real functions can be added, multiplied, and admit a scalar multiplication by real numbers, $\mathscr{F}(X)$ forms a *commutative, associative algebra*. In the "modern" approach to differential geometry the properties of this "algebra" play the basic role. This is, in a sense, a vast and deep realization of Descartes' idea that "geometric" problems can be "algebraized".

The coordinate functions $x^1, \ldots, x^n$ themselves are elements of $\mathscr{F}(X)$. Now, consider an object of the following form, which we call a *one (differential) form*

$$\theta = a_1 dx^1 + \cdots + a_n dx^n \quad , \tag{2.1}$$

with $a_1, \ldots, a_n \in \mathscr{F}(X)$.

Such a form is, of course, identifiable with the n-tuple

$$(a_1, \ldots, a_n)$$

of its components. (We shall see that this identifies one-forms with *one-covariant tensor fields in the sense of tensor analysis*.) However, as we shall see, there are great advantages to writing this particular kind of "tensor field" in this form.

From now on, we shall use the summation convention to write expressions like (2.1):

$$\theta = a_i dx^i \quad . \tag{2.2}$$

Here, the index i ranges over 1 to n; upper and lower indices (and only these!) are summed *automatically*. If more than one such index is needed, it is chosen from succeeding letters of the alphabet, e.g., j,k,... Later on, other ranges of indices are needed, e.g., from 1 to m. This is handled by using other groups of successive letters of the alphabet, e.g., a,b,...

Two expressions like (2.2) can be added

$$\theta = a_i dx^i, \qquad \theta' = b_i dx^i,$$
$$\theta + \theta' = (a_i + b_i) dx^i \quad . \tag{2.3}$$

A scalar function $f \in \mathscr{F}(X)$ and a one-form θ can be multiplied:

$$(f\theta) = (fa_i) dx^i \quad . \tag{2.4}$$

Let $\mathscr{D}^1(X)$ be the set of one-forms. This scalar multiplication (2.4)—together with the addition (2.3), makes $\mathscr{D}^1(X)$ into a *module* over the algebra $\mathscr{F}(X)$.

One can also define a map

$$d: \mathscr{F}(X) \to \mathscr{D}^1(X) \tag{2.5}$$

called *exterior derivative*

$$df = \frac{\partial f}{\partial x^i} dx^i \quad . \tag{2.6}$$

Note that d is a *linear* first order differential operator. However, d is *not* linear with respect to the $\mathcal{F}(X)$-multiplication, since

$$d(f_1 f_2) = (df_1) f_2 + f_1 df_2 \tag{2.7}$$

for $f_1, f_2 \in \mathcal{F}(X)$.

Algebraically, one says that d is R-*linear*, but not $\mathcal{F}(X)$-*linear*.

We can use d to express the "covariance" of differential forms under change of coordinates. Suppose that (y^i) are another set of elements of $\mathcal{F}(X)$ which define a "new" coordinate system for X.

Example. *Polar coordinates for* $X = R^2$

$n = 2$, $x^1 = $ "x", $x^2 = $ "y", where (x,y) are the usual Cartesian coordinates

$$y^1 = r = \sqrt{x^2 + y^2} = \sqrt{(x^1)^2 + (x^2)^2}$$

$$y^2 = \theta = \tan^{-1}(x^2/x^1) \quad .$$

So far, "dx^i" have just been algebraic symbols. Notice, however, that *they agree with* "dx^i" *as the exterior derivative of the function* x^i. *They provide a basis for* $\mathcal{D}^1(X)$ *as an* $\mathcal{F}(X)$-*module*. This completely trivial remark is enormously significant, however, since it *determines the transformation law for differential forms under change in variables* $x \to y$. Namely,

$$dx^i = \frac{\partial x^i}{\partial y^j} dy^j \quad , \tag{2.8}$$

hence,

$$\theta = a_i dx^i = \left(a_i \frac{\partial x^i}{\partial y^j} \right) dy^j \equiv a_i' du^i \quad ,$$

i.e., θ in the "new" coordinates (y) has components

$$\left(a_j \frac{\partial x^j}{\partial y^i} \right) = a_i' \quad . \tag{2.9}$$

This represents the *transformation law for a one-covariant tensor*, in the language of classical tensor analysis.

Now we define *two-differential forms*. In the coordinates (x^i), they are expressions of the form:

$$w = a_{ij} dx^i \wedge dx^j \tag{2.10}$$

with $a_{ij} = -a_{ji} \in \mathscr{F}(X)$.

Denote the set of all these objects as $\mathscr{D}^2(X)$. As the solution indicates, they are to be identified with *twice-covariant, skew-symmetric tensors*, in the sense of tensor analysis. Why we write it in this "peculiar" way—with this new symbol $\wedge$ —will now be explained.

Now, let us define *exterior multiplication*. For simplicity, we shall do so only for one-forms. It will then be a bilinear operation

$$\wedge: \mathscr{D}^1(X) \times \mathscr{D}^1(X) \to \mathscr{D}^2(X)$$

If $\theta_1, \theta_2 \in \mathscr{D}^1(X)$, their *exterior product* will be denoted as

$$\theta_1 \wedge \theta_2 \quad .$$

Here is the explicit formula: If

$$\theta_1 = a_i dx^i$$

$$\theta_2 = b_j dx^j \quad ,$$

then

$$(\theta_1 \wedge \theta_2) = \tfrac{1}{2}(a_i b_j - a_j b_i) dx^i \wedge dx^j \quad . \tag{2.11}$$

Then, we see that if

$$\theta_1 = \delta_i^k dx^i \equiv dx^k$$

$$\theta_2 = \delta_j^\ell dx^j \equiv dx^\ell \quad ,$$

(δ_1^k is the "Kronecker delta" symbol; equal to zero if $i \neq h$, to one if $i = k$.) Then

$$dx^k \wedge dx^\ell = \theta_1 \wedge \theta_2 = \tfrac{1}{2}(\delta_i^k \delta_j^\ell - \delta_j^k \delta_i^\ell) dx^i \wedge dx^j$$

$$= \tfrac{1}{2}(dx^k \wedge dx^\ell - dx^\ell \wedge dx^k)$$

i.e.,

$$\boxed{dx^k \wedge dx^\ell = -dx^\ell \wedge dx^k} \qquad (2.12)$$

This expresses the skew-symmetry of the exterior product $\wedge$. (It is also called "wedge" or "Grassmann" product. Amazingly, the recognition of the geometric value of this "non-commutative" algebra goes back to the middle 19-th century to Pfaff and Grassmann. It was extensively developed later in the 19-th century by them and by Frobenius, Goursat, E. Cartan, but then died out, to be triumphantly revived in post World War II differential geometry. It is for such reasons that I strongly believe that a strong sense of *historical* scholarship is critically important for successful work in differential geometry.)

Exterior derivative can now be extended to a map

$$d: \mathcal{D}^1(X) \to \mathcal{D}^2(X) \quad,$$

by the following explicit formula:

$$d(a_i dx^i) = \frac{\partial a_i}{\partial x^j} dx^j \wedge dx^i$$

$$= \frac{1}{2}\left(\frac{\partial a_j}{\partial x^i} - \frac{\partial a_i}{\partial x^j}\right) dx^i \wedge dx^j \qquad (2.13)$$

(Notice that it is a sort of "curl".) Similarly, $d: \mathcal{F}(X) \to \mathcal{D}^1(X)$ is a sort of "gradient". The "divergence" can also be related to exterior derivative.) In particular, if

$$\theta = df = \frac{\partial f}{\partial x^i} dx^i \quad,$$

then

$$d\theta = \frac{1}{2}\left(\frac{\partial^2 f}{\partial x^i \partial x^2} - \frac{\partial^2 f}{\partial x^i \partial x^2}\right) dx^i \wedge dx^j$$

$$= 0 ,$$

i.e.,

$$ddf \equiv 0 \qquad (2.14)$$

This is quite general -- when d is defined for differential forms of all degrees, dd will be *identically* zero. It is this property that is basic (e.g., for topological purposes), and which is obviously related to the skew-symmetry of the exterior multiplication. The following rule is also important:

$$d(f\theta) = df \wedge \theta + f d\theta$$

for $f \in \mathscr{F}(X)$, $\theta \in \mathscr{D}^1(X)$

The final foundational topic we shall discuss (because it is a key feature in the treatment of classical mechanics using differential forms) is the "covariance" or "naturality" of the basic operations on differential forms. So far we have been working "statically" with a single space. Let Y and X be another space. Consider a map

$$\phi: Y \to X \quad .$$

Suppose that $f \in \mathscr{F}(X)$, i.e., f is a map

$$f: X \to R \quad .$$

One can compose f and ϕ

$$Y \xrightarrow{\phi} X \xrightarrow{f} R$$

$$\phi^*(f) \equiv f\phi$$

For algebraic purposes, it is convenient to denote the resulting map by $\phi^*(f)$, i.e.,

$$\phi^*(f)(y) = f(\phi(y))$$

for all $y \in Y$.

ϕ^* thus defines a map

$$\mathscr{F}(X) \to \mathscr{F}(Y) \quad .$$

It satisfies the following rules

$$\phi^*(f_1 + f_2) = \phi^*(f_1) + \phi^*(f_2)$$

$$\phi^*(f_1 f_2) = \phi^*(f_1) \phi^*(f_2) \quad ,$$

i.e., ϕ^* is *a homomorphism from the algebra* $\mathscr{F}(X)$ *to the algebra* $\mathscr{F}(Y)$.

Now, $\mathscr{F}(X)$ is identified with $\mathscr{D}^0(X)$, the *zero-th degree differential forms*. Thus, we can *extend* ϕ^* to a map

$$\phi: \mathscr{D}^r(X) \to \mathscr{D}^r(Y) \quad , \qquad r = 1,2,\ldots$$

as follows

$$\phi^*(a_i dx^i) = \phi^*(a_i) d\phi^*(x^i)$$

$$\phi^*(a_{ij} dx^i \wedge dx^j) = \phi^*(a_{ij}) d\phi^*(x^i) \wedge d\phi^*(x^j)$$

and so forth. Notice that it satisfies the following algebraic rules:

$$\phi^*(\theta_1 \wedge \theta_2) = \phi^*(\theta_1) \wedge \phi^*(\theta_2)$$

$$\phi^*(d\theta) = d\phi^*(\theta)$$

Notice that ϕ^* points *backward* to ϕ

$$Y \xrightarrow{\phi} X$$

$$\mathcal{D}(Y) \xleftarrow{\phi^*} \mathcal{D}(X)$$

ϕ^* is determined by this property and its effect on the coordinate functions x^i. Suppose that

$$\phi^*(x^i) = f^i(y^1, \ldots, y^m) \quad .$$

We see that the $f^i(\ )$ are the *functions* which determine the map ϕ, i.e., ϕ assigns to the point of Y whose coordinates are $y^1, \ldots, y^n$ the point of X whose coordinates are

$$x^1 = f^1(y^1, \ldots, y^m)$$
$$\vdots$$
$$x^n = f^n(y^1, \ldots, y^m) \quad .$$

This brief review of the formalism is enough for the moment. We turn to showing how Lagrange's equations can be described in terms of differential forms.

3. LAGRANGE'S EQUATIONS IN THE LANGUAGE OF DIFFERENTIAL FORMS

In the traditional "analytical mechanics", Lagrange's equations are the basic tool. (Recall that Lagrange's book, which was the most influential of all, was called "Mécanique analytique".) There are two approaches--directly, via Newton's equations, and indirectly, via the calculus of variations or "Hamilton's principle". In this section I will briefly review the former approach (the latter has been extensively treated elsewhere in my work) then show how Lagrange's equations can be formulated very elegantly using differential forms.

Let us start with Newton's equations (for particles) in the usual form

$$\begin{aligned}
m_1 \frac{d^2 \vec{r}_1}{dt^2} &= -\frac{\partial}{\partial \vec{r}_1} V_1(\vec{r}_1,\ldots,\vec{r}_N) + \vec{F}_1(\vec{r}_1,\ldots,\vec{r}_N, \dot{\vec{r}}_1,\ldots,\dot{\vec{r}}_N) \\
&\vdots \\
m_N \frac{d^2 \vec{r}_N}{dt^2} &= -\frac{\partial}{\partial \vec{r}_N} V_N + \vec{F}_N
\end{aligned} \qquad (3.1)$$

Here $\vec{r}_1,\ldots,\vec{r}_N$ are the position coordinates of N particles in R^3. $V_1,\ldots,V_N$ are the potential functions determining the "internal forces". $\vec{F}_1,\ldots,\vec{F}_N$ (functions of the positions $\vec{r}_1,\ldots,\vec{r}_N$ and the velocities $\dot{\vec{r}}_1,\ldots,\dot{\vec{r}}_N$) are the *external forces*.

Following the classical ideas, define variables q and $\dot{q}$ called *configuration* and *velocity variables*.

$$q = (q_1,\ldots,q_{3N})$$

$$\vec{r}_1 = \begin{pmatrix} q_1 \\ q_2 \\ q_3 \end{pmatrix}, \ldots, \vec{r}_N = \begin{pmatrix} q_{3N-2} \\ q_{3N-1} \\ q_{3N} \end{pmatrix}$$

$$\dot{q} = (\dot{q}_1,\ldots,\dot{q}_{3N})$$

$$\dot{\vec{r}}_1 = \begin{pmatrix} \dot{q}_1 \\ \dot{q}_2 \\ \dot{q}_3 \end{pmatrix}, \ldots, \dot{\vec{r}}_N = \begin{pmatrix} \dot{q}_{3N-2} \\ \dot{q}_{3N-1} \\ \dot{q}_{3N} \end{pmatrix}$$

Set:

$$T = \frac{1}{2} m_1 (\dot{\vec{r}}_1)^2 + \cdots + \frac{1}{2} m_N (\dot{\vec{r}}_N)^2 \qquad (3.2)$$

$$L = T - V . \qquad (3.3)$$

L, the *Lagrangian*, then becomes a function of $(q,\dot{q})$.

$$\frac{\partial L}{\partial \vec{r}_1} = m_1 \vec{\dot{r}}_1$$

$$\frac{\partial L}{\partial \vec{r}_1} = -\frac{\partial V}{\partial \vec{r}_1}$$

$$\frac{d}{dt}\frac{\partial L}{\partial \vec{\dot{r}}_1} - \frac{\partial L}{\partial \vec{r}_1} = m_1 \vec{\ddot{r}}_1 + \frac{\partial V}{\partial \vec{r}_1}$$

$$= \text{, using Newton's Equations (3.1), } \vec{F}_1.$$

Introduce indices

$$1 \leq a, b, \ldots \leq 3N \quad,$$

and the summation convention for these indices. Then, Equations (3.1) take the form

$$\frac{d}{dt}\frac{\partial L}{\partial \dot{q}^a} - \frac{\partial L}{\partial q^a} = F_a(q, \dot{q}) \quad . \tag{3.4}$$

The F_a are just the external forces rewritten so that (3.4) is equivalent to (3.1). (For example, $F_1(q, \dot{q})$ is the first component of the "vector" $\vec{F}_1(r, \dot{r})$.)

Notice the way the indices a occur in (3.4), i.e., *downstairs*. This indicates that the forces are--in the Lagrangian framework--*differential forms*.

Let us now forget where *Lagrange's equations* (3.4) came from, and study them using techniques of the theory of differential forms. Let X be the space of variables

$$(q^a, \dot{q}^a, t) \quad .$$

Physically, X is the *state × time space*. System theoretically, it might be called the *extended state space*. Consider the following one-forms in X

$$\theta^a = dq^a - \dot{q}^a dt \quad . \tag{3.5}$$

They are called (for geometric reasons that I will not go into here) the *contact forms*. Set:

$$\theta \equiv L dt + \lambda_a \theta^a \quad . \tag{3.6}$$

The λ_a are functions on X which we shall determine in a moment. Let us calculate $d\theta$, using the rules developed in Section 2:

$$d\theta = dL \wedge dt + d\lambda_a \wedge \theta^a + \lambda_a d\theta^a \tag{3.7}$$

Now,

$$dL = \frac{\partial L}{\partial q^a} dq^a + \frac{\partial L}{\partial \dot{q}^a} d\dot{q}^a$$

$$d\theta^a = d(dq^a) - d\dot{q}^a \wedge dt$$

$$= 0 - d\dot{q}^a \wedge dt$$

Hence

$$d\theta = \left(\frac{\partial L}{\partial q^a} dq^a + \frac{\partial L}{\partial \dot{q}^a} d\dot{q}^a\right) \wedge dt + d\lambda_a \wedge d\theta^a - \lambda_a d\dot{q}^a \wedge dt$$

If we now set

$$\lambda_a = \frac{\partial L}{\partial \dot{q}^a} , \qquad (3.8)$$

notice that the terms in $d\theta$ involving $d\dot{q}$ cancel, and we are left with

$$d\theta = \frac{\partial L}{\partial q^a} dq^a \wedge dt + d\lambda_a \wedge \theta^a \qquad (3.9)$$

Use (3.5) to express dq^a in terms of θ^a, and substitute into (3.9), using also that $dt \wedge dt = 0$:

$$d\theta = \frac{\partial L}{\partial q^a} \theta^a \wedge dt + d\lambda_a \wedge \theta^a$$

$$= \left(d\left(\frac{\partial L}{\partial \dot{q}^a}\right) - \frac{\partial L}{\partial q^a} dt\right) \wedge \theta^a \quad .$$

Set:

$$\theta_a = d\left(\frac{\partial L}{\partial \dot{q}^a}\right) - \frac{\partial L}{\partial q^a} dt \qquad (3.10)$$

Then, we have what I have called the magic formula of analytical mechanics

$$\boxed{d\theta = \theta_a \wedge \theta^a} \qquad (3.11)$$

Let us recapitulate what has been done. We have taken the Lagrangian function L and constructed the one-form θ on X. (θ is called the *Cartan form* since it first appeared in E. Cartan's book, Leçons sur les Invariants Intégraux.) The contact forms θ^a are purely kinematical. However, the θ_a involve the Lagrangian L and are "dynamical". Now, set:

$$\omega_a = \theta_a - F_a dt , \qquad (3.12)$$

where the F_a are the component functions of the force, as they appear in Lagrange's equations (3.4). Also,

$$\omega_a = d\left(\frac{\partial L}{\partial \dot{q}^a}\right) - \left(\frac{\partial L}{\partial q^a} + F_a\right) dt . \qquad (3.13)$$

Now,

$$t \to \left(q(t), \dot{q}(t) = \frac{dq}{dt}, t\right)$$

be a curve in X that represents a solution of Lagrange's equations. Let T be the one-dimensional manifold whose coordinate is t. This determines a map

$$\phi: T \to X .$$

Theorem 3.1. The curve satisfies the Lagrange equations if and only if

$$\phi^*(\omega_a) = 0 = \phi^*(\theta^a) . \qquad (3.14)$$

This expresses Lagrange's equations in terms of what Cartan calls an *exterior differential* system. This is also a convenient way to express laws of physics. To document this observation for this case, I will now treat a number of standard topics with these methods.

4. QUASICOORDINATES AS DIFFERENTIAL FORMS

The notion of "quasicoordinates" is rather confusing in the classical literature. It fits rather nicely into the theory of differential forms. In fact, it is more or less identical with what E. Cartan called a *moving frame*; this is one of the key notions in his work on differential geometry.

Let Q be a space (a "configuration space") with coordinates (q^i), $1 \leq i,j \leq n$. Let M be the space of variables

$$(q^i, \dot{q}^i, t)$$

(In terms of now-standard differential-geometric notation, M is $T(Q) \times R$, where $T(Q)$ is the *tangent bundle* to Q.)

Suppose that

$$x^i = x^i(q) \tag{4.1}$$

are new coordinates for Q. One obtains coordinates for M as follows

$$x^i \; ; \; \dot{x}^i = \frac{\partial x^i}{\partial q^j} \dot{q}^j, \; t \tag{4.2}$$

Notice how the $\dot{x}^i$ are determined by the $\dot{q}^i$. Exterior differentiate (4.1):

$$dx^i = \frac{\partial x^i}{\partial q^j} dq^j \tag{4.3}$$

Let us write (4.3) as:

$$dx^i = A^i_j dq^j \tag{4.4}$$

for appropriately chosen functions $A^i_j(q)$. Then, also

$$A^i_j \dot{q}^j \; , \tag{4.5}$$

i.e., the *transformation* $\dot{q} \to \dot{x}$ *is the "same" as* $dq \to dx$.

Let us think of these relations generalized as follows:

$$\omega^i = A^i_j dq^j \tag{4.6}$$

where now (A^i_j) is an *arbitrary* $m \times m$ matrix of functions on Q whose determinant is non-zero. The ω^i are said to be a *basis for differential forms on* Q or a *moving frame* (in the sense of Cartan). Standard linear algebra implies that an arbitrary one-form θ on Q can be written in a *unique* way as a linear combination of the ω^i with coefficients in $\mathscr{F}(Q)$. Formula (4.5) now suggests that we *define* functions y^i on M as follows:

$$y^i = A^i_j \dot{q}^j \; . \tag{4.7}$$

Notice that the y^i *would be the* $\dot{x}^i$ *if the* x^i *existed*. The conditions that they exist are:

$$d\omega^i = 0 \; . \tag{4.8}$$

Proof. If (4.4) is satisfied, i.e., $\omega^i = dx^i$, then $d\omega^i = d(dx^i) = 0$ since $d^2 = 0$.

Conversely, if (4.8) is satisfied, then for each i there exists (locally) a function x^i such that

$$dx^i = \omega^i . \qquad (4.9)$$

(This is the Poincaré lemma; its proof is in all the standard books.) It remains to show that the y^i --as functions of q --form a *new coordinate system*. Now, it follows from (4.9)

$$\frac{\partial x^i}{\partial x^j} = A^i_j \equiv \text{Jacobian matrix of } y \text{ as a function of } x$$

Thus, the condition that the matrix (A^i_j) be non-singular is that the Jacobian determinant be non-zero, i.e., that the x^i form a "new" coordinate system.

Let us *not* make the assumption (4.8). In the classical literature the ω^i are essentially what are called "quasicoordinates."

Theorem 4.1. If the ω^i are defined in terms of the q by formula (4.6), with a non-singular coefficient matrix, and the y^i are defined by (4.7), then the one-forms (ω^i, dy^i, dt) form a basis for one-forms of M.

Proof. The functions $(q^i, \dot{q}^i, t)$ form a coordinate system for M. Hence, to show that the (ω^i, dy^i, dt) form a basis of one-forms for M, we must show that the one-forms dq^i, $d\dot{q}^i$ can be written in terms of them. It follows from (4.6) (and the postulated invertibility of the matrix (A^i_j)) that dq^i can be written as a linear combination of the ω^i. Hence, it only remains to deal with the $d\dot{q}^i$.

Now, relation (4.7) can be converted to yield a relation of the form:

$$\dot{q}^i = B^i_j y^j , \qquad (4.10)$$

where the $(B^i_j(q))$ are functions of q *alone*. Hence,

$$d\dot{q}^i = \text{linear combination of the } dq, dy$$
$$= \text{linear combination of the } \omega, dy$$

Q.E.D.

5. THE ROTATING RIGID BODY AND THE ROTATION GROUP

A *rotating rigid body* may be defined as a set of particles in R^3 which move so that the distance between the particles remain constant in time and so that one point is always fixed in time. To be explicit, let us suppose that there are N such particles whose positions in R^3 are vectors

$$\vec{r}_1(t), \ldots, \vec{r}_N(t) \quad,$$

and whose masses are $m_1, \ldots, m_N$.

Let $SO(3,R)$ denote the group (under matrix multiplication) of 3×3 real *orthogonal* matrices A of determinant +1, i.e.,

$$AA' = 1$$

$$\det A = 1 \quad.$$

(A' denotes transpose of a matrix.) The vectors $\vec{r}$ in R^3 are taken as column vectors

$$\vec{r} = \begin{pmatrix} x \\ y \\ z \end{pmatrix}$$

Thus, we can kinematically describe the motion as follows:

$$\vec{r}_i(t) = A(t) \cdot \vec{r}_i^0 \quad, \quad 1 \leq i \leq N \quad, \tag{5.1}$$

where $t \to A(t)$ is a curve in $SO(3,R)$ and $\vec{r}_1^0, \ldots, \vec{r}_N^0$ are a set of vectors in R^3 which are *fixed* in time.

The *kinetic energy* of the particles is then

$$T = \frac{1}{2} \sum_i m_i |\dot{\vec{r}}_i(t)|^2$$

$$= \frac{1}{2} \sum_i m_i \left| \frac{dA'}{dt}(t) \vec{r}_i^0 \right|^2$$

$$= \frac{1}{2} \sum_i m_i \left| A(t) \frac{dA'}{dt} \vec{r}_i^0 \right|^2$$

$$= \frac{1}{2} \sum_i m_i |\Omega(t) \vec{r}_i^0|^2 \tag{5.2}$$

where

$$\Omega(t) = A(t)\frac{dA'}{dt} \tag{5.3}$$

Lemma 5.1. $\Omega(t)' = -\Omega(t)$, i.e., Ω is a skew-symmetric matrix.

Proof. Orthogonality of A means that

$$AA' = 1 \tag{5.4}$$

Differentiate (5.4):

$$\frac{dA}{dt}A' + A\frac{dA'}{dt} = 0$$

$$\Omega = A(t)\frac{dA'}{dt}$$

$$\Omega' = \frac{dA}{dt}A',$$

whence skew-symmetry of Ω. Q.E.D.

We can see how Ω is related to the *angular velocity vector* $\vec{\omega}(t)$. It is characterized by the following relation:

$$\frac{d\vec{r}_i}{dt} = \vec{\omega} \times \vec{r}_i, \tag{5.5}$$

where $\times$ is the Euclidean vector product. Compare this with (5.1):

$$\vec{\omega}(t) \times \vec{r}_i(t) = \frac{dA'}{dt}\vec{r}_i^{\,0}$$

$$= \frac{dA'}{dt}A(t)\vec{r}_i(t)$$

Hence,

$$\vec{\omega}(t) \times \vec{r} = \Omega(t)\vec{r} \tag{5.6}$$
$$\text{for all } \vec{r} \in R^3.$$

(Thus, $\Omega(t)$ *is* essentially *the* angular velocity, with the standard vector algebra identification of the vector product on skew-symmetric matrices.

Theorem 5.1. There is a symmetric positive definite matrix I such that

$$T = -\text{trace}(\Omega(t) I \Omega(t)) \quad . \tag{5.7}$$

I is called the *moment of inertia matrix* of the rotating rigid body, and depends only on the vectors r_i^0 and the masses m_i.

Proof. Rewrite (5.2) using matrix algebra:

$$T = \frac{1}{2} \sum_i m_i r_i^{0\prime}(t)' (t) r^0$$

$$= -\frac{1}{2} \sum_i m_i r_i^{0\prime}(t)^2 r^0 \tag{5.8}$$

Notice that this is indeed a quadratic form in the entries of the matrix . That it can be written in the form (5.7) is straightforward matrix algebra, which is left to the reader as an exercise.

We can now reinterpret what we have done in a Lie group context. On the Lie group $SO(3,R)$ considered "abstractly", define a "kinetic energy" function

$$T = -\text{trace}(\Omega I \Omega) \tag{5.9}$$

with

$$\Omega = A\dot{A}' \quad . \tag{5.10}$$

Consider also a potential energy function $A \to V(A)$ and the Lagrangian function

$$L(A, \dot{A}) = T - V \quad .$$

Let A be coordinatized by variables

$$(\theta, \phi, \psi) \quad .$$

Set up *Lagrange's equations*

$$\boxed{\begin{aligned} \frac{d}{dt}\left(\frac{\partial L}{\partial \dot{\theta}}\right) - \frac{\partial L}{\partial \theta} &= 0 \\ \frac{d}{dt}\left(\frac{\partial L}{\partial \dot{\phi}}\right) - \frac{\partial L}{\partial \phi} &= 0 \\ \frac{d}{dt}\left(\frac{\partial L}{\partial \dot{\psi}}\right) - \frac{\partial L}{\partial \psi} &= 0 \end{aligned}} \tag{5.11}$$

Let $t \to A(t)$ be a solution of equations (5.11). Then

$$\vec{r}_i(t) = A(t)\vec{r}_i^{\,0}$$

is a solution of our original mechanics problem. This procedure then carries the *arbitrary* rotating-rigid-body mechanics problem to an analytical mechanics problem whose configuration space is $SO(3,R)$.

Let us not yet be specific about the choice of coordinates (θ,ϕ,ψ). (Traditionally, these are used as notation for "Euler angles". We shall consider this in the next section.) We can now see that the entries of the matrix Ω are "quasicoordinates". Suppose that A is a matrix function of (θ,ϕ,ψ). Thus,

$$\dot{A} = \frac{\partial A}{\partial \theta}\dot{\theta} + \frac{\partial A}{\partial \phi}\dot{\phi} + \frac{\partial A}{\partial \psi}\dot{\psi}$$

$$\Omega = A(\theta,\phi,\psi)\left(\frac{\partial A'}{\partial \theta}\dot{\theta} + \frac{\partial A'}{\partial \phi}\dot{\phi} + \frac{\partial A'}{\partial \psi}\dot{\psi}\right) \quad (5.12)$$

Ω is also a 3×3 skew-symmetric matrix, hence it can be written as follows

$$\Omega = \begin{pmatrix} 0 & \omega_1 & \omega_2 \\ -\omega_1 & 0 & \omega_3 \\ -\omega_2 & -\omega_3 & 0 \end{pmatrix} \quad (5.13)$$

with components $\omega_1, \omega_2, \omega_3$ functions of $\theta,\phi,\psi,\dot{\theta},\dot{\phi},\dot{\psi}$.

$$\omega = \begin{pmatrix} \omega_1 \\ \omega_2 \\ \omega_3 \end{pmatrix}$$

is in fact the *angular velocity*. These are not arbitrary functions, but are linear in the velocities:

$$\begin{pmatrix} \omega_1 \\ \omega_2 \\ \omega_3 \end{pmatrix} = \begin{pmatrix} a_{11}(\theta,\phi,\psi), \ldots & a_{13} \\ & \vdots & \\ a_{31} & \cdots & a_{33}(\theta,\phi,\psi,) \end{pmatrix} \begin{pmatrix} \dot{\theta} \\ \dot{\phi} \\ \dot{\psi} \end{pmatrix}$$

This obviously exhibits the $\omega_1, \omega_2, \omega_3$ as "quasicoordinates", as we have discussed in previous sections.

These "quasicoordinates" for $SO(3,R)$ are related to what in Lie group theory are called the *Cartan-Maurer one-forms*. If G is a group of $n \times n$ matrices, consider:

$$\underline{\omega}_L = A^{-1} dA$$
$$\underline{\omega}_R = dA A^{-1} \quad .$$
(5.14)

A denotes an $n \times n$ matrix which is a "variable" element of G. Notice that (5.15) provides $n \times n$ matrices whose *entries are one-differential forms*. (Alternately, and somewhat more abstractly, these may be thought of as one-forms on G which are not scalar-valued, but vector-valued, i.e., take values in the Lie algebra of G.) The subscripts L and R denote *left* and *right invariant*. In fact, if A_0 is a "constant" matrix in G,

$$(A_0 A)^{-1} d(A_0 A) = A^{-1} A_0^{-1} A_0 dA = \underline{\omega}_L$$

$$d(AA_0)(AA_0)^{-1} = dAA_0 A_0^{-1} A = \underline{\omega}_R \quad .$$

These formulas exhibit this left and right invariance.

The entries of $\underline{\omega}_L$ and $\underline{\omega}_R$ provide n^2 one-forms on G. Not all will be independent. The independent ones provide a *basis for one-forms*, i.e., *a set of quasicoordinates for* G.

Example 1. $G = GL(n,R) \equiv$ group of $n \times n$ real matrices with non-zero determinant

$$A = (a^i_j) \quad , \quad 1 \le i,j \le n$$

$$A^{-1} = (b^i_j) \quad ,$$

i.e.,

$$b^i_j a^j_k = \delta^i_k \quad .$$

$$\underline{\omega}_L = (b^i_k da^k_j)$$

$$\equiv (\omega^i_j) \quad .$$

Now,

$$d\underline{\omega}_L = -A^{-1} dA A^{-1} \wedge dA$$

$$= -\underline{\omega}_L \wedge \underline{\omega}_L \quad .$$

$$\boxed{d\omega^i_j = -\omega^i_k \wedge \omega^k_j}$$

Example 2. $G = SO(n,R) \equiv$ group of $n \times n$ real orthogonal matrices with determinant $+1$.

Since $A^{-1} = A'$, we have

$$b^i_j = a^j_i ,$$

$$\omega^i_j = \sum_k a^k_i da^k_j$$

$$\boxed{\omega^i_j + \omega^j_i = 0}$$

Hence, the form

$$\boxed{\omega^i_j , \quad i < j}$$

forms a basis for one-forms on G.

6. THE EULER ANGLES

The *Euler angles* (θ, ϕ, ψ) of a 3×3 rotation matrix A are the numbers such that

$$A = \begin{pmatrix} \cos\psi & \sin\psi & 0 \\ -\sin\psi & \cos\psi & 0 \\ 0 & 0 & 1 \end{pmatrix} \begin{pmatrix} 1 & 0 & 0 \\ 0 & \cos\theta & \sin\theta \\ 0 & -\sin\theta & \cos\theta \end{pmatrix} \begin{pmatrix} \cos\phi & \sin\phi & 0 \\ -\sin\phi & \cos\phi & 0 \\ 0 & 0 & 1 \end{pmatrix}$$

$$= \begin{pmatrix} c\phi c\psi - s\phi c\theta s\psi, & s\phi c\psi + c\phi c\theta s\psi, & s\theta s\psi \\ -c\phi s\psi - s\phi c\theta c\psi, & -s\phi s\psi + c\phi c\theta c\psi, & s\theta c\psi \\ s\phi s\theta, & -c\phi s\theta, & c\theta \end{pmatrix}$$

It is interesting that there is a general Lie-theoretic foundation for this way of construction of a coordinate system. In this section--which is basically just a side line--I will briefly go into this.

Let $\mathcal{G}$ be the Lie algebra (under commutation) of 3×3 real skew symmetric matrices. Let G be the group of $3 \times n$ real orthogonal matrices. Let

$$\text{Exp}: \mathcal{G} \to G$$

be the map

$$A \to \exp(A) \equiv 1 + A + \frac{A^2}{2!} + \cdots$$

For example, if

$$A = \begin{pmatrix} 0, & 1, & 0 \\ -1, & 0, & 0 \\ 0, & 0, & 0 \end{pmatrix}$$

Then

$$\exp(\theta A) = \begin{pmatrix} \cos\theta, & \sin\theta, & 0 \\ -\sin\theta, & \cos\theta, & 0 \\ 0, & 0, & 1 \end{pmatrix}$$

Let A_1, A_2 be two elements of $\mathscr{G}$. Label coordinates of R^3 as (θ, ϕ, ψ). Map

$$\alpha: R^3 \to G$$

as follows:

$$\alpha(\theta, \phi, \psi) = \exp(\psi A_2) \exp(\theta A_1) \exp(\phi A_2) \qquad (6.1)$$

(In the usual way of doing this,

$$\theta \to \exp(\theta A_2)$$

is the group of rotations about the z-axis.

$$\theta \to \exp(\theta A_1)$$

is the group of rotations about the x-axis.)

Set

$$\Omega = A dA' \,. \qquad (6.2)$$

In order to compute the relation between the "true" coordinates (θ, ϕ, ψ) and the "quasicoordinates", defined as the one-forms $\Omega_1, \Omega_2, \Omega_3$ such that

$$\Omega = \begin{pmatrix} 0, & \Omega_1, & \Omega_2 \\ -\Omega_1, & 0, & \Omega_3 \\ -\Omega_2, & -\Omega_3, & 0 \end{pmatrix} \qquad (6.3)$$

we must compute the "pull-back"

$$\alpha^*(\Omega) \quad .$$

This is readily done using formula (6.1):

$$d\alpha(\theta,\phi,\psi) = A_2 d\psi \alpha + \exp(\psi A_2) d\theta \, \exp(\theta A_1) A_1 \, \exp(\phi A_2) + \alpha A_2 d\phi$$

$$d\alpha' = -\alpha' A_2 d\psi - \exp(-\phi A_2) A_1 \exp(-\theta A_1) \exp(-\psi A_2) d\theta - A_2 \alpha' d\phi$$

$$\alpha d\alpha' = -A_2 d\psi - \exp(\psi A_2) \exp(\theta A_1) A_1 \exp(-\theta A_1) \exp(-\psi A_2) d\theta$$

$$-\exp(\psi A_2) \exp(\theta A_1) \exp(\phi A_2) A_2 \exp(-\phi A_2) \exp(-\theta A_1) \exp(-\psi A_2) d\phi$$

$$= -A_2 d\psi - \exp(\psi A_2) A_1 \exp(-\psi A_2) d\theta - \exp(\psi A_2) \exp(\theta A_1) A_2 \exp(-\theta A_1)$$

$$\cdot \exp(-\psi A_2) d\phi$$

In order to resolve this further, set:

$$B(\theta) = \exp(\theta A_1) A_2 \exp(-\theta A_1)$$

Then,

$$\boxed{\begin{aligned} \frac{dB}{d\theta} &= A_1 B - B A_1 = [A_1, B] \\ B(0) &= A_2 \end{aligned}} \qquad (6.4)$$

These are differential equations for the matrix function

$$\theta \to B(\theta)$$

which can be solved by power series as follows:

$$\boxed{B(\theta) = \exp(\theta \operatorname{Ad} A_1)(A_2)}$$

$$= A_2 + \theta[A_1, A_2] + \frac{\theta^2}{2!}[A_1,[A_1,A_2]] + \cdots \qquad (6.5)$$

In order to sum up these power series very explicitly, introduce the matrix A_3 such that:

$$A_3 = [A_1, A_2] \quad . \qquad (6.6)$$

Let us suppose (A_1, A_2, A_3) is a basis for the Lie algebra $\mathcal{G}$, with

$$[A_1, A_3] = A_2$$
$$[A_2, A_3] = -A_1 \tag{6.7}$$

Hence,

$$\exp(\theta A_1) A_2 \exp(-\theta A_1) = A_2 + \theta A_3 + \frac{\theta^2}{2!} A_2 + \frac{\theta^3}{3!} A_3 + \cdots$$

$$= \cos\theta\, A_2 + \sin\theta\, [A_1, A_2] \tag{6.8}$$

(6.8) can be now used to write:

$$\Omega = \alpha d\alpha'$$
$$= -A_2 d\psi - (\cos\psi\, A_1 + \sin\psi\, [A_2, A_1]) d\theta - \exp(\psi A_2)(\cos\theta\, A_1 + \sin\theta\, [A_1, A_2])$$
$$\cdot \exp(-\psi A_2) d\phi$$

$$\boxed{\begin{aligned}
= &-A_2 d\psi - (\cos\psi\, A_1 + \sin\psi\, [A_2, A_1]) d\theta - (\cos\theta\, (\cos\psi\, A_1 + \sin\psi\, [A_2, A_1]) \\
&+ \sin\theta\, (\cos\psi\, [A_1, A_2] + \sin\psi\, [A_2, [A_1, A_2]])) d\phi
\end{aligned}} \tag{6.9}$$

The "kinetic energy" of the rotating rigid body can be conveniently calculated from this formula. Convert the "differentials" into "velocities", i.e., via

$$d\theta \to \dot\theta\ ;\quad d\psi \to \dot\psi;\quad d\phi \to \dot\phi$$

Then,

$$\text{K.E.} = \text{trace}(I\Omega^2) \tag{6.10}$$

where I is the symmetric "moment of inertia" matrix. We have then reduced Ω to the form (6.9) where (6.10) can obviously be readily calculated.

7. THE BOLTZMANN-HAMEL EQUATIONS

Let Q be a manifold with coordinates

$$q^i\ ,\qquad 1 \le i,j \le n\ .$$

Let M be the space of variables

$(q^i, \dot{q}^i, t)$

(From the coordinate-free point of view, M is $T(Q) \times R$, where Q is the "tangent bundle" to Q.)

Let
$$\omega^i = A^i_j(q) dq^j \qquad (7.1)$$

be another basis for one-forms on Q. (Then, $\det(A^i_j) \neq 0$.) Set
$$y^i = A^i_j(q) \dot{q}^j \quad .$$

As we have seen,
$$\omega^i, \; dy^i, \; dt \qquad (7.2)$$

forms a basis for one-forms on M.

Let $L(q,\dot{q},t)$ be a Lagrangian function. We want to compute the *Lagrange equations in terms of this basis for one-forms on* M, i.e., to show how the curves which are solutions of
$$\frac{d}{dt} L_{\dot{q}} - L_q = F$$

affect the one-forms (7.2).

Set:
$$\theta^i = \omega^i - y^i dt \qquad (7.3)$$

$$\theta = Ldt + \lambda_i \theta^i \quad . \qquad (7.4)$$

(λ_i are functions which we shall determine as the calculation proceeds.)

We shall now adopt a special notation for the partial derivatives of L; namely, we let L_i, L'_i, L_t be the functions on M such that
$$dL = L_i \omega^i + L'_i dy^i + L_t dt \qquad (7.5)$$

We can now compute $d\theta$:
$$d\theta = (L_i \omega^i + L'_i dy^i) \wedge dt + d\lambda_i \wedge (\omega^i - y^i dt) + \lambda_i (d\omega^i - dy^i \wedge dt) \qquad (7.6)$$

We now *choose* λ_i so that the terms in (7.6) *which involve* dy^i *cancel*; this requires that:

$$\boxed{\lambda_i = L_i'} \qquad (7.7)$$

Thus,

$$\boxed{\theta = L\,dt + L_i'\theta^i} \qquad (7.8)$$

$$d\theta = L_i\omega^i \wedge dt + dL_i' \wedge \omega^i - dL_i' y^i \wedge dt + L_i' d\omega^i$$

Set:

$$\theta_i = dL_i' - L_i\,dt \quad . \qquad (7.9)$$

Hence,

$$d\theta = L_i\omega^i \wedge dt + \theta dL_i' \wedge \theta^i + L_i' d\omega^i$$

$$= L_i(\omega^i - y^i dt) \wedge dt + dL_i' \wedge \theta^i + L_i' d\omega^i$$

$$= L_i\theta^i \wedge dt + dL_i' \wedge \theta^i + L_i' d\omega^i \quad ,$$

i.e.,

$$\boxed{d\theta = \theta_i \wedge \theta^i + L_i' d\omega^i} \qquad (7.10)$$

Now, the new element in this formula is the presence of the forms $d\omega^i$. If the ω^i were the differentials of a new set of coordinates, the $d\omega^i$ would vanish. Thus, the $d\omega^i$ represent the "non-holonomic" nature of the "quasi-coordinates".

Let (a_{jk}^i) be the functions on Q such that

$$d\omega^i = a_{jk}^i \omega^j \wedge \omega^k$$

Use (7.3):

$$d\omega^i = a_{jk}^i(\theta^j + y^j dt) \wedge (\theta^k + y^k dt)$$

or

$$d\omega^i = a^i_{jk}\theta^j \wedge \theta^k + 2a^i_{jk}y^j dt \wedge \theta^k$$

Substitute this into (7.9):

$$d\theta = \theta_k \wedge \theta^k + L'_i(a^i_{jk}\theta^j + 2a^i_{jk}y^j dt) \wedge \theta^k$$

$$= (\theta_k + L'_i(a^i_{jk}\theta^j + 2a^i_{jk}y^j dt)) \wedge \theta^k$$

$$= \eta_k \wedge \theta^k \qquad (7.11)$$

with

$$\eta_k = \theta_k + L'_i(a^i_{jk}\theta^j + 2a^i_{jk}y^j dt)$$

Notice that (7.11) writes $d\theta$ in its algebraic "canonical form". Let us then copy what we did earlier, and write the following differential equations:

$$\boxed{\sigma^*(\theta^k) = 0 = \sigma^*(\eta_k - F_k dt)} \qquad (7.13)$$

where F_k are functions on M. A "solution" is then a map

$$t \to \left(q(t), \frac{dq}{dt}, t\right) = \sigma(t)$$

of $R \to M$.

Theorem 7.1 (Boltzmann-Hamel). The solutions of the differential equations (7.13) are identical with the solutions of Lagrange's equations:

$$\boxed{\frac{d}{dt}\frac{\partial L}{\partial \dot q^i} - \frac{\partial L}{\partial q^i} = A^k_i F_k} \qquad (7.14)$$

Proof. We have:

$$\theta^i = \omega^i - y^i dt$$

$$= \text{, using (7.1),}$$

$$A^i_j dq^j - A^i_j \dot{q}^j dt$$

$$= A^i_j (dq^j - \dot{q}^j dt) \qquad (7.15)$$

Thus,

$$\sigma^*(\theta^i) = 0$$

if and only if

$$\frac{dq^i}{dt} = \dot{q}^i \quad.$$

Also, using (7.5),

$$dL = L_i \omega^i + L'_i dy^i + L_t dt$$

$$= L_i A^i_j dq^j + L'_i d(A^i_j d\dot{q}^j) + L_t dt$$

$$= \frac{\partial L}{\partial q^i} dq^i + \frac{\partial L}{\partial \dot{q}^i} d\dot{q}^i + \frac{\partial L}{\partial t} dt$$

Comparing these two expressions gives

$$\boxed{L'_i A^i_j = \frac{\partial L}{\partial \dot{q}^j}} \qquad (7.16)$$

Hence, using (7.7) and (7.15),

$$\theta = Ldt + L_j A^i_j (dq^j - \dot{q}^j dt)$$

$$= \quad , \text{ using (7.16),}$$

$$Ldt + \frac{\partial L}{\partial \dot{q}^j} (dq^j - \dot{q}^j dt) \qquad (7.17)$$

(7.17) tells us that θ is the *Cartan form* constructed from L.

Work on (7.11):

$$d\theta = \eta_i \wedge A^i_j (dq^j - \dot{q}^j dt)$$

$$= (A^i_j \eta_i) \wedge dq^j - \dot{q}^j dt$$

We know from previous work that:

$$d\theta = \left(d\frac{\partial L}{\partial \dot{q}^j} - \frac{\partial L}{\partial q^i} dt\right) \wedge (dq^i - \dot{q}^i dt)$$

Hence,

$$\left(A^i_j \eta_i - d\left(\frac{\partial L}{\partial \dot{q}^j}\right) - \frac{\partial L}{\partial \dot{q}^j} dt\right) \wedge (dq^j - \dot{q}^j dt) = 0 \tag{7.18}$$

Since the one-forms $dq^u - \dot{q}^j dt$ are linearly independent, the form

$$A^i_j \eta_i - d\left(\frac{\partial L}{\partial \dot{q}^j}\right) - \frac{\partial L}{\partial q^j} dt$$

are *linear combinations* of the form $dq^j - \dot{q}^j dt$.

Suppose then that

$$\sigma: R \to M$$

is a map satisfying (7.13). Then, since $\sigma^*(dq^u - \dot{q}^j dt) = 0$, we have:

$$\sigma^*\left(d\left(\frac{\partial L}{\partial \dot{q}^j}\right) - \left(\frac{\partial L}{\partial q^j} + A^i_j F_i\right)\right) = 0 \tag{7.19}$$

These are just the Boltzmann-Hamel equations (7.14) in another notational form.

We can write the equations (7.13) in a more traditional and explicit form. Suppose

$$t \to q(t)$$

is a curve in Q which is a solution of (7.13). Set

$$y^i(t) = A^i_j(q(t)) \frac{dq^j}{dt} \tag{7.20}$$

Then,

$$\boxed{\frac{d}{dt}\left(L'_k\left(q(t), \frac{dq}{dt}\right) - L_k\left(q(t), \frac{dq}{dt}\right)\right) + 2L'_i\left(q(t), \frac{dq}{dt}\right) a^i_{jk}(q(t)) y^j(t) = F^j_k\left(q(t), \frac{dq}{dt}\right)} \tag{7.21}$$

In applications, one is often interested in situations where (7.21) becomes an equation for the $y(t)$, i.e., it does not involve $q(t)$. This is particularly likely to happen if the $a^i_{jk}(q)$ are *constant*. In the next section we shall investigate what this means differential-geometrically.

8. THE BOLZTMANN-HAMEL EQUATIONS ON A LIE GROUP

Being able to *use* the equations (7.13) depends on knowing something about the "non-holomonic" coefficients a^i_{jk} such that

$$d\omega^i = a^i_{jk} \omega^j \wedge \omega^k \tag{8.1}$$

where (ω^i) is the given basis for one-forms on Q. The simplest situation is obviously that where the (a^i_{jk}) are *constants*. It is well-known in differential geometry that this condition implies that Q is a Lie group with (a^i_{jk}) the structure constants of its Lie algebra and ω^i the left-invariant one-form on Q (the so-called *Maurer-Cartan forms*). Equations (7.21) then take the following form:

$$\frac{d}{dt} L'_k - L_k + 2L'_i a^i_{jk} y^j = F_k \quad . \tag{8.2}$$

(Recall that

$$y^i = A^i_j(q) \dot{q}^j \tag{8.3}$$

$$\omega^i = A^i_j(q) dq^j \tag{8.4}$$

$$dL = L'_i dy^i + L_i \omega^i + L_t dt \quad .) \tag{8.5}$$

Of course, the dynamical equations (8.2) take their simplest form if L and the F_k *are functions of the variables* y *alone*, i.e., if they are independent of the q. One such important case is that where:

$$L = \tfrac{1}{2} g_{ij} y^i y^j + g_i y^i \tag{8.6}$$

where g_{ij}, g_i are *constants*. Using (8.5) we see that, in this case,

$$L_i = 0 = L_t \tag{8.7}$$

$$L'_i = g_{ij} y^j + g_i \quad . \tag{8.8}$$

Hence, the equations (8.2) take the following form:

$$g_{kj} \frac{d}{dt} y^j + 2(g_{i\ell} y^\ell + g_i)(a^i_{jk} y^j) = F_k \tag{8.9}$$

For example, let us specialize these equations by making the following simplifications:

$$g_i = 0 = F_i \quad . \tag{8.10}$$

Equations (8.9) then take the following form:

$$g_{kj} \frac{d}{dt} y^j + (g_{i\ell} a^i_{jk} + g_{ij} a^i_{\ell k}) y^\ell y^j = 0 \tag{8.11}$$

Equations (8.11) are a generalization of *Euler's equations of rigid body motion*. (They are readily seen to reduce to them if the a^i_{jk} are the structure constants of the rotation group $SO(3,R)$, namely, ε^i_{jk}.) Notice their typically *quadratic* form on the y's. We can also put this into a coordinate-free form, which is useful for various purposes.

Let $\mathscr{G}$ be a real vector space of dimension n, with a basis which we label as Y_i. Define a bracket structure on $\mathscr{G}$ by the formula

$$[Y_i, Y_j] = a^k_{ij} Y_k \quad . \tag{8.12}$$

<u>Theorem 8.1</u>. Relation (8.1)—together with the hypothesis that the a^k_{ij} are *constants*—implies that formula (8.12) defines $\mathscr{G}$ as a Lie algebra, i.e., the following *Jacobi identity* is satisfied:

$$[Y_i, [Y_j, Y_k]] = [[Y_i, Y_j], Y_k] + [Y_j, [Y_i, Y_k]] \quad . \tag{8.13}$$

<u>Proof</u>. Apply the exterior derivative operation to both sides of (8.1):

$$\underline{\underline{0}} = d(d\omega^i) = d(a^i_{jk} \omega^j \wedge \omega^k)$$

$$= a^i_{jk} d\omega^j \wedge \omega^k + a^i_{jk} \omega^j \wedge d\omega^k$$

$$= a^i_{jk} (a^j_{\ell m} \omega^\ell \wedge \omega^m) \wedge \omega^k + a^i_{jk} \omega^j \wedge (a^k_{\ell m} \omega^\ell \wedge \omega^m)$$

It is now readily seen that these resulting relations on the a's are equivalent to (8.11). (Here is a more satisfying proof for the reader who understands "vector fields". Let Y_i be the vector fields on $\underline{\underline{0}}$ such that

$$\omega^i(Y_j) = \delta^i_j$$

Then,

$$d\omega^i(Y_j, Y_k) = Y_j(\omega^i(Y_k)) - Y_k(\omega^i(Y_j)) - \omega^i([Y_j, Y_k])$$
$$= a^i_{jk} ,$$

where (8.10) *follows as a relation among vector fields*. The "Jacobi identity" (8.11) is *automatically* satisfied for vector fields, so the theorem can really be proved without any calculation.)

Return to equations (8.9) and use the Lie algebra $\mathcal{G}$ to write them in a coordinate free way. Let $\mathcal{G}^d$ be the dual vector space to $\mathcal{G}$. Let Y^j be the dual basis to $\mathcal{G}^d$, i.e.,

$$Y^j(Y_i) = \delta^j_i .$$

(See [ ] for the "linear algebra" we use here.) Let $\beta: \mathcal{G} \to \mathcal{G}^d$ be the linear map defined as follows

$$\beta(Y_i) = \beta_{ij} Y^j \tag{8.14}$$

or

$$g_{ij} = (Y_j)(Y_i) \tag{8.15}$$

Set:

$$Y(t) = Y_j y^j(t) \tag{8.16}$$

$$Z = g_i Y^i \in \mathcal{G}^d \tag{8.17}$$

With these conventions, (8.18) can be rewritten as:

$$\beta(Y_k)(Y_j) \frac{d}{dt} y^j + 2(\beta(Y_\ell)(Y_i)y^\ell + Z(Y_i))Y^i([Y_j, Y_k])y^j = F_k ,$$

or

$$\beta(Y_k)\left(\frac{d}{dt} Y(t)\right) + 2(\beta(Y(t))([Y(t), Y_k]) + 2Z([Y(t), Y_k]) = F_k(y(t))$$

In order to put this into a more algebraic form, let us define what is in Lie algebra theory called *a linear representation of the Lie algebra* $\mathcal{G}$ *by linear transformations on the vector space* $\mathcal{G}^d$; namely,

$$\rho(X)(Y^d)(Y) = -X^d([X, Y]) \tag{8.18}$$

for $X, Y \in \mathcal{G}$, $Y^d \in \mathcal{G}^d$.

(In Lie algebra theory, ρ is called the *dual of the adjoint representation*.) With the aid of this formula, the basic "dynamical" equation (8.17) can be rewritten as:

$$\beta\left(\frac{d}{dt} Y(t)\right)(Y_k) - 2\rho(Y(t))(\beta(Y(t))(Y_k) - 2\rho(Y(t))(Z)(Y_k) = F_k \tag{8.19}$$

Let F be the map: $\mathscr{G} \to \mathscr{G}^d$ defined by

$$F(y_j Y^j) = F_k(y) Y^k \quad . \tag{8.20}$$

Then, Equation (8.19) takes the following basis-free and definitive form:

$$\boxed{\beta\left(\frac{d}{dt} Y(t)\right) - 2\rho(Y(t))(\beta(Y(t)) - 2\rho(Y(t))(Z) = F(Y(t))} \tag{8.21}$$

This is the *generalized Euler equation*.

Chapter 25

LINEARIZATION OF NONLINEAR SYSTEMS FROM THE POINT OF VIEW OF CONTROL THEORY

1. INTRODUCTION

Linearization of nonlinear differential equations is, of course, one of the main techniques of applied mathematics. Although there is extensive study of it for ordinary differential equations in their traditional "dynamic system" form, there does not seem to be much very explicitly in the control theory literature. (A good deal is in the "folklore" state.) In this chapter I want to develop a few topics that are of interest to me and that I believe could benefit from a systematic discussion. I have in mind essential application to aircraft control; one of my research projects for the Ames Research Center (NASA) has been to look at the "flight control" problem from the geometric point of view. I am also interested in discovering the conditions that the linearized equations of motion can be "simplified" by a clever choice of geometric "moving frames". This is what seems to happen in the classical discussion of aircraft stability and control.

2. LINEARIZATION OF A NONLINEAR INPUT SYSTEM

Here is the classic argument. Consider an "input system" of the form:

$$\frac{dx}{dt} = f(x,u) \quad . \tag{2.1}$$

Here x is the "state vector", u is the "control vector". Let

$$t \to (x_0(t), u_0(t))$$

be a "reference trajectory" which satisfies (2.1). Let τ be another real parameter. Suppose

$$t \to x(t,\tau), \; u(t,\tau) \tag{2.2}$$

are parametrized families of solutions of (2.1). Set:

$$\hat{x}(t) = \left.\frac{\partial x}{\partial \tau}\right|_{\tau=0} \quad ; \quad \hat{u}(t) = \left.\frac{\partial u}{\partial \tau}\right|_{\tau=0} \tag{2.3}$$

We can derive the differential equations satisfied by $\hat{x}, \hat{u}$. First, write down explicitly the fact that the curves (2.2) satisfy (2.1), i.e.,

$$\frac{\partial x}{\partial t} = f(x(t,\tau), u(t,\tau)) \quad . \tag{2.4}$$

Differentiate both sides of (2.4) with respect to τ, and then set $\tau = 0$:

$$\frac{d\hat{x}}{dt} = f_x(x(t),u(t))\hat{x}(t) + f_u(x(t),u(t))\hat{u}(t) \tag{2.5}$$

These are linear differential equations for $\hat{x}$, with the control variables occurring linearly. It is an input system that is the *linearization* of (2.1). Notice it is of the form:

$$\frac{d\hat{x}}{dt} = A(t)\hat{x} + B(t)\hat{u} \quad , \tag{2.6}$$

where $\hat{x}, \hat{u}$ are vectors; $A(t)$, $B(t)$ are t-dependent *linear maps* or *matrices*.

In general, the equations (2.5) and (2.6) are time dependent. This creates practical difficulties, because the most powerful control theoretic methods that we possess now are restricted to time independent linear systems.

Of course, we can extend this linearization process to an input-output system, namely,

$$\begin{aligned}\frac{dx}{dt} &= f(x,u) \\ y &= g(x) \quad .\end{aligned} \tag{2.7}$$

The linearization is then:

$$\begin{aligned}\frac{d\hat{x}}{dt} &= f_x \hat{x} + f_u \hat{u} \\ \hat{y} &= g_x \hat{x} \quad .\end{aligned} \tag{2.8}$$

3. AN INPUT SYSTEM AS AN EXTERIOR DIFFERENTIAL SYSTEM

In order to consider the linearization more "geometrically", it will be convenient to write the input system (2.1) as an *exterior differential system* in the sense of Cartan. Let X, U be manifolds of dimension n and m with coordinates

$$(x^i) \, , \qquad 1 \leq i,j \leq n$$

$$(u^a) \, , \qquad 1 \leq a,b \leq m \quad .$$

Let T be a one-dimensional manifold whose coordinate is t. Set:

LINEARIZATION

$$M = X \times U \times T \quad .$$

Thus, (x^i, u^a, t) is a coordinate system for M. Set:

$$\theta^i = dx^i - f^i(x,u)\, dt \quad , \tag{3.1}$$

where f^i are given functions on $X \times U$. The *solutions* of the input system (2.1) are then obviously determined as maps

$$\sigma : T \to M : t \to (x(t), u(t), t)$$

such that

$$\sigma^*(t) = t \; ; \qquad \sigma^*(\theta^i) = 0 \tag{3.2}$$

Let $\mathscr{P}$ be the Pfaffian system generated by the θ^i, i.e., the $\mathscr{F}(M)$-module generated by the θ^i. The *linear variational equations* are then of the form:

$$\sigma^*(V(\mathscr{P})) = 0 \quad , \tag{3.3}$$

where

$$V = A^i \frac{\partial}{\partial x^i} + B^a \frac{\partial}{\partial u^a} + \frac{\partial}{\partial t} \tag{3.4}$$

($V(\mathscr{P})$ denotes the Lie derivative of the forms in $\mathscr{P}$ by the vector field V), is a vector field on M. The A^i, B^a are functions on M, i.e., functions of x,u. To see this, let us compute:

$$V(\theta^i) = dA^i - \left(A^j \frac{\partial f^i}{\partial x^j} + B^a \frac{\partial f^i}{\partial u^a} \right) dt \tag{3.5}$$

so that equations (3.3) take the form:

$$\frac{d}{dt} A^i(t) = A^j(t) \frac{\partial f^i}{\partial x^j}(x(t), u(t)) + B^a(t) \frac{\partial f^i}{\partial u^a}(x(t), u(t)) \tag{3.6}$$

which are indeed the linear variational equations in a slightly different notational form from that used in Section 2.

4. THE LINEAR VARIATIONAL EQUATIONS IN TERMS OF A "NON-HOLONOMIC" OR "MOVING" FRAME

Continue with the notation of Section 3. Let

$$\omega^i = a^i_j(x)\, dx^j \tag{4.1}$$

be another basis for one-forms on X. It may possibly be "non-holonomic", i.e., $d\omega^i$ *may not be zero*. Suppose that their exterior derivatives take the following form:

$$d\omega^i = c^i_{jk} \omega^j \wedge \omega^k \quad , \tag{4.2}$$

where c^i_{jk} are *functions on* X. Suppose now that:

$$\theta^i = \omega^i - f^i dt \tag{4.3}$$

is the basis for $\mathscr{P}$. (The f^i are now different functions on $X \times U$; they are the "components of the control forces in terms of the new frame ω^i".)

Again, let V be a vector field on M such that $V(t) = 1$. Let $V(\theta^i)$ be the Lie derivative of the form θ^i with respect to A. Then,

$$\begin{aligned} V(\theta^i) &= V(\omega^i) - V(f^i) dt \\ &= V \lrcorner d\omega^i + d(V \lrcorner \omega^i) - V(f^i) dt \end{aligned} \tag{4.4}$$

$$= \text{, using (4.2),}$$

$$2c^i_{jk} \omega^j (V) \omega^k + d(\omega^i(V)) - V(f^i) dt \tag{4.5}$$

Now, let us suppose that:

$$df^i = f^i_j \omega^j + f^i_a du^a \tag{4.6}$$

Combine (4.5) and (4.6):

$$V(\theta^i) = 2c^i_{jk} \omega^j(V) \omega^k + d(\omega^i(V)) - (f^i_j \omega^j(V) + f^i_a V(u^a)) dt \tag{4.7}$$

Set:

$$\alpha^i(t) = \omega^i(V)(\sigma(t)) \tag{4.8}$$

$$\beta^a(t) = V(u^a)(\sigma(t)) \tag{4.9}$$

Thus,

$$\sigma^*(V(\theta^i)) = \sigma^*((2c^i_{jk} \omega^j(V) f^k + d\omega^i(V) - f^i_j \omega^j(V) - f^i_a V(u^a)) dt)$$

The linear variational equations then become:

$$\boxed{\begin{aligned}\frac{d}{dt}\alpha^i(t) &+ 2c^i_{jk}(\sigma(t))\alpha^j(t)f^k(\sigma(t)) \\ &= f^i_j(\sigma(t))\alpha^i(t) + f^i_a(\sigma(t))\beta^a(t)\end{aligned}}$$

(4.10)

These are then the definitive form of the linear variational equations. Notice that the α^i are the "state vectors", the β^a are the "controls". The second term on the left hand side of (4.10) is the new feature--a measure of the "non-holonomic" nature of the moving frame.

Chapter 26

A GENERAL ALGEBRAIC AND GEOMETRIC FORMALISM FOR PERTURBATION THEORY
OF NONLINEAR DIFFERENTIAL EQUATIONS
AND INPUT-OUTPUT SYSTEMS

1. INTRODUCTION

In previous work I have remarked on a natural setting for "perturbation theory" in terms of Lie theory. In this work I intend to develop this point of view more systematically and thoroughly. The recent excellent book by Nayfeh, *Perturbation Methods*, provides a rich source of examples and problems which I plan to assemble together in a general theory. Thus, the material in this chapter will alternate between examples chosen from the classical theory and the general formalism. It will be helpful if the reader has some general knowledge of "modern" differential geometry and Lie group theory. Later, some further general knowledge of more specialized topics, e.g., Cartan's theory of exterior differential systems and Lie algebra cohomology theory will be useful.

Restricting attention for the moment to ordinary differential equations, the "classical" differential equation theory deals with those of the form:

$$\frac{dx}{dt} = f(x,t) \tag{1.1}$$

$$x \in R^n \ .$$

"Perturbation theory" deals with the behavior of such equations which depend on one or more parameters, e.g.,

$$\frac{dx}{dt} = f(x,t,\varepsilon) \ .$$

In "modern" control theory (which I regard as one of the most creative and interesting parts of today's activity in applied mathematics) the viewpoint is expanded beyond (1.1) to include *inputs* and *outputs*. Thus, the system (1.1) is replaced by the following sort of *input-output* system:

$$\frac{dx}{dt} = f(x,u)$$
$$y = g(x) \tag{1.2}$$

This defines a mapping

(curves in u) × (initial vectors in x) → (curves in y)

that is called the *input-output* description of the differential equations. From the "perturbation theory" point of view it is natural to consider such systems depending on paramters ε.

$$\frac{dx}{dt} = f(x,u,\varepsilon)$$
$$y = g(x,\varepsilon) \quad .$$
(1.3)

Now, "perturbation theory" has never been discussed systematically in this context, and I believe that the geometric and algebraic theory to be described here can ultimately make a substantial contribution to this discussion.

Two topics of application have been my immediate motivation, one related to *mechanics*, the other to *biology*. Mechanics is, of course, *the* classical source of perturbation theory problems. As R. Brockett points out, mechanics has never systematically been studied from an input-output point of view. In biology, we encouter many questions which seem to involve complicated nonlinear oscillation questions. Often, perturbation parameters are naturally present (e.g., as "slow" and "fast" time scales) or must be put in to simplify the picture sufficiently to get answers. In particular, it seems to me that biology forces one to invent a *new* sort of "systems theory" with *periodic* functions as inputs. I believe that the classical Van der Pol-Krylov Bogoliubov perturbation theory for periodic systems is the simplest prototype example from which we will ultimately build such a systems theory.

Here is some notation of differential geometry and Lie theory that will be used frequently. Let X denote a (C^∞) manifold of dimension n. $\mathcal{F}(X)$ denotes the (commutative, associative) algebra of C^∞, real-valued functions on X. $\mathcal{V}(X)$ denotes the vector fields. Algebraically, an element $A \in \mathcal{V}(X)$ is a derivation of $\mathcal{F}(X)$-th operation on $\mathcal{F}(X)$ (and its extension to differential forms is called *Lie derivative*) $\mathcal{V}(X)$ is a Lie algebra--the commutator of two such derivations is called the *Jacobi bracket*, denoted by $[\,,\,]$. $\mathcal{D}^r(X)$ denotes the r-th degree differential forms. If (x^i), $1 \leq i,j \leq n$, is a coordinate system on X, $A \in \mathcal{V}(X)$ is of the form

$$A = A^i \frac{\partial}{\partial x^i} \quad ,$$

with $A^i \in \mathcal{F}(X)$.

The *orbit curve* of $t \to x(t)$ satisfies

$$\frac{d}{dt} f(x(t)) = A(f)(x(t)) \quad ,$$

or

$$\frac{d}{dx} x^i(t) = A^i(x(t))$$

or

$$\frac{d}{dt} x(t) = A(x(t))$$

If $t \to x(t)$ is such an orbit curve with $x(0) = 0$,

$$x(t) = \exp(tA)(x(0)) \quad ,$$

Then, for $f \in \mathscr{F}(X)$, $x \in X$,

$$\frac{d}{dt} f(\exp(tA)(x)) = A(f)(\exp(tA)(x))$$

or

$$\frac{d}{dt} \exp(tA)^*(f) = \exp(tA)^*(A(f))$$
$$= A(\exp(tA)^*(f))$$

Thus,

$$\exp(tA)^*(f) = f + tA(f) + t^2 \frac{A^2 f}{2!} + \cdots$$

The right hand side of this formula is called a *Lie series*, and plays a key role in recent work on applied perturbation theory.

I would like to thank Brian Doolin and Hyman Hartman, with whom I have had extensive discussions about the "systems theoretic" and "geometric" aspects of mechanics and biology, respectively.

2. VAN DER POL'S EQUATION

Let us begin with one of the most traditional perturbation theory situations, van der Pol's equation. We follow the treatment in Nayfeh's book, *Perturbation Methods*, page 164. Consider an input-output system of the form

$$\frac{dx}{dt} = f(x,u)$$
$$y = g(x) \quad , \tag{2.1}$$

where the relation between input u and output y is given as follows:

$$y_{tt} + \omega_0^2 y = \varepsilon(1 - y^2) y_t + u \tag{2.2}$$

Subscripts denote partial derivatives. ω_0 is given as constant. ε is a small parameter. k and λ are *parameters for the input*.

Let us look for solutions (2.1) - (2.3) of the form:

$$y = a(t,\varepsilon)e^{i\lambda t} + \bar{a}e^{-i\lambda t} \qquad (2.3)$$

The basic van der Pol techniques is to assume expansions in t for a,b such that:

$$a_t = \alpha(t,\varepsilon)\varepsilon + \text{(higher order terms in } \varepsilon)$$

$$b_t = \beta(t,\varepsilon)\varepsilon + \text{(higher order terms in } \varepsilon)$$

$$a_{tt} = \gamma(t,\varepsilon)\varepsilon^2 + \cdots$$

$$b_{tt} = \gamma(t,\varepsilon)\varepsilon^2 + \cdots$$

Now,

$$y^2 = a^2 e^{2i\lambda t} + \bar{a}^2 e^{-2i\lambda t} + 2|a|^2$$

$$\varepsilon(1-y^2)y_t = \varepsilon(1-y^2)i\lambda(ae^{i\lambda t} - \bar{a}e^{-i\lambda t}) + \cdots$$

$$= i\lambda\varepsilon(1 - 2|a|^2 + a^2 e^{2i\lambda t} + |a|^2 e^{-2i\lambda t})ae^{i\lambda t} - \bar{a}e^{-i\lambda t}$$

$$= i\lambda\varepsilon(e^{i\lambda t}(a(1-2|a|^2) - a^2\bar{a}) + e^{-i\lambda t}(a|a|^2 - (1-2|a|^2)\bar{a})$$

$$+ e^{3i\lambda t}a^3 - |a|^3 e^{-3i\lambda t})$$

$$= i\lambda\varepsilon((e^{i\lambda t}(a(1-3|a|^2) + e^{-i\lambda t}(\bar{a}(3|a|^2 - 1) + a^3 e^{3i\lambda t}$$

$$- |a|^3 e^{-3i\lambda t}) + \cdots$$

Now, let us make the further approximation that the a^3 is negligible:

$$\varepsilon(1-y^2)y_t = i\lambda\varepsilon(1-3|a|^2)(ae^{i\lambda t} + \bar{a}e^{-i\lambda t}) + \cdots$$

Now, substitute all of these approximations into the differential equation (4.2):

$$(-\lambda^2 a + 2i\lambda a_t)e^{i\lambda t} + (-\lambda^2\bar{a} - 2i\lambda\bar{a}_t)e^{-i\lambda t} + \omega_0(ae^{i\lambda t} + \bar{a}e^{-i\lambda t})$$
$$\qquad (2.4)$$
$$= i\lambda\varepsilon(1-3|a|^2)(ae^{i\lambda t} + \bar{a}a e^{-i\lambda t}) + u + \cdots$$

So far the input function $u(t)$ has been arbitrary. Let us suppose it too is a combination of $e^{i\lambda t}$, $e^{-i\lambda t}$:

$$u = v(t)e^{i\lambda t} + \overline{v(t)}\, e^{-i\lambda t} \qquad (2.5)$$

Thus, we can compare coefficients in (4.4) and (4.5), to come up with the following differential equation:

$$2i\lambda a_t - a\lambda^2 + a\omega_0^2 = i\lambda\varepsilon a(1 - 3|a|^2) + v(t) \quad . \qquad (2.6)$$

At first sight, we have not accomplished anything, just converted the nonlinear input-output system (2.2) into the input-output system (2.6). The value in this conversion is the interpretation solving (2.6) for constant input $v(t) \equiv v_0$ and constant $a(t) \equiv a_0$. Substituted back into (2.2), it means

$$y(t) \sim a_0 e^{i\lambda t} + \bar{a}_0 e^{-i\lambda t} \quad , \qquad (2.7)$$

i.e., that $t \to y(t)$ is a *periodic* solution of the van der Pol equation. This is then a very clever transformation--searching for the *periodic* solutions of one differential equation in terms of the *stationary* solution of another. Both are nonlinear problems, but, of course, much more is known about the latter problems, and there are powerful topological and geometric techniques available for studying them.

We can see immediately what the stationary solutions of (2.6) must be:

$$-a_0\lambda^2 + a_0\omega_0 = i\lambda\varepsilon a_0(1 - 3|a_0|^2) + v_0 \qquad (2.8)$$

Of special interest are their stability properties, which can be studied by linearizing (2.6) about the stationary solution. This study is made in all the treatises on nonlinear oscillators, e.g., those by Stoker and Minorsky. Another interesting question is whether there actually *are* periodic solutions of (2.2) corresponding to the stationary solutions of (2.6), i.e., whether these considerations are a good guide to the "qualitative" properties of the van der Pol equation. The most illuminating discussion with which I am familiar is in the book by Hale.

Finally, notice that the most important systems-theoretic "moral" to this classical business is that the input-output system (2.6) is in some sense an "approximation" to the original system (2.2). Understanding the general features of such "approximations" is one of my main goals. Notice that this does not have to be an "approximation" in a literal sense--we are mainly interested in what the approximating system can say about the *periodic* solutions of the original system.

3. **THE METHOD OF "MULTIPLE TIME-SCALE" FOR THE VAN DER POL INPUT-OUTPUT**

Consider again an input-output system of the form:

$$\frac{d^2y}{dt^2} + y = t(1-y^2)\frac{dy}{dt} + \varepsilon u(t,\varepsilon) \tag{3.1}$$

ε is a parameter. Let us look for a solution to this equation of the form:

$$y(t,\varepsilon) = f(t,\varepsilon t, \varepsilon), \tag{3.2}$$

where f depends on three variables, t, τ, ε. Then (denoting partial derivatives by subscripts),

$$y_t = f_t + \varepsilon f_\tau$$

$$y_{tt} = f_{tt} + 2\varepsilon f_{t\tau} + \varepsilon^2 f_{\tau\tau} \quad .$$

We now ask that y be a solution of (5.1) *up to terms in* ε^2. Hence,

$$f_{tt} + 2\varepsilon f_{t\tau} + f = \varepsilon(1-f^2)(f_t) + \varepsilon u + \cdots \tag{3.3}$$

($\cdots$ denotes terms in ε^2 or higher.) Let us next throw away terms in $\varepsilon, \varepsilon^2, \ldots,$ and equate, obtaining

$$f_{tt} + f = 0$$

or

$$f(t,\tau,\varepsilon) = a(\tau,\varepsilon)e^{it} + cc \tag{3.4}$$

("c,c" means "complex conjugate".) Let us now substitute (3.4) back into (3.3) and throw away terms in ε^2:

$$2ia_\tau e^{it} + cc = (1-f^2)(iae^{it} + cc) + U \tag{3.5}$$

$$f^2 = a^2 e^{it} + 2|a|^2 + \bar{a}^2 e^{-2it}$$

$$(1-f^2)(iae^{it}+cc) = (1-2|a|^2)iae^{it} - ia^3 e^{3it} - ia|a|^2 e^{-it} + cc$$

$$= ((1-2|a|^2)ia + ia|a|^2)e^{it} - a^3 ie^{3it} + cc$$

$$= (1-|a|^2)iae^{it} - ia^3 e^{3it} + cc \quad .$$

Suppose also that:

$$u(t,\varepsilon) = v_0(\varepsilon) + v_1(\varepsilon)e^{it} + \cdots + cc \quad .$$

Then, (3.5) becomes:

$$2ia_\tau e^{it} + cc = (1 - |a|^2) ia e^{it} - ia^3 e^{3it} + cc$$
$$+ v_0(\varepsilon) + v_1(\varepsilon) e^{it} + \cdots + cc \quad .$$

(Here, ... stands for *higher harmonics* in t.) This *suggests* that we impose a differential equation on $a(\tau)$:

$$2ia_\tau = (1 - |a|^2) ia + v_1(\varepsilon) \quad . \tag{3.6}$$

This is essentially the same equation we derived in Section 4 using the van der Pol ideas. Hence, this viewpoint is somewhat different, the differential equation is a *consequence of throwing away functions of* (t,τ,ε) *which are linear combinations of* $f(t)\varepsilon^n e^{\pm imt}$, $n,m \geq 2$.

4. AN ABSTRACT PERTURBATION THEORY FOR CURVES IN LIE ALGEBRAS

"Perturbation theory" has a significant Lie-theoretic foundation. This point of view is potentially very valuable for systems theory and I now want to go into it more systematically.

Let $\mathcal{G}$ be a real Lie algebra. To avoid analytic subtleties, we will, for the moment, suppose $\mathcal{G}$ is finite dimensional. Let G be the connected Lie group whose Lie algebra is $\mathcal{G}$. G acts on $\mathcal{G}$ via the *adjoint representation*

$$(g,A) \to \text{Ad } g(A)$$

for $g \in G$, $A \in \mathcal{G}$

If G is a linear group, i.e., a group of $n \times n$ matrices, then

$$\text{Ad } g(A) = gAg^{-1} \quad .$$

In the abstract situation, $\mathcal{G}$ is identified with the set of one-parameter subgroups

$$t \to \exp(tA)$$

of G, and Ad $g(A)$ is the one-parameter subgroup

$$t \to g(\exp(tA))g^{-1} \quad .$$

Thus, in the "abstract" case, we have as a *definition*

$$g \exp(tA) g^{-1} = \exp(t \text{ Ad } g(A)) \quad ,$$

while in the matrix case, this identity is a result of a *theorem*.

Suppose ε is a real parameter (and "small"), and

$$\varepsilon \to A(\varepsilon) = A_0 + A_1\varepsilon + A_2\varepsilon^2 + \cdots$$

is a curve in $\mathcal{G}$. In the clearest and simplest most naive form, we may think of the *perturbation problem* as that of finding a curve $\varepsilon \to g(\varepsilon)$ in G such that

$$\mathrm{Ad}(g(\varepsilon))(A(\varepsilon)) = A_0 \tag{4.1}$$

for all ε.

To a differential/algebraic geometer-Lie theorist, this looks like the following problem:

> Consider Ad G acting on $\mathcal{G}$ as an action of a transformation group. Study the orbit space
>
> $$\mathrm{Ad}\, G \backslash \mathcal{G} \quad.$$
>
> (4.1) then says something about the "structure" of the orbit space, and particularly the projection of the curve $\varepsilon \to A(\varepsilon)$ in this orbit space.

The most straightforward way of computing $g(\varepsilon)$ is to use the exponential map:

$$g(\varepsilon) = \exp(B_0 + \varepsilon B_1 + \varepsilon^2 B_2 + \cdots)$$

(This is related to what the practical perturbation theorists call the *Lie series* method.) One obtains in this way a set of (messy) equations for $B_0, B_1, \ldots \in \mathcal{G}$. This is not too useful, except perhaps for low orders in ε.

Here is another technique: Find $B_1, B_2, \ldots \in \mathcal{G}$ such that:

$$\mathrm{Ad}\,\exp(B_1)(A(\varepsilon)) = A_0 + \varepsilon^2 A_2' + \cdots \equiv A'(\varepsilon)$$

$$\mathrm{Ad}\,\exp(B_2)(A'(\varepsilon)) \equiv A''(\varepsilon) = A_0 + \varepsilon^3 A_3'' + \cdots$$

In other words, a sequence

$$g_1 = \exp(B_1) \quad,$$

$$g_2 = \exp(B_2)$$

of elements of G is to be found so that

$$\text{Ad}(g_n \cdots g)(A(\varepsilon)) = A_0 + \cdots + A_{n+1}^{(n)} \varepsilon^{n+1} + \cdots$$

The sequence of elements of G is used iteratively to get rid of higher and higher powers of ε.

Chapter 27

LIE THEORY, THE METHOD OF "VARIATION OF PARAMETERS",
VOLTERRA SERIES, PERTURBATION THEORY
AND THE VOLTERRA PATH INTEGRAL

1. INTRODUCTION

The "method of variation of parameters" has done yeoman service for a wide spectrum of problems of applied mathematics ranging from analytical and celestial mechanics (where it is called "Lagrange's method"), quantum mechanics, (where it is associated with the slogan "interaction picture") and systems theory (where it is ubiquitous in the linear theory, and intimately involved, as we shall see in this chapter, in the more recent "Volterra series" approach to nonlinear systems). In this chapter I will first describe an abstract version of the theory, then proceed to discuss its ramifications.

2. FLOWS ON MANIFOLDS AND THEIR INFINITESIMAL GENERATORS

For the moment, consider a "manifold" as a C^∞, finite dimensional, paracompact object, although undoubtedly there is interest in extending the framework to more general things.

Let X be a manifold. $\mathscr{F}(X)$ denotes the (commutative, associative algebra) of C^∞, real-valued functions on X. A *vector field* is an R-linear map $V: \mathscr{F}(X) \to \mathscr{F}(X)$ which is a *derivation*, i.e.,

$$V(f_1 f_2) = V(f_1) f_2 + f_1 V(f_2)$$

for $f_1, f_2 \in \mathscr{F}(X)$.

$\mathscr{V}(X)$ denotes the set of all such vector fields. $\mathscr{V}(X)$ admits two sorts of algebraic structure--it is an $\mathscr{F}(X)$-module ($f_0 V$ is the derivation $f \to f_0 V(f)$) and a real Lie algebra (the Jacobi bracket $[V_1, V_2]$ is the commutator $V_1 V_2 - V_2 V_1$).

If $\phi: X \to Y$ is a map between manifolds,

$$\phi^*: \mathscr{F}(Y) \to \mathscr{F}(X)$$

is the dual map on functions, i.e.,

$$\phi^*(f)(x) = f(\phi(x))$$

for $f \in \mathscr{F}(Y)$; $x \in X$

353

It is, of course, an algebra homomorphism. A map $\phi: X \to Y$ is a *diffeomorphism* if ϕ^{-1} exists *as a* C^∞ *map*. (The example $x \to x^3$ of a map $R^3 \to R^3$ shows that ϕ^{-1} may exist, but not be C^∞.)

A *flow* on X is a one-parameter family $t \to \phi_t: X \to X$, $0 \le t \le \infty$, of diffeomorphisms satisfying the following conditions:

$$\phi_0 = \text{identity map}$$

The map $(x,t) \to \phi_t(x)$ of $[0,\infty) \times X \to X$ is C^∞.

Let such a flow be denoted as $\underline{\phi}$. For each t, $f \in \mathcal{F}(X)$, set:

$$V_t(f) = \phi_t^{-1*}\left(\frac{\partial}{\partial t}(\phi_t^*(f))\right) . \tag{2.1}$$

Note that V_t is a derivation of $\mathcal{F}(X)$, i.e., $V_t \in \mathcal{V}(X)$. Thus, we obtain a curve

$$t \to V_t \quad \text{in} \quad \mathcal{V}(X) , \tag{2.2}$$

which we denote as $\underline{V}$, or $\underline{V}(\phi)$. It is called the *infinitesimal generator of the flow* $\underline{\phi}$. The core of Lie theory is the study of this correspondence between flows and their infinitesimal generators .

<u>Exercise.</u> Show that $\underline{\phi}$ is a *semigroup*, i.e.,

$$\phi_{t_1+t_2} = \phi_{t_1}(\phi_{t_2})$$

if and only if V_t is independent of t.

We shall now investigate the relation between these ideas and *ordinary differential equations*. A curve $t \to x(t)$ in X is an *orbit* of the flow $\underline{\phi} = \{\phi_t\}$ if

$$x(t) = \phi_t(x(0)) \tag{2.3}$$

for all t .

Let us see what this means in terms of the infinitesimal generator $\underline{V} = \{V_t\}$. For $f \in \mathcal{F}(X)$,

$$\frac{d}{dt} f(x(t)) = \frac{d}{dt} f(\phi_t(x(0)))$$

$$= \left(\frac{d}{dt} \phi_t^*(f)\right)(x(0))$$

LIE AND PERTURBATION THEORY   355

$$= \text{, using (2.1),}$$

$$\phi_t^*(V_t(f))(x(0))$$

$$= V_t(f)(\phi_t(x(0))$$

$$= V_t(f)(x(t)) \qquad (2.4)$$

Classically, this means that $t \to x(t)$ satisfies the ordinary differential equation

$$\frac{dx}{dt} = V_t(x)(t) \qquad . \qquad (2.5)$$

For example, if $X = R^n$ with Cartesian coordinates (x^i), $1 \leq i, j \leq n$ (summation convention), and

$$V_t(x)\underline{V}_t = v^i(x,t) \frac{\partial}{\partial x^i} \quad ,$$

then (2.5) takes the form:

$$\frac{dx^i}{dt} = v^i(x(t),t) \qquad . \qquad (2.6)$$

<u>Exercise</u>. Show that, conversely, every solution of these differential equations is an orbit of the flow.

3.   THE PERTURBATION-VARIATION OF CONSTANTS FORMULA

Suppose that ϕ is a flow whose infinitesimal generator is a curve in $\mathcal{V}(X)$ of the form:

$$t \to A_t + B_t = V_t \quad ,$$

where $t \to A_t, B_t$ are curves in $\mathcal{V}(X)$. Suppose $\underline{\alpha}$ is the flow whose infinitesimal generator is $\underline{A}$. Set:

$$\beta_t = \alpha_t^{-1} \phi_t \qquad . \qquad (3.1)$$

β_t is a flow. Let us compute its infinitesimal generator.

$$\beta_t^{-1*} \frac{\partial}{\partial t} \beta_t^*(f) = \alpha_t^* \alpha_t^{-1*} \frac{\partial}{\partial t} \phi_t^*(\alpha_t^{-1*} f)$$

$$= \alpha_t^*(A_t + B_t(\alpha_t^{-1*}(f)) - \alpha_t^* \alpha_t^{-1*} \frac{\partial}{\partial t}(\alpha_t^*)\alpha_t^{-1*}(f)$$

$$= \alpha_t^*(A_t + B_t)\alpha_t^{-1*}(f) - \alpha_t^* A_t \alpha_t^{-1*}(f)$$

$$= \alpha_t^* B_t \alpha_t^{-1*}(f) \quad , \tag{3.2}$$

i.e., we have proved the following main result.

<u>Theorem 3.1.</u> The infinitesimal generator of the flow β_t defined by formula (3.1) is the curve $t \to \alpha_t^* B_t \alpha_t^{-1*}$ in $\mathscr{V}(X)$.

This result--as trivial as it is--is of great importance for the practical problem of solving differential equations. Invert (3.1) and write it in the form

$$\phi_t = \alpha_t \beta_t \quad . \tag{3.3}$$

Then, if $t \to x(t) = \phi_t(x(0))$ is an orbit curve of the flow whose infinitesimal generator is $A_t + B_t$,

$$x(t) = \alpha_t(\beta_t(x(0))) \quad . \tag{3.4}$$

This formula shows how the orbit curves of $A_t + B_t$ are "known" if those of $A_t + B_t$, and the curve

$$t \to \alpha_t^* B_t \alpha_t^{-1*} \equiv C_t \tag{3.5}$$

in $\mathscr{V}(X)$. We shall see later on that many classical approximation and/or perturbation methods are based on this observation, i.e., substitute an "approximate" orbit curve $t \to \hat{x}(t)$ of C_t and, with α_t known exactly,

$$t \to \alpha_t(\hat{x}(t))$$

will be an "appropriate" orbit curve of $A_t + B_t$. We can use "Picard iteration" to carry out this approximation.

4. THE PICARD APPROXIMATIONS

Suppose X is R^n. Adopt the usual vectorial notation $x, y, \ldots$ for points of R^n. A "vector field" is then just a map $v: R^n \to R^n$. (The solutions of

$$\frac{dx}{dt} = v(x)$$

are the orbit curves of the vector field.)

Continue with the notation of Section 3. Suppose that

$$C_t = \alpha_t^* B_t \alpha_t^{-1*}$$

is of the form:

$$C_t = v(x,t) \quad ,$$

i.e., the orbit curves of C_t are the solutions of

$$\frac{dx}{dt} = \cdot v(x,t) \quad . \qquad (4.5)$$

The *Picard approximations* of (4.5) are the curves $x_j(t)$, $j = 0,1,2,$ such that

$$x_0(t) = x(0) \quad .$$

$$x_j(t) = \int_0^t v(x_{j-1}(t) - t) \, dt$$

As is well-known, these will converge if t is sufficiently small. Thus, the curves

$$t \to \alpha_t(x_m(t)) = y_n(t)$$

will be approximations to the solutions of

$$\frac{dy}{dt} = (A_t + B_t)(y(t)) \quad .$$

Later on, in connection with Volterra series, we will encounter a case where these equations are of the form

$$\frac{dx}{dt} = w(x,t)'u(t) \qquad (4.6)$$

where $x \in R^n$, with a map of $R^n \times R \to R^n$, $u \to u(t)$ is given a curve in R^n. ' denotes "transpose". Let $x_j(t)$, $j = 0,1,2,\ldots$ be the Picard iterates with, say,

$$x_0(t) = 0 \; ; \quad x_j(0) = 0 \quad .$$

Then,

$$x_{j+1}(t) = \int_0^t w(x_j(t),t)'u(t) \, dt \quad . \qquad (4.7)$$

Thus,

$$(x_{j+1}(t) - x_j(t)) \leq \int_0^t ((w(x_j(t),t) \cdot u(t))$$
$$= w(x_{j-1}(t),t) \cdot u(t)) \, dt \qquad (4.8)$$

Let t vary over the interval

$$0 \leq t \leq a$$

and let K be a Lipchitz constant for w over this interval, i.e.,

$$|w(x,t) - w(x_1,t)| \leq K|x - x_1|$$

Thus,

$$|x_{j+1}(t) - x_j(t)| \leq K \sup_{0 \leq t \leq a} |x_j(t) - x_{j-1}(t)| \int_0^a (u(t)) \, dt \qquad (4.9)$$

Thus, the convergence of the Picard iterate will be *uniform in its* L_1-norm for the "input" $t \to u(t)$. This observation plays a key role in the "Volterra series" work of Gilbert, Brockett, Krener,... If all the data is *real analytic*, it implies that the solution of (4.6), considered as a function of the curve $t \to u(t)$, which is in the Banach space of L, is *analytic*.

Another way to see this is the use the "method of majorants" due to Cauchy, and extensively developed in the 19-th century. There, everything reduces to the study of a *scalar* "majorant" *scalar* equation of the form

$$\frac{dx}{dt} = \frac{1}{1 - bx} u(t) \quad .$$

This can, of course, be solved explicitly:

$$\int dx \, (1 - bx) = \int u(t) \, dt \quad ,$$

$$x - \frac{bx^2}{2} = u(t) \, dt$$

$$x(t) = \frac{-1 \pm \sqrt{1 - 2b^2 \int_0^t u(t) \, dt}}{b}$$

The analyticity of this as a functional of u (when $\|u\|_1 = \int_0^a (u(t)) \, dt$ is sufficiently small) now follows, hence, by Cauchy's method, the analyticity of other equations which this one "majorizes".

LIE AND PERTURBATION THEORY   359

In determining the properties of these approximations, it will be important to know something about the vector field C_t. Here is a technique for doing this.

5. THE DIFFERENTIAL EQUATIONS FOR THE CURVE $t \to C_t = \alpha_t^* B_t \alpha_t^{-1*}$

$$\frac{d}{dt} C_t = \alpha_t^* A_t B_t \alpha_t^{-1*} + \alpha_t^* \frac{d}{dt} B_t \alpha_t^{-1*} - \alpha_t^* B_t \alpha_t^{-1*} \frac{d}{dt} \alpha_t^* \alpha_t^{-1*}$$

$$= \alpha_t^* [A_t, B_t] \alpha_t^{-1*} + \alpha_t^* \frac{d}{dt} B_t \alpha_t^{-1*}$$

$$= [\alpha_t^* A_t \alpha_t^{-1*}, C_t] + \alpha_t^* \frac{d}{dt} (B_t) \alpha_t^{-1*} \qquad (5.1)$$

Let us examine the special case that:

$$\frac{d}{dt}(A_t) = 0 = \frac{d}{dt}(B_t) \quad .$$

Then,

$$A_t \equiv A, \; B_t \equiv B \in \mathcal{V}(X)$$

The differential equation for α_t becomes:

$$\frac{d\alpha_t^*}{dt} = \alpha_t^* A \; ,$$

or

$$\alpha_t^* = \exp(tA)$$

$$= 1 + tA + \frac{t^2 A^2}{2!} + \cdots \qquad (5.2)$$

(5.1) becomes:

$$\frac{dC_t}{dt} = [A, C] \qquad (5.3)$$

$$C_0 = B \quad .$$

Hence the solution of (5.3) is:

$$C_t = \exp(\text{Ad } tA)(B) \qquad (5.4)$$

$$= B + t[A,B] + \frac{t^2}{2!}[A,[A,B]] + \cdots$$

In order to make more geometric sense of this, introduce natural Lie-theoretic notation. If $A \in \mathscr{V}(X)$, $t \to \exp(tA)$ denotes the one-parameter group of diffeomorphisms X generated by A. (Of course, it may not exist *globally*; for simplicity, assume we deal with vector fields that do generate global groups.) The problem solved by these formulas is then basically to calculate the orbit curve $x(t) = \exp(t(A+B))(x(0))$. This is done by setting:

$$y(t) = \exp(-tA)x(t) \quad . \tag{5.5}$$

Then,

$$\frac{dy}{dt} = \exp(\text{Ad } tA)(B)(y(t)) \quad . \tag{5.6}$$

(5.6) constitutes a time-dependent ordinary differential equation for $y(t)$. (If $B = 0$, note that y = constant. This reflects what is classically called "variation of constants".

6. THE VOLTERRA SERIES AND THE FEYNMAN-DYSON EXPANSION

In the mathematical systems theory literature, a recent development has been a description of input-output relations of nonlinear systems in the form:

$$y(t) = \int_0^t k_1(t,t_1) u(t_1) \, dt$$

$$+ \int_0^t \int_0^t k_2(t,t_1,t_2) u(t_1) u(t_2) \, dt_1 dt_2 \tag{6.1}$$

$$+ \cdots$$

We will now show how this is related to an abstract expansion first made by Volterra, and extensively developed by physicists in quantum field theory [hence sometimes called the *Feynman-Dyson expansion*.

To recapitulate notation, suppose that X is a manifold with vector fields $A, B_1, \ldots, B_m$ on X. Let $t \to (u_1(t), \ldots, u_m(t)) = u(t)$ be a curve in R^m. Associate with this curve as "input", the "state" curve $t \to \dot{x}(t)$ which is a solution of

$$\frac{dx}{dt} = (A + u_1(t)B_1 + \cdots + u_m(t)B_m)(x(t)) \quad , \tag{6.2}$$

and the "output"

$$y = f(x(t)) \quad , \tag{6.3}$$

LIE AND PERTURBATION THEORY   361

where $f \in \mathcal{F}(X)$. Set:

$$B_t = \varepsilon u_1(t) B_1 + \cdots \varepsilon u_m(t) B_m \quad .$$

Then $t \to B_t$ is a curve in $\mathcal{V}(X)$ (with ε a small, real parameter) so the equations for the state are as follows:

$$x(t) = \phi_t(x(0)) \quad , \tag{6.4}$$

$$\phi_t^{-1*} \frac{d}{dt} \phi_t^* = A + B_t \quad . \tag{6.5}$$

Set:

$$\beta_t = \exp(-tA) \phi_t \quad . \tag{6.6}$$

Then,

$$\beta_t^{-1*} \frac{d}{dt} \beta_t^* = \exp(\text{Ad } tA)(B_t)$$
$$= \varepsilon(u_1(t) \exp(\text{Ad } tA)(B_1) + \cdots + u_m(t) \exp(\text{Ad } tA)(B_m)) \tag{6.7}$$

Now,

$$y(t) = f(x(t))$$
$$= \phi_t^*(f)(x(0))$$
$$= \beta_t^* \exp(tA)^*(f)(x(0)) \quad . \tag{6.8}$$

The right hand side of (6.8) depends on ε in a C^∞ way. ("Analytic" if the data is analytic.) *The Taylor series in powers of ε is the Volterra series of the input-output system.*

We can find this Taylor series by either Picard iteration or direct power series expansion in ε. Let us do the latter.

$$\beta_t^* = \sum_{j=0}^{\infty} \gamma_j(t) \varepsilon^j \tag{6.9}$$

Then,

$$\frac{d\beta_t^*}{dt} = \sum_{j=0}^{\infty} \frac{d\gamma_j}{dt} \varepsilon^j$$
$$= \left(\sum_{j=0}^{\infty} \gamma_j \varepsilon^{j+1} \right) \text{Ad } \exp(tA)(u_1 B_1 + \cdots + u_m B_m)$$

We see that $d\gamma_0/dt = 0$ (in fact, γ_0 = identity), hence,

$$\sum_{j=0}^{\infty} \frac{d\gamma_{j+1}}{dt} \varepsilon^{j+1} = \sum_{j=0}^{\infty} (\gamma_j \varepsilon^{j+1}) \text{ Ad exp}(tA)(u_1 B_1 + \cdots + u_m B_m) \quad,$$

or

$$\frac{d\gamma_{j+1}}{dt} = \gamma_j \text{ Ad exp}(tA)(u_1 B_1 + \cdots + u_m B_m)$$

$$\gamma_{j+1}(t) = \int_0^t \gamma_j(s) \text{ Ad exp}(sA)(i_1(s)B_1 + \cdots + u_m(s)B_m) \, ds \quad (6.10)$$

Use vectorial notation. Write

$$u(s) = \begin{pmatrix} u_1(s) \\ \vdots \\ u_m(s) \end{pmatrix}$$

$$B = \begin{pmatrix} B_1 \\ \vdots \\ B_m \end{pmatrix}$$

Then,

$$\gamma_1(t) = \int_0^t \text{Ad exp}(sA) u(s)'B \, ds \quad (6.11)$$

Similarly,

$$\gamma_2(t) = \int_0^t \gamma_1(s_1) \text{ Ad exp}(s_1 A) u(s_1)'B \, ds_1$$

$$= \int_0^t \int_0^{s_1} u(s)'(\text{Ad exp}(sA)(B)) u(s_1)' \text{ Ad exp}(s_1 A)(B) \, ds \, ds_1$$

$$(6.12)$$

etc.

This expansion makes explicit the "Volterra" nature of the dependence of the output $t \to y(t)$

$$\frac{dx}{dt} = f(x) + ug(x) \quad (6.13a)$$

$$y = h(x) \tag{6.13b}$$

on the input $t \to u(t)$. However, the convergence properties are not at all obvious. (This is also the great defect of the formalism for quantum-field theory.) However, in case all the data in (6.13) is *real-analytic*, it can be shown, using the argument given in the previous section (and that are just another form of those given by Krener and Brockett), that the output as a functional of the input, is *analytic*, with the L_1-norm put on the input. Thus, the "Volterra-Feynman-Dyson" expansion *must* converge (for $0 \leq t \leq a$ sufficiently small, $\int_0^a (u(t))\, dt$ sufficiently small) because it provides a "formal" power series solution, which must be identical with the convergent one. Unfortunately, the analogous understanding of the analytical properties in the quantum field theory case (where A,B are essentially vector fields on *infinite dimensional manifolds*) is lacking!

7. THE "VARIATION OF CONSTANTS" IN ANALYTICAL MECHANISMS

Let (X,ω) be a *symplectic manifold*. This means that X is a differentiable manifold of even dimension, and ω is a closed two-form of maximal rank, i.e., for $V \in \mathcal{V}(X)$

$$\omega(V, \mathcal{V}(X)) = 0 \Rightarrow V = 0 \quad.$$

Given $h \in \mathcal{F}(X)$, there is then a vector field $V_h \in \mathcal{V}(X)$ such that

$$dh = V_h \lrcorner \omega \quad. \tag{7.1}$$

The orbits of V_h are then *the solutions of the Hamilton equations* with *Hamiltonian* h. A diffeomorphism $\phi: X \to X$ is *canonical* if:

$$\phi^*(\omega) = \omega \quad. \tag{7.2}$$

If (7.2) is satisfied, we have:

$$\phi^* V_f \phi^{*-1} = V_{\phi^*(f)} \quad. \tag{7.3}$$

Proof of (2.3):

$$V_f(f_1) = \{f, f_1\} \equiv \text{Poisson bracket} \quad.$$

ϕ^* is an automorphism of Poisson bracket, i.t.,

$$\phi^*\{f, f_1\} = \{\phi^*(f), \phi^*(f_1)\}$$

or

$$\phi^* V_f \phi^{-1*}(f_1) = \{\phi^*(f), \phi^*(\phi^{-1*}(f_1))\}$$

$$= \{\phi^*(f), f_1\}$$

$$= V_{\phi^*(f)}(f_1) \quad ,$$

whence (7.3).

Frequently, in mechanics one encounters the following situation:

$$h = h_0 + h_1 \quad . \tag{7.4}$$

h_0 is the Hamiltonian of a system for which the solutions of the Hamiltonian equations are "known", and f_1 is a "perturbation" of it. Apply the formalism of the previous sections with

$$A = V_{h_0} \quad , \qquad B = V_{h_1} \quad . \tag{7.5}$$

Set

$$\beta_t = \exp(-tA)\, \exp(t(A+B)) \quad .$$

Then

$$\frac{d}{dt}\beta_t^* = \exp(t(A+B))^*(A+B)\, \exp(-tA)^* - \exp(t(A+B))^* A\, \exp(-tA)^*$$

$$= \exp(t(A+B))^* B\, \exp(-tA)^*$$

$$= \beta_t^*(\exp(tA)^* B\, \exp(-tA)^*) \tag{7.6}$$

This provides a time-dependent differential equation for β_t.

Now, we know that $A \equiv V_{h_0}$ is canonical, hence, using (7.3), we know that

$$\exp(tA)^* B\, \exp(-tA)^* = V_{\exp(tA)^*(h_1)} \quad . \tag{7.7}$$

Carrying out the computations is facilitated by a suitable choice of coordinate system for X. Suppose

$$(q^i, p_i) \quad , \qquad 1 \leq i, j \leq n \quad ,$$

are local coordinates for X chosen so that

h_0 is a function $h_0(p)$ of the p_i above .

$$\omega = dp_i \wedge dq^i \quad .$$

Then,

$$A = V_{h_0} = \frac{\partial h_0}{\partial p_i} \frac{\partial}{\partial q^i} \quad . \tag{7.8}$$

The p_i are constants along the orbit curves of A. Thus, $A(p^i) = 0$,

$$\exp(tA)^*(p_i) = p_i \tag{7.9}$$

$$\exp(tA)^*(q^i) = q^i + tV_{h_0}(q^i) + \frac{t^2}{2!} V_{h_0} V_{h_0}(q^i) + \cdots$$

$$= q^i + \frac{\partial h_0}{\partial p^i} \tag{7.10}$$

$$\exp(tA)^*(h_1) = h_1\left(p_i, q^i + t\frac{\partial h}{\partial p^i}\right) \tag{7.11}$$

Suppose that

$$dh_0 = h_0^i dp_i = \frac{\partial h_0}{\partial p^i} dp_i$$

$$dh_1 = h_{1,i} dq^i + h_1^i dp_i$$

$$h_0^{ij} = \frac{\partial^2 h_0}{\partial p_i \partial p_j}$$

Then,

$$\exp(tA)^*(dh_1) = \exp(tA)^*(h_{1,j})(dq^j + t h_0^{ij} dp_i) + \exp(tA)^* h_1^i dp_i \quad .$$

Thus,

$$V_{\exp(tA)^*(h_1)} = \exp(tA)^*(h_{1,i}) \frac{\partial}{\partial q^i} - (\exp(tA)^*(h_1^i))$$

$$+ t \exp(tA)^*(h_{1,j}) h_0^{ij} h_0^i) \frac{\partial}{\partial p^i} \quad .$$

The ordinary differential equations for β_t are then:

$$\frac{dq^i}{dt} = \frac{\partial h_1}{\partial p_i}\left(q^i + t\frac{\partial h_0}{\partial p^i}, p\right) \tag{7.9a}$$

$$\frac{dp_i}{dt} = -\frac{\partial h_1}{\partial q^i}\left(q + t\frac{\partial h_0}{\partial p}, p\right) - t\frac{\partial h_1}{\partial q^j}\left(q + t\frac{\partial h_0}{\partial p}, p\right)\frac{\partial^2 h_0}{\partial p^i \partial p^j} \quad (7.9b)$$

<u>Theorem 7.1</u>. Let β_t be the canonical flow

$$t \to \exp(-tA)\exp(t(A+B))$$

on the symplectic manifold. Then, in the coordinates (p,q) for which the zero-th order problem (with Hamiltonian h_0) is "solved", the orbit curves of the flow β_t are the solutions of the equations (7.9).

Often, h_1 is "small" relative to h_0. One then wants a perturbation approximation to Equations (7.9). It is simplest to begin over again with this problem.

8. THE FIRST ORDER APPROXIMATION TO THE ANALYTICAL MECHANICS PERTURBATION PROBLEM

Let us begin with the abstract situation. Let A, B be vector fields in a manifold X. Set

$$\beta_t = \exp(-At)\exp(t(A+\varepsilon B)) \quad .$$

Then we know that:

$$\frac{\partial}{\partial t}\beta_t^* = \beta_t^* \varepsilon (\exp(tA)^* B \exp(-tA)^*) \quad . \quad (8.1)$$

The first approximation in ε is then

$$\beta_t^* = 1 + \varepsilon \int_0^t \exp(sA)^* B \exp(-sA)^* \, ds \quad (8.2)$$

Thus, if (using the symplectic manifold solutions of Section 7)

$$h = h_0 + \varepsilon h_1 \quad ,$$

$$V_h = V_{h_0} + \varepsilon V_{h_1}$$

$$A = V_{h_0} \quad , \quad B = V_{h_1}$$

$$\exp(sA)^* B \exp(-sA)^* = V_{\exp(sA)^*(h_1)}$$

LIE AND PERTURBATION THEORY 367

$$\int_0^t \exp(sA)^*(h_1) \, ds = A^{-1}(\exp(tA)^*(h_1) - h_1)$$

$$= A^{-1}\, tA(h_1) + \frac{t^2 A^2(h_1)}{2!} + \cdots$$

$$= th_1 + \frac{t^2}{2!} A(h_1) + \cdots$$

$$\equiv \hat{h}_1 \qquad (8.3)$$

<u>Remark</u>. Note the "averaging" process involved in the first approximation. This is a first clue of the relation of this method to that of Krylov and Bogoliubiv, called the "method of averaging"

Thus,

$$\beta_t^* = 1 + \varepsilon V_{\hat{h}_1} \qquad (8.4)$$

$$\exp(t(A + \varepsilon B))^* = (1 + \varepsilon V_{\hat{h}_1})(\exp(tA)^*)$$

$$= \exp(tA)^* + \varepsilon V_{\hat{h}_1} \exp(tA)^* \, . \qquad (8.5)$$

9. PERTURBATION OF THE HARMONIC OSCILLATOR

The "harmonic oscillator" (with one degree of freedom) is the simplest physical system to illustrate these generalities. Suppose

$$X = R^2, \text{ with coordinates } (x,y) \, .$$

$$h = h_0(x,y) + h_1(x,y)$$

$$= \frac{1}{2}(x^2 + y^2) + h_1(x,y)$$

$$A = V_{h_0} = y \frac{\partial}{\partial x} - x \frac{\partial}{\partial y}$$

$$\exp(tA)^*(x) = \cos tx + \sin ty$$

$$\exp(tA)^*(y) = -\sin tx + \cos ty$$

For example, suppose that

$$h_1(x,y) = x^4 .$$

(The corresponding differential equation is called *Duffing's equation*. It is one of the simplest nonlinear differential equations. Physicists call it the *anharmonic oscillator*.) Then

$$\exp(tA)^*(h_1) = (\cos tx + \sin ty)^4 .$$

The orbit curves of $_t$ are then the solutions of:

$$\frac{dx}{dt} = 4(\cos tx + \sin ty)^3 \sin t$$

$$\frac{dy}{dt} = -4(\cos tx + \sin ty)^3 \cos t \tag{9.1}$$

These equations are *exact*.

<u>Exercise</u>. Show that the "first approximation" derived in the previous section is essentially just the first Picard approximation of (9.1), i.e.,

$$\hat{x}(t) = 4 \int_0^t (\cos sx_0 + \sin sy_0)^3 \sin s \, ds + x(0)$$

$$\hat{y}(t) = -4 \int_0^t (\cos sx_0 + \sin sx_0)^3 \cos s \, ds + x(0) . \tag{9.2}$$

Work out these integrals, and then write down the approximation to the orbits of V_h.

10. THE APPROACH TO PERTURBATION THEORY VIA LAGRANGE'S ORIGINAL METHOD AND THE HAMILTON-JACOBI EQUATION

Let (X,ω) be a symplectic manifold, let T be a one-dimensional manifold, with coordinate labeled "time", t. Let $h: X \times T \to R$ be a real-valued C^∞ function on the product. For $t \in T$, let

$$h_t : X \to R$$

be the element of $\mathcal{F}(X)$ obtained by holding the t-coordinate fixed. Let

$$V_{h_t} \in \mathcal{V}(X)$$

be the vector field on X such that

$$dh_t = V_{h_t} \lrcorner \omega .$$

LIE AND PERTURBATION THEORY

As t varies, $t \to V_{h_t}$ is a curve in $\mathcal{V}(X)$. It determines (modulo the usual global problems, which we ignore) a flow on X called the *Hamiltonian flow*, with h the *Hamiltonian*. The orbit curves of this flow are the curves $t \to x(t)$ in X which satisfy the *Hamilton equations* with h *the Hamiltonian*. Thus, if (p_i, q^i), $1 \leq i \leq n$, are canonical coordinates on X, i.e.,

$$\omega = dp_i \wedge dq^i \quad ,$$

and if h is a function $h(p,q,t)$ in these coordinates, then these curves are the solutions of

$$\frac{dq^i}{dt} = \frac{\partial h}{\partial p_i} \; ; \qquad \frac{dp_i}{dt} = -\frac{\partial h}{\partial q^i} \quad . \tag{10.1}$$

Form

$$M = X \times T$$

and let

$$\omega_h = \omega - dh \wedge dt \quad , \tag{10.2}$$

a closed two-form on M. It is a basic observation in Cartan's book *Leçons sur les invariants integraux* that the solutions of (10.1), considered as a curve

$$t \to (p(t), q(t), t)$$

in M, are *characteristic curves* of the "augmented" form ω_h. Thus, if A is the vector field on M such that

$$A \lrcorner \omega_h = 0$$
$$A(t) = 1 \tag{10.3}$$

(and it is readily verified that A is uniquely determined by these conditions) then the orbit curves of A are the "augmented" solutions of (10.1).

Our aim is to show (again following Cartan, and briefly discussed in my *Differential Geometry and the Calculus of Variations*) that this observation leads to a geometric perturbation scheme, that is basically a modernized version of the method of Lagrange and Jacobi.

Suppose that the Hamiltonian h is split into two parts:

$$h = h_0 + h_1$$

with h_0 the "unperturbed energy", h_1 the "perturbing system". Trivially,

$$\omega_h = (\omega + dh_0 \wedge dt) + (dh_1 \wedge dt) \quad . \tag{10.4}$$

Thus, if we find a new coordinate system (α_i, β^i, t) for M such that

$$\omega - dh_0 \wedge dt = d\alpha_i \wedge d\beta^i \tag{10.5}$$

and express h_1 as a function $h_1(\alpha, \beta, t)$ of these new variables, then

$$\omega_h = d\alpha_i \wedge d\beta^i - dh_1 \wedge dt \quad .$$

The equations of the curves (10.1) now take the form:

$$\frac{d\beta^i}{dt} = \frac{\partial h_1}{\partial \alpha_i} \quad ; \quad \frac{d\alpha_i}{dt} = -\frac{\partial h_1}{\partial \beta^i} \quad . \tag{10.6}$$

Notice that the effect of h_0 has been absorbed *completely* in the (α, β), they are *constants* of motion if $h_1 = 0$, hence their derivatives may be expected to be "small" if h_1 is "small" relative to h_0. This is the essence of Lagrange's method.

In practice, one usually finds these new, marvelous coordinates by finding a "complete solution" of the Hamilton-Jacobi equation. Let us briefly recall how this goes. The Hamilton-Jacobi equation itself is

$$\frac{\partial S}{\partial t} + h_0\left(q, \frac{\partial S}{\partial q}\right) = 0 \quad . \tag{10.7}$$

Find a solution which depends on n additional parameters α_i

$$S(q, t, \alpha)$$

In the space of variables

$$(q, p, \alpha, \beta, t) \quad ,$$

define a submanifold by the following equations:

$$p = \frac{\partial S}{\partial q} \quad ; \quad \beta = \frac{\partial S}{\partial \alpha} \quad . \tag{10.8}$$

Consider the two-form

$$\Omega = dp_i \wedge dq^i - dh_0 \wedge dt - d\alpha_i \wedge d\beta^i \quad . \tag{10.9}$$

Restrict Ω to the submanifold determined by condition (10.8). The result is

LIE AND PERTURBATION THEORY

$$\Omega \text{ restricted} = d\left(\frac{\partial S}{\partial q^i} dq^i\right) - d\left(h_0\left(q, \frac{\partial S}{\partial q}\right)\right) dt + d\left(\frac{\partial S}{\partial \alpha_i} d\alpha_i\right)$$

$$= -d\left(\frac{\partial S}{\partial t} dt\right) + d\left(\frac{\partial S}{\partial t} dt\right)$$

$$= \underline{\underline{0}} \qquad (10.10)$$

Suppose there is a mapping

$$\phi: (q,p,t) \to (\alpha,\beta)$$

whose graph is the submanifold. Then, relating (10.10) means that

$$\phi^*(d\alpha_i \wedge d\beta^i) = dp_i \wedge dq^i - dh_0 \wedge dt$$

Another way of putting this is to assume that the (α_i, β^i, t) are *new* coordinates for M, and then, in these new coordinates,

$$dp_i \wedge dq^i - dh_0 \wedge dt = d\alpha_i \wedge d\beta^i ,$$

which is the relation needed to prove the Lagrange method.

Here is another way of looking at this. Suppose

$$h = h_0 + h_1 \varepsilon + h_2 \varepsilon^2 + \cdots$$

with ε a small parameter. Let

$$\phi_0 : M \to M$$

be a diffeomorphism such that:

$$\phi_0^*(\omega - dh_0 \wedge dt) = \omega \qquad (10.11)$$

(In practice, ϕ_0 will only be defined locally, but I will put that complication to the side.) Thus,

$$\phi_0^*(\omega - dh \wedge dt) = \omega - \varepsilon(dh_1 + \varepsilon dh_2 + \cdots) \wedge dt .$$

Let $\phi_1 : M \to M$ be a diffeomorphism such that

$$\phi_1^*(\omega - \varepsilon dh_1) = \omega \qquad (10.12)$$

Thus,

$$\phi_1^* \phi_0^*(\omega - dh \wedge dt) = \omega - \varepsilon^2 dh_2 \wedge dt - \cdots$$

Continuing in this way, we see that formally, if

$$\lim_{j \to \infty} \phi_0 \phi_1 \cdots \phi_j = \phi \,, \qquad (10.13)$$

then

$$\phi^*(\omega - dh \wedge dt) = \omega \,,$$

i.e., ϕ will be *the* solution to the problem. In practice what seems to happen is that the limit (10.13) does *not* exist (except perhaps in a "formal power series" sense), but cutting it off at the N-th stage

$$\phi_N \sim \phi$$

provides a *useful* approximation to the complete solution. This technique was extensively used in the 19-th century celestial mechanics literature (and inspired much of Poincaré's work, attempting to "justify" the method) and a variant was used by Kolmogoroff and Arnold in their famous work in the stability of Hamiltonian systems (and the solar system).

11. THE VOLTERRA PATH INTEGRAL

Let us now return to the general theory. We have seen that the solution

$$\frac{dx}{dt} = f(x) + \varepsilon u(t) g(x,t)$$

can be expanded in powers of ε. However, this expansion is not quite "natural" from the point of view of group theory. Volterra also has developed the formalism to give a better answer to this question; it is called the *Volterra path integral*. The best exposition in the modern literature (as for so much else) is in Gantmacher's *Theory of Matrices*.

The starting point for the development of the ideas is the following simple linear scalar equation:

$$\frac{dx}{dt} = a(t) x(t) \qquad (11.1)$$

Its solution (with, say, $x(0) = x_0$) can be written down explicitly

$$\frac{dx}{x} = a(t) \, dt$$

$$\log x = \int a(t) \, dt$$

$$x = x_0 e^{\int_0^t a(s)\, ds} \qquad (11.2)$$

Now, let us use the Riemannian approximation of $\int_0^t a(s)\, ds$, namely:

$$\int_0^t a(s)\, ds \sim \sum_{j=0}^{N-1} a\left(\frac{jt}{N}\right)\frac{t}{N} \qquad (11.3)$$

Substitute this approximation into (11.2)

$$x(t) = \prod_{j=0}^{N-1} e^{a\left(\frac{jt}{N}\right)\frac{t}{N}} x_0 \quad . \qquad (11.4)$$

Notice that (11.4) is much improved from the group point of view, over the approximations one would obtain using Picard iteration, since it *directly* involves the underlying group--in this case, the multiplicative group of real numbers acting on the real numbers. Now, the special feature here is that this is an *abelian* group. It should be intuitively clear that one way to take possible non-abelian groups into account is to *time-order* the approximation, i.e.,

$$x(t) \sim \cdots e^{a\left(\frac{jt}{N}\right)\frac{t}{N}} e^{a\left(\frac{t}{N}\right)\frac{t}{N}} e^{a(0)\frac{t}{N}} x_0 \quad . \qquad (11.5)$$

We can now set up to give the Lie-theoretic generalization. Suppose G is a Lie group which acts on a manifold X. Consider the Lie algebra $\mathscr{G}$ "infinitesimally" as a Lie algebra of vector fields on X. Suppose $x \to A(t)$ is a curve in $\mathscr{G}$. Our problem is to find an approximation to the orbit curves of the flow generated by $t \to A(t)$, i.e., to the solutions of:

$$\frac{dx}{dt} = A(t)(x(t)) \quad . \qquad (11.6)$$

Proceed by generalizing (11.5). For $A \in \mathscr{G}$, let

$$t \to \exp(tA)$$

be the one-parameter group of diffeomorphisms of X generated by A. Then, set:

$$x_N(t) = \exp\left(A\left(\frac{(N-1)t}{N}\right)\frac{t}{N}\right) \cdots \exp\left(A(0)\frac{t}{N}\right) x_0 \qquad (11.7)$$

$x_N(t)$, when N is large, is taken as the approximation to $x(t)$, solution of (11.6).

Let us see the form this takes for the "perturbation" situation.

$$y(t) = \exp(t(A + \varepsilon B))y(0) \tag{11.8}$$

$$x(t) = \exp(-tA)y(t) \tag{11.9}$$

$$\frac{dx}{dt} = \varepsilon \, \text{Ad} \exp(tA)(B)(x(t)) \tag{11.10}$$

The approximation $x_N(t)$ to (11.10), following the Volterra path-integral formula (11.7) is then

$$x_N(t) = \exp\left(A \frac{(N-1)t}{N}\right) \exp\left(\varepsilon B \frac{t}{N}\right) \exp\left(-A \frac{(N-1)t}{N}\right)$$

$$\cdots \exp\left(\varepsilon B \frac{t}{N}\right) x(0) \tag{11.11}$$

This is not exactly an easy formula to work with, but it does indicate more clearly the Lie-theoretic nature of the approximation.

Chapter 28

SOME POTENTIAL USES OF ALGEBRAIC GEOMETRY IN
SYSTEMS THEORY AND NUMERICAL ANALYSIS

1. INTRODUCTION

Systems Theory is no doubt the most heavily mathematized branch of engineering--and possibly of all of science. There are basically two reasons for this: the reliance on computers as the most powerful and useful component of the discipline, and the orientation towards studying unified features of the "systems" that occur in a wide spectrum of engineering and societal problems, which forces the sort of abstraction and generalization which is typical of modern mathematics.

Now, it is difficult to define Systems Theory precisely, since it sprawls over such a wide area. I tend to see it as the attempt to apply scientific methodology to the areas that have grown up outside of the traditional scientific and engineering disciplines. It is "interdisciplinary", par excellence. For example, building a radio set is traditional electrical engineering, designing a communications network for which the radio is one component might be systems theory. Designing a thermostat is not (necessarily) systems theory, investigating the general principles by which wide classes of regulatory and control devices work *is*.

This trend towards mathematical thought as the unifying glue of Systems Theory took hold during the 1960's. It would be nice and pat to think that this was because mathematics was ideally suited to this role, and because there had been an effective collaboration between mathematicians and engineers. In fact, neither of these statements is really true--rather, it snuck in the back door, spurred on by certain relatively spectacular popularizations and applications (e.g., the Maximum Principle, Dynamic Programming, the Kalman Filter, etc.). The sociology of science in the 1960's played a role too, particularly in determining the chaotic and spasmodic way that mathematics was developed and applied. (I have in mind the laissez-faire scientific atmosphere, with plenty of money available for ideas that had some plausible claim to be ultimately applicable.) There is now a reaction against this--in part deserved and needed--but there is also an acute danger that the useful component will also be destroyed! Further, it is probably only now, with the imminent dramatic and qualitative change induced by computer miniturization and decentralization, and the emphasis on the much more complicated and difficult "large scale" systems, that we are in a position to really utilize mathematical thought at anywhere close to its potential.

Of course, by "utilizing mathematics" in Systems Theory, I do not simply mean an emphasis on certain mathematical topics that (whether they are called "pure" or "applied") have an incidental relevance. A random glance at the relevant journals seems to indicate that, in fact, this is the interpretation that in practice is most often put on it! A certain amount of this is probably inevitable and desirable, particularly in order to develop the scholarly context in which applications can be understood. It is, of course, a matter of degree, especially because this sort of thing might crowd out the more interactive and innovative work, particularly the application of mathematical ideas to new applied areas.

The aim of this essay is to suggest applications of algebraic and geometric methodology to Systems Theory. Both mathematical disciplines have already been successfully applied. The theory of linear systems (see Section 2) and associated engineering areas like network theory, estimation and filtering, etc. have long utilized material of considerable algebraic sophistication. For example, I would be willing to bet that every fact in Gantmacher's magnificent treatise "Matrix Theory" has been (or could be!) applied in this context. Basically, this book is a compendium of 19th century algebra, and it seems that the Old Boys had a much better feeling for concreteness and "practicality" than today's algebraists. (Of course, there is another, more positive, moral to be drawn--it just takes a long time to apply new mathematics in a non-trivial way. Funding agencies, please note!) There has been some work trying to apply 20th century algebraic ideas to Systems Theory--mainly, to give him due credit, under Rudy Kalman's impulsion--but this has had a certain "softness" that I must criticize. One must probably try to avoid using all of the fancy stuff that is available until one has a firmer grasp of the specific context where it might be applicable and were the older, but *more concrete*, and therefore, other things being equal, *more desirable* methods are *definitely* less powerful. The curse of modern mathematics is the tendency to generalize and invent fancy terminology in areas where the motivation and intuition is not all that interesting or compelling! Of course, one traditional role of the applications in *pure* mathematics is to provide just that extra input that saves the subject from sterility.

Now, it is my personal opinion that modern algebraic geometry often has that typical sterility, at least in comparison with the classical material. Often, the only motivation used to choose problems is the *hope* that the material will have some relevance, some day, to relatively concrete problems in classical mathematics, e.g., number theory. In fact, the growth of *technology* offers magnificent problems--turning one's back on these in disdain (which usually only masks ignorance) is a classic example of something or other.

In this Essay I hope to suggest more specific and powerful ways that modern algebraic geometry can be useful in Systems Theory. For me, this involves putting more emphasis on *geometry* than on *algebra*. This, too, is really a 19th century influence. What I would like to do is to see the magnificent work of Abel, Riemann, Picard, Poincaré, Enriques, Castelnuovo, Severi, Lefschetz,... applied. Lefschetz once said that in his day it was "algebraic GEOMETRY", today it is "ALGEBRAIC geometry". (He makes it clear, in his own inimitable way, what he thinks of this!) It is also not without significance that Lefschetz is also one of the few first rank mathematicians of the 20th century who thought it important to devote his time to applications.

2. ALGEBRA IN LINEAR SYSTEMS THEORY: THE IDENTIFICATION PROBLEM

One of the great accomplishments of "mathematical" Systems Theory in the 1960's was completely working out the *linear, finite-dimensional* case. Luckily this material is very useful and applicable; knowledge of it *in detail* is essential for anyone seriously interested in the subject. There are several excellent treatises--I like those by Anderson and Moore, Brockett, Desoer, and Kalman-Arbib-Falb. I will not try to duplicate this material, but will only discuss here several algebraic features and insights.

Consider such a system, written as:

$$\frac{dx}{dt} = Ax + Bu$$
$$y = Cx \ . \tag{2.1}$$

Here, u, x, y are vectors with, say, complex components, and A, B, C are matrices of the appropriate size. x is called the *state vector*; u the *input vector*; y the *output vector*.

The usual procedure is to pass from the "time domain" description (2.1) of the system to the "frequency domain"

$$sx = Ax + Bu$$
$$y = Cs \ , \tag{2.2}$$

via the Laplace transform

$$x(t) \to \int_0^\infty e^{-st} x(t) \, dx \ .$$

Of course, the classic useful feature of Laplace transform--converting "differential" into "algebraic" equations--appears here. (Assume, for simplicity, that the solution of x(t) of (2.1) has zero initial condition, so that we do not have to bother with carrying along the initial conditions in (2.2).) Equations (2.2) can be solved explicitly, of course.

$$x = (sI - A)^{-1} Bu \ .$$

or

$$y = C(sI - A)^{-1}Bu \qquad (2.3)$$
$$= T(s)u ,$$

where

$$T(s) = C(sI - A)^{-1}B . \qquad (2.4)$$

$T(s)$ is a matrix-valued function of s. It is called the *transfer matrix* (or *frequency response*), and the "input-output" properties (in electrical circuit theory, these are the "steady state" properties) of the system are obviously determined by it.

Mathematically, $T(s)$ is an interesting object. Its entries are *rational functions of the complex variables*. Now, in fact Gantmacher's "Theory of matrices" is primarily about such objects--the close links between it and modern systems theory should thus be no surprise!

"Algebraic geometry" applies to such objects also, but only as a very special case. It really is mainly concerned with *rational functions of many complex variables*. I will now give an example of how this topic might appear very naturally in linear systems theory.

Given the transfer matrix $T(s)$, it can be expanded in a Laurent series:

$$T(s) = s^{-1}(\alpha_0 + \alpha_1 s^{-1} + \alpha_2 s^{-2} + \cdots) .$$

The coefficients $\alpha_0, \alpha_1, \ldots$ are *constant* matrices. For an integer N, form the partitioned matrix

$$h_N = \begin{pmatrix} \alpha_0 \alpha_1 & \cdots & \alpha_{N-1} \\ \alpha_1 \alpha_2 & \cdots & \\ \vdots & & \\ \alpha_{N-1} \alpha_N & \cdots & \end{pmatrix} \qquad (2.5)$$

It is called a *Hankel matrix*. It turns out that its properties (for N sufficiently large, but fixed) determine the input-output properties of the system as well as the transfer functions. This is important, at least from a mathematical point of view, because the *Hankel matrices only depend on a finite number of parameters*, while the transfer functions at *first sight* depend on an infinite number, i.e., the coefficients of the rational functions that appear in the matrix entries. In fact, the Hankel matrices form a "parameterization" for the transfer functions.

Let us examine in more detail how the Hankel matrix h_N depends on the system (2.1). The system itself can be parameterized by the triple

(A,B,C)

of matrices. Let Σ denote the set of such matrices, i.e., the set of such systems. h_N thus depends on Σ. We have explicitly:

$$h_N(A,B,C) = \begin{pmatrix} CB & CAB & \cdots \\ CAB & CA^2B & \cdots \\ \vdots & \vdots & \end{pmatrix} \qquad (2.6)$$

Let us formalize this relation more precisely. Suppose (A,B,C) is described by m parameters. The space of m complex variables we denote as

$$\mathbb{C}^m \quad .$$

Suppose h_N is an $(r \times s)$ matrix. Denote the $r \times s$ matrix by

$$M(r,s) \quad .$$

Thus, the assignment

$$(A,B,C) \to h_N(A,B,C)$$

defines a map

$$h_N \colon \Sigma \equiv \mathbb{C}^m \to M(r,s) \quad .$$

This map is given explicitly, via (2.6), in terms of *polynomials* on the variables that make up $\mathbb{C}^m$.

Let

$$H(r,s) \subset M(r,s)$$

denote the subset of $M(r,s)$ consisting of the matrices that have the special "Hankel" form (2.5). $H(r,s)$ is a *linear subspace* of the vector space $M(r,s)$. We have

$$h_N(\mathbb{C}^m) \subset H(r,s) \quad . \qquad (2.7)$$

Let n be the number of components of the state vector x. Let

$$M(r,s;n)$$

denote the subset of the r,s matrices *which have rank* n.

Now, *a matrix has rank* n *if and only if all* $(n+1) \times (n+1)$ *subdeterminants vanish, while at least one* $n \times n$ *subdeterminant is nonzero*. Hence,

$$M(r,s;0) \cup M(r,s;1) \cup \cdots \cup M(r,s;n) \equiv \text{set of matrices where rank is } \leq n \quad ,$$

is a subset of $M(r,s)$ *which is defined by setting a certain number of polynomial functions equal to zero.* It is a closed subset of $M(r,s)$ (with the usual topology on the space of matrices) of a special sort. In algebraic geometry it is called a *Zariski closed subset*. (I will explain what this terminology means later on.)

Its intersection with H_N is a *Zariski closed subset* of the space of Hankel matrices. Let Σ_{co} denote the subset of Σ consisting of the *controllable, observable systems*. It is *known* that

$$h_N(\Sigma_{co}) = H_N \cap M(r,s;n) \quad . \tag{2.8}$$

(This is the content of the basic *realization theorem* of linear systems theory. See "Interdisciplinary Mathematics", Vol. 8.)

Now, $h_N(\Sigma_{co})$ is also the complement in

$$H_N \cap (M(r,s;n) \cup \cdots \cup M(r,s;0))$$

of a Zariski closed subset. It is then a *Zariski open subset*.

Further, h_N is a mapping that is described by *polynomial functions*; it is called a polynomial map. We are then dealing with a special sort of map, defined by polynomials, between special sorts of spaces, defined by *algebraic relations*. The general theory of such things is the content of *modern algebraic geometry*.

I should hasten to point out (before the reader becomes put off by this sort of jargon) that the study of the map h_N is an essential topic for modern systems theory—even of the most applied sort—called the *identification problem*. Input-output relations are often given by measurement, and one wants to construct a system (or "identify the state vector") to realize these relations. In the sort of complicated systems encountered in today's applications, there is just no way to deny that algebraic geometry *must* be relevant.

There are also many other problems of modern systems theory where the outlook of algebraic geometry is useful and probably even essential. (The case described above is the favorite example used by Rudy Kalman.) My favorite example is the "pole placement via feedback problem" on which Clyde Martin and I have done extensive work.

3. SOME BASIC IDEAS OF ALGEBRAIC GEOMETRY

"Algebraic geometry", as presently understood, goes back to Descartes—the very familiar material (to anyone who has taken calculus) describing how "geometric" figures are described by "algebraic" equations. For example, in the plane, with coordinates (x,y), a line is given by a *linear* equation

$$ax + by = c \quad ,$$

USES OF ALGEBRAIC GEOMETRY

a circle by

$$(x-a)^2 + (y-b)^2 = c^2 \ .$$

A general "quadric" (i.e., a circle, parabola, ellipse, or hyperbola) by a "quadratic equation"

$$ax^2 + by^2 + cxy + d = 0 \ .$$

Here is the general pattern: a set of points in the plane, i.e., in R^2, has associated with it a *polynomial* in (x,y) with the property that a point lies in the set if and only if the polynomial is zero on that point. Now, not *every* set of points in the plane has this property. For example, the set:

$$\left\{ (x,y) \in R^2 : \ y = x \sin \frac{1}{x} \right\}$$

does not

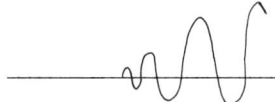

We will say that a subset of R^2 is *algebraic* if it does have this property. The goal of algebraic geometry is to study the *geometric* properties of algebraic sets, with special attention to such properties that are closely related to their algebraic nature. For example, consider a "curve", i.e., a point-set defined by a single polynomial equation

$$f(x,y) = 0 \ .$$

The "curvature" of this curve is a "geometric" property, but the study of curvature is more profitably studied in the wider context of *differential* geometry, and is not usually considered to be within the domain of algebraic geometry. The study of the intersection points of this curve with, say, a line *is* an "algebraic" property, and is "algebraic geometry" par excellence.

Let us try to formalize this process of assigning polynomials to sets of points. For technical purposes, it is most convenient to work with the *complex numbers*, instead of the reals. (This is because the complexes are *algebraically closed*, so that sets of points defined by polynomials will be rich enough to prove general properties. One can *attempt* to do algebraic geometry over the reals--and it is a very important and useful topic--but often one cannot prove sufficiently general and non-trivial results to make it useful.)

$\mathbb{C}^n$ denotes the set of n-tuples

$$x = (x_1, \ldots, x_n)$$

of complex numbers. $PF(\mathbb{C}^n)$ denotes the set of all complex-valued functions

$$f: \mathbb{C}^n \to \mathbb{C}$$

which are given by polynomials. Thus,

$$f(x) = a_{(m_1,\ldots,m_n)} x_1^{m_1} \cdots x_n^{m_n} + \cdots$$

Definition. A subset $A \subset \mathbb{C}^n$ is said to be an *algebraic subset* if there is a set $f_1,\ldots,f_n \in PF(\mathbb{C}^n)$ such that:

$$A = \{x \in \mathbb{C}^n : f_1(x) = 0 = \cdots = f_r(x)\} \tag{3.1}$$

(By convention, the empty set and $\mathbb{C}^n$ itself is also algebraic. We will call them the *trivial* algebraic subsets.)

What characteristic properties do algebraic sets have? We shall see various examples of such properties as we go along. Here are two such. (The proofs of these facts can be found in [1-3].

Theorem 3.1. The union and intersection of a finite number of algebraic sets is algebraic. If $A_1 \supset A_2 \supset \cdots$ is a nested sequence of algebraic sets, then, for *some* integer n,

$$A_n = A_{n+1} = \cdots .$$

(The proof of the first statement is easy. The second is equivalent to the famous Hilbert theorem asserting that every ideal of polynomials in n variables has a finite basis.)

Thus, we endow $\mathbb{C}^n$ with a type of mathematical structure--the collection of algebraic subsets. In modern mathematics one has learned that it is important to not only study mathematical structures, but their "homomorphisms" also, i.e., "structure preserving maps". This we shall now do for the algebraic subset-structure. We shall pursue this by analogy with the way "continuous map" (which is a "structure preserving map" for topological structures) is defined in topology.

Let us say that a subset of $\mathbb{C}^n$ is a *Zariski-open* (Z-open, for short) if it is the *complement* of an algebraic set. To be consistent with the terminology in topology, we also say that a subset is Z-*closed* if its complement is Z-open. (This is, of course, just an algebraic set by another name, but it is often convenient to use this language.)

USES OF ALGEBRAIC GEOMETRY

Now, $\mathbb{C}^n$ is, as a real vector space, R^{2n}. In particular, it is a metric space and has the usual topology with which everyone who knows calculus is very familiar. In this topology, a set is *open* if each point of the set contains a ball of sufficiently small radius. One sees readily that the Z-open sets are open in the usual sense.

However, the Z-open sets are much fewer in number. In particular, they are not sufficient to *separate points*, i.e., given two points $x, y \in \mathbb{C}^n$, there are *not* two Z-open sets O_1, O_2 such that

$$x \in O_1 \quad , \quad y \in O_2$$

$$O_1 \cap O_2 = \text{empty set} \quad .$$

(In fact, any two non-trivial Z-open subsets must intersect.) Thus, of one defines a "topology" for $\mathbb{C}^n$ by *defining* the Z-open sets as *the* open sets (the so-called *Zariski topology*) *some* of the intuitive point-set topology ideas (developed, say, in calculus) carry over, some do not. In fact, I will avoid this topic here--when I refer to a topological concept, it will refer to the "usual", analysis, topology on $\mathbb{C}^n$. (Modern books on algebraic geometry *do* carry along the Z-topology, since they want to work over more general scalar fields than the complex numbers.)

These ideas can be "relativized" by restricting them to a given subset S of $\mathbb{C}^n$. Recall that a subset $S' \subset S$ is *open* or *closed* if it is the intersection with S of an open or closed subset of $\mathbb{C}^n$. Similarly, let us say that S' is Z-open or closed if it is the intersection of S with a Z-open or closed subset of $\mathbb{C}^n$.

Here is an intuitive idea that will be very important to our work.

> Z-open subsets are very *big*,
> Z-closed subsets are very *small*.

After all, a Z-closed subset is a subset determined by a *finite number of polynomial equations*. "Almost all" of the points do *not* satisfy these equations. (In the technical measure theory sense, Z-closed sets have "measure zero". However, it is important for applications to realize that thinking of Z-closed sets as "sparse" makes much more intuitive sense than thinking of "measure zero" sets in this way. For example, the points of $\mathbb{C}^n$ whose coordinates are all *rational* numbers is of measure zero, but it is in no way Z-closed! In the next section, we shall see that this way of thinking also makes sense in terms of dimension-- Z-closed subsets of $\mathbb{C}^n$ have *lower dimension*.

In calculus, a map

$$\phi: \mathbb{C}^m \to \mathbb{C}^m$$

is continuous if the following condition is satisfied:

> For every open subset of $O \subset \mathbb{C}^m$,
> $\phi^{-1}(O)$ is an open subset of $\mathbb{C}^m$.

We can generalize this by saying that ϕ is Z-*continuous* if $\phi^{-1}(O)$ is Z-open whenever O is Z-open. We can also relativize it by saying that a map

$$\phi: S \to S'$$

between subsets $S \subset \mathbb{C}^n$, $S' \subset \mathbb{C}^m$ is Z-continuous if $\phi^{-1}(O)$ is Z-open whenever $O \subset S'$ is Z-open. (Notice that the usual definition of "continuity" in terms of "limits" would not be useful for Z-purposes, because "limits" make no sense. This is an important illustration of how the typical "abstract nonsense" of modern mathematics is often very convenient, particularly as an aid to thinking in new situations.)

Similarly, we can say that a map $\mathbb{C}^n \to \mathbb{C}^n$ is a Z-*homeomorphism* if it and its inverse are Z-continuous. We see that these maps have an important geometric meaning—*they are the maps which map algebraic sets into algebraic sets in a one-one way*, i.e., they are the natural "isomorphisms" of "geometric" algebraic geometry. Just as topology is the study of the equivalence of spaces under homomorphisms, "geometric" algebraic geometry should, in principle, be the study of equivalence under Z-homeomorphism. Unfortunately, things are not handled this way (probably because it is too hard!) and only special types of Z-continuous maps are considered. In fact, for our purposes, two types are sufficient—the *polynomial* and *rational maps*—both of which are obviously Z-continuous.

<u>Definition</u>. A map $\phi: \mathbb{C}^n \to \mathbb{C}^m$ is a *polynomial map* if there are polynomial functions $f_1, \ldots, f_m$ on $\mathbb{C}^n$ such that:

$$\phi(x) = (f_1(x), \ldots, f_m(x))$$

for $x \in \mathbb{C}^n$.

<u>Definition</u>. Let $O \subset \mathbb{C}^n$ be a Z-open subset. A map $\phi: O \to \mathbb{C}^m$ is said to be *rational* if there are polynomials $f, f_1, \ldots, f_m$ such that:

$$\phi(x) = \left(\frac{f_1(x)}{f(x)}, \ldots, \frac{f_m(x)}{f(x)} \right) \tag{3.2}$$

$$f(x) \neq 0, \quad \text{for } x \in O.$$

These concepts can be "relativized"—one defines a map $\phi: S \to S'$ between subsets $S \subset \mathbb{C}^n$, $S' \subset \mathbb{C}^m$ as *polynomial* if it is the restriction to S of a polynomial map $\mathbb{C}^n \to \mathbb{C}^m$. Similarly, one can define (with a bit more care necessary to avoid various subtle complications) define *rational maps* between subsets. However, we shall only systematically consider *polynomial* maps. There is a trick (due to C. Chevalley) which allows rational maps to be formulated in terms of polynomial ones. In fact, suppose a rational map is defined by formula (3.2). Introduce another variable λ, and consider (x,λ) as a point of $\mathbb{C}^{n+1}$. Let

$$A \subset \mathbb{C}^{n-1}$$

be the subset of (x,λ) such that

$$\lambda f(x) = 1 \quad.$$

Set:

$$\phi'(x,\lambda) = (\lambda f_1(x), \ldots, \lambda f_m(x)) \quad.$$

ϕ' is then a polynomial map: $\mathbb{C}^{n+1} \to \mathbb{C}^m$. Restrict it to A. Now, A can be identified with O, i.e., identify $x \in O$ with $(x, f(x)^{-1})$. With this identification, ϕ' becomes equal to ϕ.

4. CALCULUS PROPERTIES OF ALGEBRAIC SETS

As we have emphasized, "algebraic sets" are just the formulation in a general setting of the usual curves and surfaces that everyone is familiar with (and that play such a key role in applications). In geometry, we are used to introducing ideas of calculus by defining "tangent lines", "tangent planes", ... to curves, surfaces, ... In this section we shall do this for general algebraic sets.

First, we begin with material that is the Z-version of the notion of "connectivity" of sets in $\mathbb{C}^n$, as it is encountered in calculus.

<u>Definition</u>. An algebraic set $A \subset \mathbb{C}^n$ is said to be *irreducible* if it *cannot* be written as the union

$$A = A_1 \cup A_2$$

of two non-empty algebraic subsets.

<u>Remark</u>. This can also be conveniently and usefully put into an algebraic form. Recall that we have defined $PF(\mathbb{C}^n)$, the polynomial map $\mathbb{C}^n \to \mathbb{C}$. They can be added and multiplied, i.e., they form a *ring*. Given $A \subset \mathbb{C}^n$, an algebraic set,

$$I_A = \{f \in PF(\mathbb{C}^n) : f(A) = 0\} \tag{4.1}$$

I_A is an *ideal* of the ring $PF(\mathbb{C}^n)$, i.e.,

$$I_A + I_A \subset I_A$$

$$PF(\mathbb{C}^n) I_A \subset I_A \quad .$$

I_A is called the *ideal of the set* A.

Theorem 4.1. A is irreducible, in the sense of the definition given above, if and only if

$$\boxed{I_A \text{ is a prime ideal}} \quad ,$$

i.e.,

$$\boxed{f_1 f_2 \in I_A \Rightarrow f_1 \in I_A \text{ or } f_2 \in I_A}$$

(Equivalently, $PF(\mathbb{C}^n)/I_A$ is an *integral domain*.)

Examples. Let (x,y) be coordinates for $\mathbb{C}^2$.

a) $\quad f = x^2 - y^2$

$\quad A = \{(x,y) : f(x,y) = 0\}$

Since $f = (x+y)(x-y)$, set

$$A_1 = \{(x,y) : x+y = 0\}$$

$$A_2 = \{(x,y) : x-y = 0\}$$

Then,

$$A = A_1 \cup A_2 \quad ,$$

i.e., A *is not irreducible*. Set:

$$f_1 = (x+y) \quad , \quad f_2 = (x-y) \quad .$$

We see that

USES OF ALGEBRAIC GEOMETRY

$$f_1, f_2 \notin I_A \quad,$$

yet

$$f_1 f_2 \in I_A \quad,$$

i.e., I_A is *not* prime.

b) $\quad f = x^2 - y^2 - 1$

$\qquad A = \{(x,y) : f(x,y) = 0\}$

We will show that A is irreducible. Let

$$f_1, f_2 \in \mathrm{PF}(\mathbb{C}^2) \quad,$$

with

$$f_1 f_2 \in I_A \quad.$$

Set:

$$x(\theta) = \cos\theta \quad, \quad y(\theta) = i \sin\theta \quad.$$

Then, $f(x(\theta), y(\theta)) = 0$, for all $\theta \in \mathbb{C}$, i.e., $(x(\theta), y(\theta)) \in A$ for all θ. Then,

$$f_1(x(\theta), y(\theta)) \, f_2(x(\theta), y(\theta)) = 0$$

hence, for *each* θ,

$$f_1(x(\theta), y(\theta)) = 0$$

or

$$f_2(x(\theta), y(\theta)) = 0 \quad.$$

In particular

$$f_1(x(\theta), y(\theta)) = 0$$

for a *continuum of values of* θ, hence,

$$f_1(x(\theta), y(\theta)) = 0$$

for *all* $\theta \in \mathbb{C}$.

But, as θ runs through $\mathbb{C}$, $(x(\theta), y(\theta))$ runs through A, hence, $f_1 \in I_A$, i.e., I_A is a prime ideal, and A is *irreducible*.

Exercise. Find a similar, but "rational" parameterization for A, hence, show that the argument generalizes to other scalar fields.

Remark. There is an interesting general fact underlying this argument that is often quite useful in proving irreducibility, particularly in situations where A is acted on by a *Lie group*. Suppose that A is an algebraic subset of $\mathbb{C}^n$, that M is a *connected* complex analytic manifold, and that

$$\phi: M \to \mathbb{C}^n$$

is a complex-analytic map such that

$$\phi(M) = A \ .$$

Then, A is *irreducible*. (In the example, $M = \mathbb{C}$, $\phi(\theta) = (\cos\theta, i\sin\theta)$.) In particular, notice that this proves that the subset

$$\{(x,y) \in \mathbb{C}^2 : x^2 - y^2 = 0\}$$

cannot be made into a *connected* complex manifold, although as a subset of $\mathbb{C}^2 \equiv \mathbb{R}^4$ it is a connected set.

Now, let us turn to the main topic of this section, the definition of "tangent vector" to an algebraic set.

Definition. Let $A \subset \mathbb{C}^n$ be an irreducible algebraic set, and let $x_0 \in A$. Let I_A be the ideal of A as defined above. A *tangent vector to* A *at* x_0 is an element $x \in \mathbb{C}^n$ such that:

$$\frac{d}{dt} f(x_0 + tx)\Big|_{t=0} = 0 \tag{4.1}$$

for all $f \in I_A$.

The set of tangent vectors at x_0 forms a vector space (a sub-vector space of $\mathbb{C}^n$), denoted by

$$A_{x_0} \ .$$

Example. Let us see the relation to the usual "tangent plane to a surface". Let (x,y,z) be the usual coordinates of $\mathbb{C}^3$, and let

$$f(x,y,z) = 0 \tag{4.2}$$

be the equation to a surface. Suppose this surface is irreducible, and that f generates the ideal I_A.

USES OF ALGEBRAIC GEOMETRY 389

In classical "analytic geometry" the "tangent plane" to the surface (4.2) at (x_0, y_0, z_0) is defined as follows: Let

$$df = \frac{\partial f}{\partial x} dx + \frac{\partial f}{\partial y} dy + \frac{\partial f}{\partial y} dz$$

be the "differential" of f. The "tangent plane" then consists of the $(x,y,z) \in \mathbb{C}^3$ such that

$$df((x,y,z) - (x_0, y_0, z_0)) = 0 ,$$

i.e.,

$$\frac{\partial f}{\partial x}(x_0, y_0, z_0)(x - x_0) + \cdots + \frac{\partial f}{\partial z}( )(z - z_0) = 0 \qquad (4.4)$$

Set:

$$u = x - x_0 , \quad v = (y - y_0) , \quad w = (z - z_0) .$$

Then (4.4) means that:

$$\frac{d}{dt} f((x_0, y_0, z_0) + t(u,v,w)) = 0 ,$$

i.e.,

$$(u,v,w) \in A_{(x_0, y_0, z_0)} .$$

Thus, the "tangent plane" is the plane of $\mathbb{C}^3$ going through (x_0, y_0, z_0) which is *parallel* to the plane passing through $(0,0,0)$ determined by $A_{(x_0, y_0, z_0)}$.

For the rest of this section, A denotes an *irreducible algebraic* subset of $\mathbb{C}^n$.

Definition. A point $x_0 \in A$ is a *simple point* of A if:

$$\text{dimension } A_{x_0} \leq \text{dimension } A_x$$

for all $x \in A$.

The dimension of A_{x_0} at such a simple point (which is obviously the same for *all* simple points) is called the *dimension of* A. If x_0 is not a simple point, it is called a *singular point*.

Warning. This "dimension" is the *complex* dimension which is twice the "real" dimension. This often leads to confusion. Thus, one speaks of a "Riemann *surface*" (i.e., "surface" implying two dimensions), which is an algebraic set of *one* complex dimension. One also speaks of an "algebraic *surface*"; now, the "two" associated with the word "surface" really means two complex dimensions. Thus, a "Riemann surface" (or "algebraic *curve*") is defined by an equation

$$f(x,y) = 0 \;,$$

while an "algebraic surface" is defined by

$$f(x,y,z) = 0 \;.$$

Examples of simple and singular points.

Again, go to $\mathbb{C}^2$.

$$f(x,y) = x^2 + y^2 - a^2 \;,$$

"a" is a constant.

$$A = \{(x,y) : f(x,y) = 0\} \;.$$

$$A_{(x_0,y_0)} = \left\{(x,y) : \frac{\partial f}{\partial x}(x_0,y_0)x + \frac{\partial f}{\partial y}(x_0,y_0)y = 0\right\} \;,$$

or

$$A_{(x_0,y_0)} = \{(x,y) : x_0 x + y_0 y = 0\} \tag{4.5}$$

Now, so long as $(x_0,y_0) \neq 0$,

$$\dim A_{(x_0,y_0)} = 1 \;.$$

Hence,

(x_0,y_0) *is a singular point*

$x_0 \neq 0$

$x_0 = 0 = y_0$.

Thus, if $a = 0$, $(0,0)$ *is the only singular point.* (A is then a *cone*; the singular point is at the *vertex*.) If $a \neq 0$, there are no singular points (A is a *circle*).

There are now available a number of extremely useful results about the regular and singular points of algebraic sets. We shall state and use them, again without proof.

Theorem 4.1. Let $A \subset B \subset \mathbb{C}^n$ be algebraic sets, B irreducible. Then

$$\text{dimension } A < \dim B$$

or

$$A = B \quad .$$

Theorem 4.2. Let A be an irreducible algebraic set. Then A_s is a Z-closed subset of A of lower dimension.

Theorem 4.3. Let A be an irreducible algebraic subset of $\mathbb{C}^n$ whose dimension is m. Then, $A - A_s$ is a connected complex manifold of complex dimension m, i.e., for each point $x \in A - A_s$ there is an open subset $O \subset \mathbb{C}^n$ and an open subset $O' \subset \mathbb{C}^m$, and a map

$$\phi : O \to O'$$

such that:

a) ϕ is a complex analytic map

b) ϕ restricted to $\text{On}(A - A_s)$ is one-one, and onto, a map we denote by ϕ^{-1}.

c) $\phi'^{-1} : O' \to A - A_s \subset \mathbb{C}^n$ is a complex analytic map.

Remark. These conditions say that *neighborhoods of regular points can be parameterized by complex analytic coordinates*. This enables us to apply methods of complex function theory to the study of algebraic sets. This topic is often called *transcendental algebraic geometry*. The name--which is somewhat peculiar and contradictory--comes about because one studies *algebraic* relations with "transcendental", i.e., "non-algebraic", tools. Often one can prove things with amazing power and simplicity (once one knows the relevant machinery!) in this way, which would be much more difficult (or impossible) using "purely" algebraic methods. Since, in systems theory, one often meets sets defined by complicated algebraic conditions, I believe that this "transcendental" side of algebraic geometry will prove to be the most useful.

5. THE ALMOST-ONTO PROPERTY

Now we are ready to state a theorem which is most useful in applications. Let

$$\phi : \mathbb{C}^n \to \mathbb{C}^m$$

be a polynomial map and let

$$A \subset \mathbb{C}^n \; ; \quad A' \subset \mathbb{C}^m$$

be algebraic subsets such that:

$$\phi(A) \subset A' \quad .$$

Definition. ϕ is said to be *almost onto* A' if there is a Z-*open* subset O of A' such that

$$\phi(A) \supset O \quad .$$

Here is the meaning of this. "O Z-open" means that there is an algebraic subset $B \subset \mathbb{C}^m$ such that

$$A' - O = B \cap A' \quad . \tag{5.2}$$

Combining (5.1) and (5.2), we see that $y \in A'$ *and* if $y \notin B$, then there is an $x \in A$ such that

$$y = \phi(x) \quad , \tag{5.3}$$

i.e., the equation

$$" \; y - \phi(x) = 0 \; " \tag{5.4}$$

is solvable for x, given y. Thus, the set of $y \in A'$ *such that* the equation (5.4) is *not* solvable for x *at most* lies in an algebraic subset of A' of *lower dimension*.

This is a weaker property than "onto-ness", but it will turn out to be much more common and much simpler to prove. Often it suffices for applied purposes. (Another paraphrase: *For a "generic" choice of* y, *there exists an* x.)

Theorem 5.1. Suppose that $\phi: A \to A'$ is a polynomial map, and that A' is an irreducible algebraic subset of $\mathbb{C}^m$. Suppose also that the following condition is satisfied:

$$\boxed{\text{Every polynomial function on } \mathbb{C}^m \text{ which vanishes on } \phi(A) \text{ also vanishes on } A'.} \tag{5.5}$$

Then ϕ is almost onto A'.

For the *Proof*, see "Interdisciplinary Mathematics", Volume 13. Especially notice that the complex numbers are essential here--it is not true over the reals. (Think of the map $\phi(x) = x^2$ of $R \to R$.)

Of course, the usefulness of condition (5.5) as a criterion for almost-onto-ness depends on finding ways to show that it is verified. It is here the "tangent space" material considered in Section 4 comes into its own. Let

$$\phi: \mathbb{C}^n \to \mathbb{C}^m$$

be a polynomial map. The (Fréchet) *differential* is a map

$$d\phi: \mathbb{C}^n \times \mathbb{C}^n \to \mathbb{C}^m .$$

defined as follows:

$$d\phi(x_0, x) = \frac{d}{dt} \phi(x_0 + tx) \Big|_{t=0} \quad (5.6)$$

For fixed x_0, $x \to d\phi(x_1, x)$ is a linear map $\mathbb{C}^n \to \mathbb{C}^m$.
Now, suppose that

$$A \subset \mathbb{C}^n , \quad A' \subset \mathbb{C}^m$$

are algebraic subsets such that

$$\phi(A) \subset A' . \quad (5.7)$$

Recall that their tangent spaces

$$A_x , \quad A'_x$$

are linear subspaces of $\mathbb{C}^n$ and $\mathbb{C}^m$.

Theorem 5.2. If (5.7) is satisfied, i.e., ϕ maps A into A', then

$$d\phi(x_0, A_{x_0}) \subset A'_{\phi(x_0)}$$

for all $x_0 \in A$.

Theorem 5.3. Let (5.7) be satisfied. Suppose that x_0, x'_0 are *simple points* of A, A' such that

$$\phi(x_0) = x'_0 .$$

Suppose further that

$$d\phi(x_0, A_{x_0}) = A'_{x'_0} . \quad (5.8)$$

Then, x'_0 contains an open subset $O' \subset \mathbb{C}^m$ such that:

$\phi(A) \supset A' \cap O'$.

<u>Theorem 5.4</u>. With the hypotheses of Theorem 5.3, suppose in *addition* that A' is irreducible. Then, condition (5.5) is satisfied, hence ϕ is almost onto A', i.e.,

$A' - \phi(A)$

is contained in a Z-closed subset of A' of lower dimension.

To prove Theorem 5.4, one must show that condition (5.5) is satisfied. Suppose that $f: \mathbb{C}^m \to \mathbb{C}$ is a polynomial function that vanishes on $\phi(A)$. Consider $A' - A'_s$, a connected complex manifold. By Theorem 5.3, $\phi(A)$ contains an open subset of it, hence f vanishes on an open subset of $A' - A'_s$, which implies that f vanishes identically on A' (since f is an analytic function).

Thus, after all this statement of general terminology and facts (and I apologize to the reader for asking him to take them on faith--they are just too complicated and long to go into here) we come to an extremely useful result. A condition--which only involves *linear* conditions on the first partial derivatives of ϕ at one point of A --that a polynomial map be almost onto. This theorem can also be extended--for example, using the "Chevalley trick" mentioned above--to *rational* maps. Indications of how these results are useful in Systems Theory have already been extensively given in the series of papers that Martin and I have written, "Applications of Algebraic Geometry to Systems Theory". Since I do not want to repeat this material here in detail, I will give another example of how the "almost onto" technique can be used in mathematical engineering. (I assume that the reader is aware that "numerical analysis" is now virtually a part of "mathematical engineering".)

6. THE METHOD OF GAUSSIAN ELIMINATION AS AN EXAMPLE OF AN ALMOST ONTO MAP

"Gaussian elimination" is a method for solving linear equations. (It is basically the method one learns in high school.) It has an interesting matrix-theoretic background, e.g., see Gantmacher , Chapter 2. In modern computer science-numerical analysis, it is important as a classical prototype of what one thinks of now as "algorithms" to effectively do computations on a computer. The problem of "solving linear equations" is *in principle* solved by "linear algebra", e.g., the usual proof of the existence of the inverse of a matrix, but Gaussian elimination offers an effective and constructive implementation of this existence theorem.

For simplicity, let us think of the problem of solving n linear equations in n unknowns, written in matrix form as follows:

USES OF ALGEBRAIC GEOMETRY

$$\alpha x = y \quad , \tag{6.1}$$

with $x \in \mathbb{C}^n$, $y \in \mathbb{C}^n$, α an $n \times n$ matrix.

Gantmacher shows that the Gaussian elimination process essentially involves writing α as a product

$$\alpha = \beta\gamma \tag{6.2}$$

where β is an $(n \times n)$ matrix in *lower* triangular form, γ is an $(n \times n)$ matrix in upper triangular form. We can look at this from various algebro-geometric points of view. Here is the simplest. Let

$M(n,n;U)$

$M(n,n;L)$

be the space of upper and lower triangular $n \times n$ matrices. Let

$M(n,n)$

denote the space of all $(n \times n)$ matrices. We can then define a map

$$\phi: M(n,n;U) \times M(n,n;L) \to M(n,n)$$

as follows:

$$\phi(\beta,\gamma) = \beta\gamma \tag{6.3}$$

Now, $M(n,n;U)$ and $M(n,n;L)$ are vector spaces of dimension $n(n+1)/2$, i.e., are $\mathbb{C}^{n(n+1)/2}$. ϕ is obviously a *polynomial* map

$$: \mathbb{C}^{n(n+1)} \to \mathbb{C}^{n^2} \quad .$$

We shall now prove that its Frechet derivative is *onto* at one point, hence the map itself is almost onto.

(Of course, the Gauss algorithm looks at this from the opposite point of view--given α, find β,γ such that

$$\alpha = \beta\gamma \quad . \tag{6.4}$$

This interpretation is most interesting from the "computational complexity-computer science" point of view. What (6.4) does is to decompose the object α into objects β,γ that are less "complex", in some sense. In fact, many of the results now being investigated by Computer Scientists under the name "computational complexity" are of this type, and I expect algebro-geometric methods to make a fundamental contribution here.)

In order to prove that the map ϕ defined by (6.3) is almost onto, it is convenient to go to a more general setting. Let $\underset{\sim}{A}$ denote an *algebra* over the complex numbers, which is--as a complex vector space--*finite dimensional*, and has a unit element 1. This means that one is given a product

$$(\alpha_1, \alpha_2) \to \alpha_1 \alpha_2$$

$\phi : \underset{\sim}{A} \times \underset{\sim}{A} \to \underset{\sim}{A}$ with the usual bilinear properties. (However, we do not need to assume that $\underset{\sim}{A}$ is *commutative* or *associative* as an algebra.) We have

$$1\alpha = \alpha 1 = \alpha$$

for $A \in \underset{\sim}{A}$.

Now, let

$$\underset{\sim}{A}', \underset{\sim}{A}'' \subset \underset{\sim}{A}$$

be algebraic subsets. Define a map

$$\phi : \underset{\sim}{A}' \times \underset{\sim}{A}'' \to \underset{\sim}{A}$$

as follows:

$$\phi(\alpha_1, \alpha_2) = \alpha_1 \alpha_2 \tag{6.5}$$

for $\alpha_1 \in A'$, $\alpha_2 \in A''$.

We can now compute the Frechet differential of ϕ:

$$d\phi(\alpha_1, \alpha_2)(\delta_1, \delta_2) = \frac{d}{dt}(\alpha_1 + t\delta_1)(\alpha_2 + t\delta_2)\big|_{t=0}$$

$$= \delta_1 \alpha_2 + \alpha_1 \delta_2 \quad . \tag{6.6}$$

Theorem 6.1. Suppose that the following conditions are satisfied:

a) $1 \in \underset{\sim}{A}' \cap \underset{\sim}{A}''$

b) 1 is a simple point of the algebraic subsets $\underset{\sim}{A}', \underset{\sim}{A}''$

c) Every element of $\underset{\sim}{A}$ can be written (in at least one way) as a sum of elements of $\underset{\sim}{A}'_1$ and $\underset{\sim}{A}''_2$.

Then, ϕ is almost onto, i.e., almost every element of $\underset{\sim}{A}$ can be written as a product of elements of $\underset{\sim}{A}', \underset{\sim}{A}''$.

USES OF ALGEBRAIC GEOMETRY

Proof. Let us use formula (6.6) to compute the Fréchet differential of ϕ at the point $(1,1)$ of $\underset{\sim}{A}' \times \underset{\sim}{A}''$. It is:

$$d\phi(1,1)(\delta_1, \delta_2) = \delta_1 + \delta_2$$

for $\delta_1 \in \underset{\sim}{A}'_1$, $\delta_2 \in \underset{\sim}{A}''_2$.

(Recall that $\underset{\sim}{A}'_1$ denote the *tangent* space to $\underset{\sim}{A}'$ at the point 1.) Condition c) then says that this *linear* map is onto, finishing the proof.

We can now specialize to derive one consequence of the Gauss algorithm:

Almost all $(n \times n)$ matrices can be written as products of upper and lower triangulars.

(Take $\underset{\sim}{A} = M(n,n)$; $\underset{\sim}{A}' = M(n,n,U)$; $\underset{\sim}{A}'' = (n,n,L)$.) We can generalize the argument to cover the "finer structure" of the Gauss algorithm. Again, let $\underset{\sim}{A}$ be a finite dimensional algebra with a unit element "1". Now, assume that it is *associative*. (It would be possible to generalize to non-associative situations, with a bit more detail.) Let

$$\underset{\sim}{A}', \underset{\sim}{A}'', \ldots, \underset{\sim}{A}^{(n)} \tag{6.7}$$

be algebraic subsets such that:

a) Each $\underset{\sim}{A}', \ldots, \underset{\sim}{A}^{(n)}$ contains the unit element "1".

b) "1" is a simple element of each of the subsets (6.7).

c) Each element of A can be written as a sum of elements of the tangent spaces $\underset{\sim}{A}'_1, \underset{\sim}{A}''_1, \ldots, \underset{\sim}{A}^{(n)}_1$.

With these conditions, we can define a map

$$\phi: \underset{\sim}{A}' \times \cdots \times \underset{\sim}{A}^{(n)} \to \underset{\sim}{A}$$

as follows:

$$\phi(\alpha', \alpha'', \ldots) = \alpha' \alpha'' \ldots \tag{6.8}$$

Condition (a) guarantees that the differential of ϕ is *onto* at the element $(1,1,\ldots)$; algebraic geometry then implies (if the scalar field is the complex numbers) that ϕ is almost onto, i.e., almost all elements of $\underset{\sim}{A}$ can be written as a product of the form given on the right hand side of (6.8).

Let us now return to the Gauss algorithm, applied to $n \times n$ matrices. $\underset{\sim}{A}$ is then the algebra of $(m \times m)$ complex matrices. For each pair of indices

$$(j,i) , \qquad 1 \leq j < i \leq m$$

let $\underset{\sim}{A}^{(j,i)}$ be the subset of $\underset{\sim}{A}$ consisting of the matrices of the form

$$\alpha^{(ji)} = \begin{pmatrix} 1 & \overset{(j)}{0} & \overset{(i)}{0} & 1 \\ \vdots & \vdots & \vdots & \vdots \\ 0 & \cdots a \cdots & 1 & \cdots 0 \\ \vdots & \vdots & \vdots & \vdots \\ 0 & 0 & 0 & 1 \end{pmatrix}$$

a is an arbitrary complex number.

If α is an arbitrary $m \times m$ matrix,

$$\alpha^{(ji)}\alpha$$

differs from α by the following operation:

The i-th row of $\alpha^{(ji)}\alpha$ is the sum of the i-th row of α and the j-th row of α multiplied by the scalar a.

We will call $\alpha^{(ji)}$ an *elementary row matrix*. The Gauss algorithm then expresses almost all matrices α in the form:

$$\alpha = \alpha'\alpha'' \cdots \alpha^{(n)}\beta \quad , \tag{6.9}$$

where β is an upper triangular, $\alpha', \alpha'', \ldots, \alpha^{(n)}$ are elementary row matrices, *each taken from some* $\underset{\sim}{A}^{(ij)}$.

We can now abstract from this an interesting general framework.

7. A GENERAL GROUP-THEORETIC AND GEOMETRIC FRAMEWORK FOR THE GAUSS ALGORITHM

Let M be a space and let G be a group which acts as a transformation group on M. (See "Interdisciplinary Mathematics", Volume 1, for the general terminology to be used here.)

For $g \in G$, $p \in M$,

$$gp$$

denotes the *transform of* p *by* g. Let

$$M' \subset M \quad ,$$

$$G' \subset G$$

be subsets. (G' is *not necessarily a subgroup*.) Define a map

$$\phi' : G' \times M' \to M$$

USES OF ALGEBRAIC GEOMETRY

as follows

$$\phi'(g',p') = g'p' \tag{7.1}$$

for $g' \in G'$, $g' \in M'$.

Suppose that M is an algebraic set, and ϕ' is almost onto.

Now, let

$$G'' \subset G',$$

$$M'' \subset M'$$

be subsets such that

$$G''M' \subset M'.$$

Define a map

$$\phi'': G'' \times M'' \to M'$$

as follows

$$\phi'(g'',p'') = g''p''.$$

Assume it is *almost onto*. Continue in this way. For $p \in M$, write

$$p = \phi'(g',p') = g'p'.$$

Write

$$p' = \phi''(g'',p'') = g''p'',$$

i.e.,

$$p = g''g'p''.$$

Continuing in this way, we see that we have decomposed "almost all" $p \in M$ in the form

$$p = g \cdots g'p. \tag{7.2}$$

To obtain the Gauss algorithm, take:

$$M = M(n,n),$$

$$G = GL(n,c),$$

acting in the *left*.

I believe that many of the more recently developed algorithms for numerical analysis (and there has been a proliferation since the wide spread use of of computers) also fall within this pattern. Of course, this procedure leads to algorithms which come to an end after a finite number of steps (since a decreasing sequence $M \supset M' \supset M'' \supset \cdots$ of algebraic sets is constructued). More general algorithms--for example, for numerical solutions of differential equations, optimization, etc. involve construction of an "approximating" sequence. There are certain general features of these algorithms which have the same geometric and algebraic flavor; I plan to work further in this direction.

GROUP VI

SURVEYS AND DEVELOPMENT OF NEW GEOMETRIC METHODOLOGY FOR APPLICATIONS

INTRODUCTION

Much of my research effort goes into development of new geometric *methodology* for a wide spectrum of applied problems. This has, in fact, been my main interest since I began work (in 1959) on the geometry of analytical mechanics and control theory. Unfortunately, this work must continually be justified and "sold", even though it has had considerable influence and has grown impressively since 1959. (Geometry always seems to lose out in *both* in the "pure" and "applied" bureaucracy. The trouble is that it takes a long time to illustrate its importance, and the subjects with a more immediate short-term payoff seem always to gain the spotlight.)

In this group, I gather together various essays on such methodological points. Some of this is material which has appeared already in previous volumes, perhaps rethought, and some is new. For example, the Chapter 35 is in a completely new area--*Lie group theory* and *numerical analysis*. I have long believed that Lie methodology has much to offer numerical analysis (for example, both involve variants of classical matrix theroy), but I have not found an appropriate setting to illustrate my ideas. I believe this topic, the QR algorithm, is an ideal one for this purpose and that it *introduces tremendous new insight* into the applied problem. I plan much more work later on; this is only the first tidbit.

SCATTERING THEORY FOR THE
RICCATI EQUATION

1. INTRODUCTION

In the 19-the century, *the* major mathematical topic was differential equations. Of course, this was quite a different subject than it is today; the mathematical emphasis was much more on *geometry*, and there was only the faintest beginning of our era's dominant approach, functional analysis. The culmination of this work was the burst of activity around the turn of the century by such men as Darboux [1], Goursat [2-5], E. Cartan [6-8], Klein, Lie, Riquier, Bäcklund, Bianchi, Ricci, Vessiot, Von Weber, and Engel. (I put Cartan in with this group since his work in this area was basically finished about 1910, and is in a sense a marvelous synthesis of the work of his predecessors--in fact, it is the chief route by which one can now penetrate into the work of the others.) Then, an abrupt end, as in geology with the death of the dinosaurs! Its only resonance in recent times has been in the work of a group of scholars centered around D.C. Spencer (Goldschmidt, Guillemin, Kumpera, Kuranishi, Malgrange,...), whose concerns have been more technical and specialized, i.e., more "modern".

This work would have remained neglected in its historical isolation but for a spectacular recent development in pure and applied mathematics--the discovery of an interrelation between linear and nonlinear partial differential equations called the inverse scattering technique and the associated theory of solitons, Bäcklund transformations, periodic solutions of Hill's equation, Korteweg-de Vries, etc. Although, to a certain extent, this material is still at the level of ingenious special tricks, it has gone far enough to be clear that there is a beautiful general theory underlying it all, which might lead to a great breakthrough in the study of nonlinear partial differential equations and, in turn, lead to important progress in "nonlinear" science and engineering. I believe that this material will turn out to be one of the classics of 20-th century mathematics, comparable, say, to the role that the theory of elliptic functions played in the 19-th century.

Frank Estabrook and Hugo Wahlquist (physicists and applied mathematicians at the Jet Propulsion Laboratory) have been the most successful [9-11] in suggesting the route to follow toward this general theory. They emphasize the general geometric nature of three concepts: *Prolongation, Bäcklund transformation and Bianchi superposition, and generalized conservation law.* Their ideas are based directly on Cartan's theory of *exterior differential systems* [8] which is, in a sense, the unification of this 19-th century work mentioned above. My own work (partially done in collaboration with Estabrook and Wahlquist) has involved [12-15] bringing out even more directly the relation to another part of Cartan's work, the theory of connections, and developing new relations to "Lie theory". In this note I want to illustrate these developments by presenting a new geometric version of the physicists' "scattering theory".

2. SCATTERING THEORY FOR THE RICCATI EQUATION

Let $\mathcal{G}$ be the Lie algebra of 2×2 real matrices of trace zero. Let $G = SL(2,R)$ be the Lie group of 2×2 real matrices of determinant one. (As the notation indicates, $\mathcal{G}$ is the Lie algebra of G.) Let x be a real variable $-\infty < x < \infty$.

Suppose

$$x \to A(x) = \begin{pmatrix} A_{11}(x) & A_{12}(x) \\ A_{21}(x) & -A_{11}(x) \end{pmatrix}$$

is a curve in $\mathcal{G}$. Let

$$z = \begin{pmatrix} z_1 \\ z_2 \end{pmatrix}$$

denote a vector in R^2. Consider the linear differential equations

$$z_x = A(x)z . \quad (2.1)$$

They can be solved as

$$z(x) = g(x)z(0) , \quad (2.2)$$

with

$$g_x = Ag , \quad g(0) = 1 , \quad (2.3)$$

i.e., $x \to g(x)$ is a curve in the Lie group $SL(2,R)$ whose infinitesimal generator is A. (As usual in the theory of nonlinear waves, derivatives are denoted by letter subscripts.)

We can convert (2.1) into a Riccati equation for a scalar y in the following way: set

SCATTERING THEORY

$$y(x) = \frac{z_1(x)}{z_2(x)} \quad . \tag{2.4}$$

Then,

$$y_x = \frac{z_2 z_{1,x} - z_1 z_{2,x}}{z_2^2}$$

$$= \frac{z_2(A_{11}z_1 + A_{12}z_2) - z_1(A_{21}z_1 - A_{11}z_2)}{z_2^2}$$

$$= 2A_{11}y + A_{12} - A_{21}y^2 \quad .$$

Thus, y satisfies the following Riccati equation:

$$y_x = a + by + cy^2 \quad , \tag{2.5}$$

with

$$a = A_{12} ; \quad b = 2A_{11} ; \quad c = -A_{21} \quad . \tag{2.6}$$

Now, suppose that

$$\lim_{x \to \pm\infty} A(x) = A^0 = \begin{pmatrix} A_{11}^0 & A_{12}^0 \\ A_{21}^0 & -A_{11}^0 \end{pmatrix} \tag{2.7}$$

Set:

$$a^0 = A_{12}^0 ; \quad b^0 = 2A_{11}^0 ; \quad c^0 = -A_{21}^0 \quad .$$

Let us *suppose* that the equation

$$a^0 + b^0 r + c^0 r^2 = 0$$

has two roots, r_1, r_2. Let us also suppose that there are four solutions y_1, y_2, y_3, y_4 of (2.5) such that:

$$\lim_{x \to \infty} \begin{pmatrix} y_1 \\ y_3 \end{pmatrix} = \begin{pmatrix} r_1 \\ r_2 \end{pmatrix} \quad , \quad \lim_{x \to -\infty} \begin{pmatrix} y_2 \\ y_4 \end{pmatrix} = \begin{pmatrix} r_1 \\ r_2 \end{pmatrix} \tag{2.8}$$

Set:

$$S = \frac{y_1 - y_2}{y_1 - y_3} \frac{y_3 - y_4}{y_2 - y_4} \qquad (2.10)$$

It is called the *invariant scattering function*. (It is a function of the curve $x \to A(x)$.) Notice the classical property of the cross-ratio appearing on the right hand side of (2.10) is that it is invariant under *linear fractional transformations* on the y's. In particular, S *is independent of* x. We shall see later that S is the absolute value squared of the "reflection coefficient", in the sense of traditional scattering theory. What is desired is an understanding of how S changes as the curve $x \to A(x)$ is changed. Suppose then that

$$x \to A'(x)$$

is another curve in $\mathcal{G}$. Use it to form the Riccati equation

$$y'_x = a' + b'y' + c'y'^2 \qquad . \qquad (2.11)$$

<u>Theorem 2.1</u>. Suppose that there is a curve

$$x \to g(x) = \begin{pmatrix} g_{11} & g_{12} \\ g_{21} & g_{22} \end{pmatrix}$$

in $SL(2,R)$ with the following property: The following formula

$$y'(x) = \frac{g_{11}(x) y(x) + g_{12}(x)}{g_{21}(x) y(x) + g_{22}(x)}$$

maps a curve $x \to y(x)$ which satisfies (2.5) into a curve $x \to y'(x)$ which satisfies (2.11). Suppose further that

$$\lim_{x \to +\infty} \begin{pmatrix} y_1 \\ y_3 \end{pmatrix} = \begin{pmatrix} r_1 \\ r_2 \end{pmatrix}$$

$$\lim_{x \to -\infty} \begin{pmatrix} y_2 \\ y_4 \end{pmatrix} = \begin{pmatrix} r_1 \\ r_2 \end{pmatrix}$$

Then

$$S = S' \qquad . \qquad (2.13)$$

SCATTERING THEORY   407

The <u>Proof</u> of Theorem 2.1 is a "trivial" consequence of the invariance of the cross-ratio function under linear fractional transformations of the type (2.12). It then transfers the question to one of deciding when the "general solutions" of the Riccati equations (2.5) and (2.11) are related via (2.12). This, in turn, is reflected in a statement about the "orbit" of the "gauge group" of all "curves" $x \to g(x)$ in $SL(2,R)$ *acting in the* "space" of Riccati equations. (In [16], Chapter 1, I have described why this is a very general setting for Lie's ideas about the relations between groups and differential equations.) What seems to happen in certain cases is that the conditions for the existence of such a curve $x \to g(x)$ can be described in terms of a *system of ordinary differential equations between the functions* (a,b,c,a',b',c'). These differential equations *define the "Bäcklund transformation"*. Let us now see how these general ideas work in the main case.

3. SCATTERING FOR THE SCHRÖDINGER-STURM-LIOUVILLE EQUATION AND THE BÄCKLUND TRANSFORMATION FOR THE KORTEWEG-DE VRIES EQUATION

Consider the following differential equation:

$$\psi_{xx} + u\psi = \lambda\psi \quad . \tag{3.1}$$

$x \to u(x)$ is a smooth function which vanishes sufficiently rapidly as $x \to \pm\infty$. Define the associated Riccati equation:

$$y = \psi_x/\psi$$
$$y_x = \lambda - u - y^2 \quad . \tag{3.2}$$

Let us also be given another equation of this type:

$$\psi'_{xx} + u'\psi' = \psi' \tag{3.3}$$

$$y' = \psi'_x/\psi$$

Thus, we have"

$$a = \lambda - u \: ; \qquad b = 0 \: ; \qquad c = -1$$
$$a^0 = \lambda \: ; \qquad b^0 = 0 \: ; \qquad c^0 = -1$$
$$r_i = \pm\sqrt{\lambda} \: ; \qquad \text{for } i = 1,2 \quad .$$

Let us see how the "invariant scattering function" S is defined in terms of the usual "scattering data" [17]: Fix $\lambda < 0$. Let ψ_+, ψ_- be solutions of (3.1) such that:

$$\lim_{x \to \pm\infty} \left(\psi_\pm(x) - e^{\sqrt{\lambda}\, x} \right) = 0 \tag{3.4}$$

Let $\overline{\psi_\pm(x)}$ denote their complex conjugate. They are also solutions of (3.1) and

$$\lim_{x \to \pm\infty} \left(\overline{\psi_\pm(x)} - e^{-\sqrt{\lambda}\, x} \right) = 0 \tag{3.5}$$

(This type of solution--determined by "boundary conditions at infinity--is generally called a *Jost function* in the physics literature.) The relation between the Jost function at $+\infty$ and $-\infty$ is then determined by a relation of the following form:

$$\psi_+ = a\psi_- + b\overline{\psi_-} ,$$

where a,b are functions of λ. If ψ, ψ' are solutions of (3.1), set

$$W(\psi,\psi') = \psi_x \psi' - \psi \psi'_x \tag{3.6}$$

$$\equiv \text{the } Wronskian .$$

Then,

$$W(\psi_+, \overline{\psi_+}) = 2\sqrt{\lambda} = W(\psi_-, \overline{\psi_-}) . \tag{3.7}$$

Hence,

$$a = W \frac{(\psi_+, \overline{\psi_-})}{2\sqrt{\lambda}}$$

$$b = \frac{(\psi_+, \psi_-)}{2\sqrt{\lambda}} \tag{3.8}$$

Definition. Set

$$R(\lambda) = \frac{b(\lambda)}{a(\lambda)} . \tag{3.9}$$

This is the *reflection coefficient* for the linear differential equation (3.1).

Now set

$$y_1 = \frac{\psi_{+,x}}{\psi_+} , \quad y_2 = \frac{\psi_{-,x}}{\psi_-} , \quad y_3 = \frac{\overline{\psi_{-,x}}}{\overline{\psi_-}} , \quad y_4 = \frac{\overline{\psi_{+,x}}}{\overline{\psi_+}} \tag{3.10}$$

SCATTERING THEORY 409

y_1, y_2, y_3, y_4 are solutions of the Riccati equation (3.2). They have the asymptotic behavior as $x \to \pm\infty$ needed (following the ideas of Section 2) to define the invariant scattering function

$$S(\lambda) = \frac{y_1 - y_2}{y_1 - y_3} \frac{y_4 - y_3}{y_4 - y_2} . \qquad (3.11)$$

Theorem 3.1.

$$S(\lambda) = |R(\lambda)|^2 . \qquad (3.12)$$

Proof. Using (3.8) - (3.9) we have:

$$R = \frac{b}{a} = \frac{W(\psi_+, \psi_-)}{W(\psi_+, \overline{\psi_-})}$$

$$= \frac{\psi_{+,x}\psi_- - \psi_+\psi_{-,x}}{\psi_{+,x}\overline{\psi_-} - \psi_+\overline{\psi}_{-,x}}$$

$$= \text{, using (3.10)}$$

$$\frac{y_1 - y_2}{y_1 - y_3} \frac{\psi_+ \overline{\psi_-}}{\psi_+ \psi_-}$$

Hence,

$$\overline{R} = \frac{y_4 - y_3}{y_4 - y_2} \frac{\psi_-}{\overline{\psi_-}} ,$$

hence

$$R\overline{R} = \text{right hand side of (3.11)} . \blacksquare$$

Definition. The Riccati differential equations

$$\begin{aligned} y_x &= \lambda - u - y^2 \\ y'_x &= \lambda - u' - y'^2 \end{aligned} \qquad (3.13)$$

are related via a *Bäcklund transformation of Wahlquist-Estabrook type* if there are a pair (w, w') of functions of x, and a constant λ' such that the following conditions are satisfied:

$$u = -w_x \; ; \qquad u' = -w'_x \; ;$$
$$(w+w')_x = (w-w')^2 - \lambda \; .$$
(3.14)

We recognize (3.14) as part of the formulas associated with the Korteweg-de Vries Bäcklund transformation [10]. That the solutions of (3.13) are connected via relations of the form (2.12) is now a consequence of the Bianchi-type superposition formula, first proved by Wahlquist and Estabrook [10].

Thus we see that as *a consequence of general principles* (which are broadly generalizable!) that absolute value of the reflection coefficient. In fact, it is known that the reflection coefficient itself is invariant--I do not know if this is also a "general" fact, or is more specially tied to the specific dynamical situation. I suspect it is the former.

4. CHANGES IN THE SCATTERING FUNCTION INDUCED BY A LIE ALGEBRA-VALUED ONE-FORM OF CURVATURE ZERO

Let us return to the general setting of Section 2. Introduce a "time" variable t. Suppose given a pair $(A(x,t), B(x,t))$ of maps $R^2 \to \mathcal{G}$. It defines a $\mathcal{G}$-valued one-form:

$$\eta = A \otimes dx + B \otimes dt$$
(4.1)

on R^2. Its *curvature* is then the following $\mathcal{G}$-valued two-form:

$$\Omega = dA \wedge dx + dB \wedge dt + \frac{1}{2} [\eta, \eta]$$
$$= (A_t - B_x - [A,B]) \otimes dt \wedge dx \; .$$
(4.2)

Suppose that

$$\Omega = 0 \; .$$
(4.3)

Then the following vectorial equations:

$$z_x = A(x,t) z$$
$$z_t = B(x,t) z$$
(4.4)

is *completely integrable* in the *classical* sense. ("Complete integrability" is a term that has been used in a very confusing--and incorrect--way in the recent literature.) Their solution can be written in the following form:

SCATTERING THEORY

$$z(x,t) = g(x,t)(z(0,0)) \qquad (4.5)$$

where $(x,t) \to g(x,t)$ is a surface in G such that:

$$\begin{aligned} g_x &= gA \\ g_t &= gB \end{aligned} \qquad (4.6)$$

Now set

$$y = \frac{z_1}{z_2} . \qquad (4.7)$$

It satisfies the following Riccati equations:

$$y_x = a + by + cy^2 \qquad (4.8)$$

$$y_t = \alpha + \beta y + \gamma y^2 , \qquad (4.9)$$

where the coefficients $(a,b,c,\alpha,\beta,\gamma)$ are functions of x and t constructed from A and B.

We can now form the invariant scattering function S using Equation (4.8). It will depend on t. Its evolution in t is via a linear fractional transformation under which S (as a cross-ratio) is invariant. Hence:

$$\boxed{S \text{ is independent of } t.} \qquad (4.10)$$

Thus, we see that "deformations" determined by (4.3) and (4.10) *leave the invariant scattering function unchanged*. Thus we see that there are two ways of changing the curve $x \to A(x)$ in $\mathcal{G}$ so as to leave unchanged the invariant scattering function--via the "discrete" Bäcklund transformation and via the "continuous" deformations generated by equations of type (4.3) - (4.4).

Remark. I have been sloppy about the precise asymptotic conditions required for (4.4) in order to leave S unchanged. What must happen is that the equations (4.4) must be such as to imply that formally

$$\frac{\partial}{\partial t}(y(\pm\infty,t)) = 0 . \qquad (4.11)$$

In the examples with which I am familiar, this is accomplished by requiring that:

$$\lim_{x \to \pm\infty} B(x,t) = \text{constant} \times \left(\lim_{x \to \pm\infty} A(x,t) \right) . \qquad (4.12)$$

At this point is established a close relation to the standard material concerning Korteweg-de Vries-Schrödinger obtained via the standard inverse scattering technique. If the potential function $x \to u(x)$ changes via a solution $x \to u(x,t)$ of the Korteweg-de Vries equation, it is known [17,18] that the reflection coefficient $A \to R(\lambda)$ changes via the rule

$$R(\lambda,t) = e^{\sqrt{\lambda}^3 t} R(\lambda) \quad . \tag{4.13}$$

In particular, $(R(\lambda,t)) = (R(\lambda))$. Again, the rule (4.12) is a more precise result which may follow from a sharpening of the tools developed here.

5. PROLONGATIONS AND GENERALIZED CONSERVATION LAWS FOR EXTERIOR DIFFERENTIAL SYSTEMS

Where do relations like (4.3) come from? Geometrically, (4.3) says that a *curvature* is zero. A comprehensive geometric genesis for relations of this sort has been provided by Estabrook and Wahlquist [11] and I will now briefly develop their ideas in the form presented in my work [12-15,19].

Let X be a manifold. An *exterior differential system* for X -- abbreviated to ED -- is a collection of differential forms with the following properties:

> ED is an ideal in the Grassmann algebra formed by the differential forms in X

$$d(ED) \subset ED \quad .$$

(ED is also called a *differential ideal*.)

Let X be a manifold with such an ED. A submanifold map $\alpha: Z \to X$ is an *integral manifold* of

$$\alpha^*(ED) = 0 \quad .$$

Definition. Let ED, ED' be exterior differential systems on manifolds X, X'. A map $\pi: X \to X$ is called a *prolongation* if

$$\pi^*(ED') \subset ED \quad .$$

Thus, such a prolongation maps integral manifolds of ED into integral submanifolds of ED'. In view of the standard relation between exterior differential systems and partial differential equations, this is one way of defining the notion of "geometric homomorphism" between solutions of different differential equations.

Definition. Let (ED,X), (ED',X') be exterior differential systems. A *Bäcklund transformation* between them is another system (ED'',X''), together with maps

$$\beta: X'' \to X \; ; \qquad \beta': X'' \to X'$$

such that

$$\beta^*(ED) \subset ED'' \supset \beta'^*(ED') \quad .$$

Such a Bäcklund transformation provides a "correspondence" between integral submanifolds of ED and ED' (thus, a "correspondence" or "relation" between solutions of the underlying differential equations).

6. ESTABROOK-WAHLQUIST PROLONGATIONS AND GENERALIZED CONSERVATION LAWS

Start off with ED as an exterior differential system on the manifold X. Let (ED',X') be another exterior system, and let

$$\pi: X' \to X$$

be a prolongation map. It is said to be an *Estabrook-Wahlquist prolongation* if the following conditions are satisfied:

a) π is a submersion map, i.e., $\pi_*: T(X') \to T(X)$ is onto.

b) There is a set of one-forms $\mathscr{P}$ on X' with the following properties:
   i) $\mathscr{P}$ is a module over the ring $F(X')$ of C^∞ functions on X',
   ii) A tangent vector $v \in T(X')$ is tangent to the fibers of π if and only if $\mathscr{P}(v) = 0$. (In other words, $\mathscr{P}$ defines an Ehresmann connection for the fiber space $\pi: X' \to X$.)
   iii) ED' is the Grassmann algebra ideal generated by $\pi^*(ED)$ and $\mathscr{P}$.

Condition (iii) is the geometric kicker. It says that over the submanifolds

$$\phi: Z \to X$$

such that $\phi^*(ED) = 0$, the connection defined by $\mathscr{P}$ is *flat*. Another way of putting this is to say that the *curvature* form of the connection defined by $\mathscr{P}$ lie in the ED.

In view of the relations between "connections" and "Lie-algebra valued one-forms", it is convenient to make further definitions. Let $\mathscr{G}$ be a real Lie algebra (possibly infinite dimensional). A $\mathscr{G}$-*valued one-form* on X is a linear mapping

$$\eta : T(X) \to \mathcal{G} \quad .$$

Its *curvature* is a $\mathcal{G}$-valued two-form on Ω defined by the following formula:

$$\Omega = d\eta + \frac{1}{2} [\eta, \eta] \quad . \tag{6.1}$$

Definition. η is called a *generalized conservation law* for the system ED if Ω lies in ED $\otimes$ $\mathcal{G}$. (Identify $\mathcal{G}$-valued differential forms in X with elements of $\mathcal{D}(X) \otimes \mathcal{G}$, where $\mathcal{D}(X) = all$ differential forms on X and the tensor product $\otimes$ is that of *real* vector spaces.

Such an η defines connections. Let Y be a space on which $\mathcal{G}$ acts as a Lie algebra of vector fields. Set:

$$X' = X \times Y \quad .$$

We can use η to define a connection in the following way. Suppose (y^a), $1 \leq a,b \leq m$, is a coordinate system for Y, and (x^i), $1 \leq i,j \leq n$, is a coordinate system for X. Suppose

$$\frac{\partial}{\partial x^i} = \Gamma_i^a(y) \frac{\partial}{\partial y^a} \quad . \tag{6.2}$$

Set:

$$\theta^a = dy^a - \Gamma_i^a dx^i \quad . \tag{6.3}$$

These are one-forms on $X \times Y = X'$. They generate $\mathcal{P}$, and the connection. It is readily seen that the connection is flat, i.e., the Pfaffian system $\mathcal{P}$ is completely integrable, if and only if the curvature form Ω vanishes. If η is a generalized conservation law for ED, it is readily verified that

$$X' = X \times Y \to X \quad ,$$

the Cartesian projection map, is an Estabrook-Wahlquist prolongation in the sense defined above.

Let us now specialize to the case of two-dimensional integral submanifolds of ED, i.e., $Z = R^2$ parameterized by x and t, and $\mathcal{G}$ the Lie algebra of 2×2 real matrices of trace zero. Let

$$\phi : Z \to X$$

be an integral submanifold. Suppose $\eta : T(X) \to \mathcal{G}$ is a generalized conservation law. Then

$$\phi^*(\eta) = A(x,t) dx + B(x,t) dt \quad ,$$

where $A(x,t)$, $B(x,t)$ are 2×2 matrix valued functions of (x,t). The condition $\phi^*(\Omega) = 0$ then gives the relation

$$A_t - B_x = [A,B]$$

that we encountered in the previous section in terms of scattering theory.

For the Korteweg-de Vries equation, X is the space of variables (x,t,u,u_x,u_{xx}). Estabrook and Wahlquist have determined [11] *all* "generalized conservation laws" associated with *one* way of writing Korteweg-de Vries as an exterior differential system. This leads to an interesting *infinite dimensional* Lie algebra; it is generated (as a Lie algebra) by seven elements, satisfying structure relations given by formula (3.7) of their paper [11]. The standard "inverse scattering" equations associated with the Korteweg-de Vries equation is obtained by looking for Lie algebra homomorphisms from this Lie algebra to the Lie algebra of $SL(2,R)$. It is also clear from the work of Wahlquist and Estabrook that the notions of "Bäcklund transformation" and "superposition formulas" are closely tied with the Lie algebra. We would very much like to know a more intrinsic definition of this Lie algebra, as well as general techniques for computing it for other partial differential equations! (The more extensive work toward generalization has been done by H. Morris [20-24], J. Corones [25,26], R. Dodd and J. Gibbon [27,28].) Another interesting question is the possible *physical* meaning of these "generalized conservation laws".

7. GENERAL REMARKS

One of the obvious deductions from the history of mathematics is that study of scientific phenomena is often the most fruitful and stimulating influence on the *long-run* development of mathematics. In the "short run" we have had, say in the last thirty years, a swamping of this effect by the flood of new material generated *within* mathematics, particularly motivated by the more abstract, general (and, I believe, sterile) techniques. Mathematics *as a profession* has had more than its share of lumps in the last ten years; ironically, at the same time that the influence of mathematical thought has spread in ever-wider circles and mathematical research is thriving. At the beginning of this period of "professional" decline, appeared the basic paper [17] from which the subject touched on here developed, slowly at first, then recently in a rush of new results and techniques by some of the world's leading mathematicians. One of the marvelous aspects of this subject is that it involves a whole spectrum of ideas, techniques, concepts, and branches of mathematics. Key contributions have come from the sort of detailed, concrete calculations in which physicists and applied mathematicians specialize, from general abstract reasoning that is the bulwark of "modern" pure mathematics, and from computer calculation. In addition to the obvious branches of mathematics (functional analysis, differential equations, numerical analysis), increasingly the subject leads into algebraic geometry, differential geometry and Lie-theory--indeed, there are suggestions of the development of vast new areas of these disciplines.

Let us also hope that the subject begins its contact with the study of nonlinear phenomena in *science*. After all, for all the propaganda in the last ten years for the development of "new" sorts of mathematics for use in biology, economics, computer science, etc., I feel that we may have lost touch with another bit of historical reality--that mathematics only can find optimally congenial ground in the sciences and technology that depend on *Newton's laws*, i.e., *differential equations* (most of which are *nonlinear*). I do not want to say dogmatically that *no* applications outside of this framework are possible-- only that the further we get from the sciences that have a "Newtonian" flavor, the more and more difficult it becomes to find meaningful and meaty material for mathematicians to work on *which has sufficient relevance to the subject to justify the work "applied"!* (I might mention also how well "systems theory"-- the other applied subject for which I have great enthusiasm--fits in here as a *bridge between the "Newtonian" and "non-Newtonian" disciplines*. Thus, it has its roots in electrical and aerospace engineering--which one would classify as basically "Newtonian"--but reaches out to other subjects like economics, biology, etc.) In the subject of "nonlinear waves" we have entirely new and exciting possibilities of description of physical, technological, and possibly even chemical-biological phenomena. Of course, a major effort will be required to break the barrier which seems to separate us from an effective theory of equations with *two or more space dimensions*. My personal belief is that an ultimate answer to it might be *Lie's infinite dimensional groups*. Elementary particle physics also offers vast possibilities--however, there is a formidable barrier in the physicist's "pragmatic" attitude toward mathematics, which inhibits him from learning new techniques unless he can see an *immediate* payoff. At any rate, many of the papers written by physicists on "solitons" (and there have been a flood, though it has now abated) have often been wrong-headed, treating a "soliton" as a slogan rather than as a major new mathematical technique which they must master before they write papers about them!

However, I believe it is the subject of differential equations itself that will recieve the greatest benefit from these new ideas. In the last twenty years there has been an overwhelming emphasis on technique. These improved techniques have mostly been sterile in terms of their usefulness in applications. Here is a new field with numerous interesting practical possibilities, with almost infinite potential "payoff", and which obviously offers tremendous challenges to the "technicians".

Bibliography

1. G. Darboux, <u>Theorié genérale des surfaces</u>, Chelsea Pub. Co., New York.

2. E. Goursat, <u>Lecons sur le problème de Pfaff</u>, Herman, Paris, 1922.

3. E. Goursat, <u>Lecons sur l'intégration des équations partielles du première ordre</u>, Herman, Paris, 1921.

4. E. Goursat, <u>Lecons sur l'intégration des équations aux dérivés partielles du second ordre...</u>, Herman, Paris, 1898.

5. E. Goursat, <u>A Course in Mathematical Analysis</u>, Ginn and Co., Boston, 1904.

6. E. Cartan, <u>Oeuvres Complètes</u>, Part 2, Gauthier-Villars, Paris.

7. E. Cartan, <u>Lecons sur les invariants intégraux</u>, Herman, Paris.

8. E. Cartan, <u>Les systèmes différentielles et leurs applications géométriques</u>, Herman, Paris, 1940.

9. B.K. Harrison and F.B. Estabrook, "Geometric Approach to Invariance Groups and Solution of Partial Differential Equations", <u>J. Math. Phys.</u> 12 (1971), 653-656.

10. H.D. Wahlquist and F.B. Estabrook, "Bäcklund Transformations for Solution of the Korteweg-de Vries Equation", <u>Phys. Rev. Lett.</u> 31 (1973), 1386-90.

11. H.D. Wahlquist and F.B. Estabrook, "Prolongation Structures of Nonlinear Evolution Equations", <u>J. Math. Phys.</u> 16 (1975), 1-7.

12. R. Hermann, "The Pseudopotentials of Estabrook and Wahlquist, the Geometry of Solitons and the Theory of Connections", <u>Phys. Rev. Lett.</u> 36 (1976), 835.

13. R. Hermann, "The Inverse Scattering Technique of Soliton Theory, Lie Algebras, the Quantum Mechanical Poisson-Moyal Bracket, and the Rigid Rotating Body", <u>Phys. Rev. Lett.</u> 37 (1961), 1591.

14. R. Hermann, <u>The Geometry of Nonlinear Differential Equations, Bäcklund Transformations, and Solitons</u>, Part A (1976), Part B (1977), and Part C (in preparation). Interdisciplinary Mathematics, Math Sci Press, Brookline, Mass.

15. R. Hermann, <u>Toda Lattices, Cosymplectic Manifolds, Bäcklund Transformations and Kinks</u>, Part A, Interdisciplinary Mathematics, Vol. 15, Math Sci Press, Brookline, Mass., 1977.

16. R. Hermann, <u>Geometric Structure of Systems Control Theory and Physics</u>, Part A, Interdisciplinary Mathematics, Vol. 9, Math Sci Press, Brookline, Mass., 1975.

17. C.S. Gardner, J.M. Greene, M.D. Kruskal, and R. Muira, <u>Phys. Rev. Lett.</u> 19 (1967), 1095-1097.

18. V.E. Zakharov and L.D. Faddeev, "Korteweg-de Vries Equation: A Completely Integrable Hamiltonian System", <u>Functional Analysis and its Application</u> 5 (1971), 280-287.

19. R. Hermann (ed.), <u>The Ames Research Center (NASA) 1976 Conference on the Geometry of Nonlinear Waves</u>, Lie Groups: History, Frontiers and Applications, Vol. 6, Math Sci Press, Brookline, Mass., 1977.

20. H.C. Morris, "Prolongation Structures and a Generalized Inverse Scattering Problem", J. Math. Phys. 17 (1976), 1867-1869.

21. H.C. Morris, "Prolongation Structures and Nonlinear Evolution Equations in Two Spatial Dimensions", J. Math. Phys. 17 (1976), 1870-1872.

22. H.C. Morris, "A Prolongation Structure for the AKNS System and its Generalizations", J. Math. Phys. 18 (1977), 533-536.

23. H.C. Morris, "A Generalized Prolongation Structure and the Bäcklund Transformation of the Anticommuting Massive Thiring Model", preprint, 1977.

24. H.C. Morris, articles in Ref. 19 above.

25. J. Corones, "Solitons and Simple Pseudopotentials", J. Math. Phys. 17 (1976), 756-759.

26. J. Corones, "Solitons, Pseudopotentials and Certain Lie Algebras", J. Math. Phys. 18 (1977), 163-164.

27. R.K. Dodd and J.D. Gibbon, "The Prolongation Structure of a Class of Nonlinear Evolution Equations", preprint, 1977.

28. R.K. Dobb and J.D. Gibbon, "The Prolongation Structure of Some Higher Order Korteweg-de Vries Equations", preprint.

EXTERIOR SYSTEMS, INPUT-OUTPUT SYSTEMS, AND VECTOR BUNDLES

1. INTRODUCTION

I have been working off and on for twenty years on a program of application and adaptation of the differential geometry of Elie Cartan to applied problems, particularly in systems theory and physics. This work is now reaching full fruition, and I want to outline here some of the general ideas and indicate how they apply to more specific problems of current interest.

Now, most "applied mathematicians", whether they are trained as such or move into it from "pure", have a background in *analysis*. Similarly, the mathematical background of most engineers and physicists is in analysis. Modern geometry is extraordinarily diverse and complicated compared to the classical material and it is very difficult to learn on one's own. All of this has combined to hinder the development of this side of science and it is only now approaching its full glory.

Cartan's masterpieces are *Leçons sur les invariants intégraux* and *Les systèmes différentielles extérieures et leurs applications géométriques*. They are (along with the more well-known 19th century work of Gauss, Riemann, and Poincaré) the highest sort of "applied mathematics". Cartan's work is really about the *geometric structure* of differential equations, and I believe that it could play the same sort of role in the science and technology of our day that 19th century mathematics did in its time. Unfortunately, most of today's professional "applied mathematicians" do not seem interested in such possibilities, but seem content to cultivate certain technical specialities (e.g., combinatorics and numerical analysis), which were overlooked in the general progress of pure mathematics. I believe that one of the historic roles of Systems Theory could be to provide the framework for a renewed interaction between the most vital elements of "pure" mathematics and the science (taken in the broadest, generic sense) of our day.

2. EXTERIOR DIFFERENTIAL SYSTEMS

Since Newton, the ultimate aim of "physical mathematics" has been to describe phenomena in terms of differential equations. This is no less important for the areas which have most recently been mathematized and which are not dominated by ideas inherited from physics. In fact, one way of thinking of Systems Theory is that is is mainly concerned with mathematical models (usually involving differential equations) of phenomena for which the precise description is not given by Nature, as they are traditionally in physics. (However, physics often finds itself up against models that are too complicated to understand, or that are imperfectly understood on a fundamental level, as in elementary particle physics. Systems ideas could be very useful here, but physicists do not know it, and seem to have lost the talent for learning ideas which exist outside their own closed circles.)

At any rate, differential equations will continue to be important. Now, this subject, as mathematics, is vast, but lately has been dominated (as a "pure" subject) by functional analysts. There is an alternate historical tradition, which one might call "geometric". Cartan's book, *Les systèmes différentielles extérieures et leurs applications géométriques* is the closest thing I know of to a unified statement of what this subject means. His motivation was mainly the problems of differential geometry; the applications I envision lead to a different emphasis. Thus, the more advanced material in the book is (for the moment) not useful for applied work; what is important is the overall spirit, the way of working with and thinking about differential equations in a systematic, geometric way. I will now outline some of the ideas from a very general point of view; later we will see how these ideas are used in Systems Theory in more specific ways.

Let X be a manifold (C^∞, paracompact, finite dimensional, unless mentioned otherwise). $\mathscr{F}(X)$ denotes the commutative associative algebra of real-valued C^∞ functions on X. $\mathscr{V}(X)$ denotes the *vector fields*, i.e., the derivations of $\mathscr{F}(X)$

$$\mathscr{D}(X) = \mathscr{D}^0(X) (\equiv \mathscr{F}(X)) \oplus \mathscr{D}^1(X) \oplus \cdots$$

denotes the graded, associative (but not commutative) algebraic differential forms on X with exterior multiplication $\wedge$ and exterior differentiation

$$d: \mathscr{D}^r(X) \to \mathscr{D}^{r+1}(X) \quad .$$

If $T(X)$, the *tangent bundle*, is the vector bundle over X whose cross-sections are $\mathscr{V}(X)$, $T^d(X)$ is the *cotangent bundle*, the *dual bundle* to $T(X)$ and the vector bundle whose cross-sections are $\mathscr{D}^1(X)$. If

$$\phi: Y \to X$$

is a C^∞ map between manifolds, it induces a *pull-back map*

$$\phi^*: \mathcal{D}(X) \to \mathcal{D}(Y)$$

on differential forms such that:

$$\phi^*(\theta_1 \wedge \theta_2) = \phi^*(\theta_1) \wedge \phi^*(\theta_2)$$

$$\phi^*(d\theta) = d\phi^*(\theta)$$
(2.1)

for $\theta, \theta_1, \theta_2 \in \mathcal{D}(X)$.

These "naturality" properties under *arbitrary* maps are one of the main sources of the strength of Cartan's methods. He methodically translated everything into differential form language (and he began his work before the rise of tensor analysis) and thus achieved this very wide "covariance" *automatically*. *Classical tensor analysis* deals with a much more restricted "covariance" under only *diffeomorphisms*. I have always wondered why physicists make such a big deal about this sort of covariance, but then have never been interested in and/or motivated to go beyond to learn Cartan's kind.

The map ϕ induces a map which is dual to ϕ^*,

$$\phi_*: T(Y) \to T(X) \quad ,$$

on the tangent bundles. This map is most convenient for the purposes of *differential topology*. For example, here are key concepts:

a) ϕ is a *submanifold map* (or defines Y as a submanifold of X) if $\phi_*: T(Y) \to T(X)$ is one-one.

b) ϕ is a *submersion map* (or defines Y as a *fiber space* over X) if $\phi_*(T(Y)) = T(X)$.

Definition. An exterior differential system $\mathcal{E}$ on a manifold X is a set of differential forms such that:

a) $\mathcal{E}$ is a linear subspace of $\mathcal{D}(X)$ as a real vector space,

b) $\mathcal{E}$ is an ideal in the Grassmann algebra $\mathcal{D}(X)$, i.e.,

$$\theta_1 \wedge \theta \in \mathcal{E}$$

for $\theta_1 \in \mathcal{D}(X)$, $\theta \in \mathcal{E}$

c) $d\mathcal{E} \subset \mathcal{E}$, i.e., $\mathcal{E}$ is a "differential ideal".

In algebraic topology one learns to separate the "algebraic" from the "geometric", *but to build a bridge between them*. This attitude is very much present in Cartan's work. The exterior systems described above are the "algebraic" gadgets. The "geometric" ones are the *integral submanifolds*, i.e., the maps

$$\phi: Y \to X$$

such that:

  a) ϕ is a **submanifold** map
  b) $\phi^*(\mathscr{E}) = 0$.

One can extend this to define *integral maps* which only satisfy condition (b).

The category theorists have taught us to think of constructing "morphisms" along with "objects". The morphisms for exterior differential systems may be defined as follows.

<u>Definition</u>. Let $(X, \mathscr{E})$, $(X', \mathscr{E}')$ be exterior differential systems. A map

$$\phi: X \to X'$$

is a *homomorphism* (of exterior systems) if

$$\phi^*(\mathscr{E}') \subset \mathscr{E} \quad .$$

Thus, if

$$\alpha: Y \to X$$

is an integral map of $(X, \mathscr{E})$, the composite

$$\phi\alpha: Y \to X'$$

is an integral map of $(X, \mathscr{E}')$. A homomorphism thus induces a map

$$(\text{integral maps of } \mathscr{E}) \to (\text{integral maps of } \mathscr{E}') \quad .$$

This concept is really the ultimate version of the concept of "symmetry" in science and engineering. Prolongations [9,23] are also a special case.

A homomorphism $\phi: X \to X'$ such that ϕ^{-1} exists is an *isomorphism* or *equivalence*. The *equivalence problem* is to find effective criteria for deciding whether such equivalences exist and describing their properties (e.g., whether they depend on a finite or infinite number of parameters). This concept played a key role in the work both of Cartan and Lie and there are many examples worked out in their Collected Works. Unfortunately, little has been done since their day in terms of the general theory. (Differential geometers have worked extensively on developing further properties of the situations where it was known that the equivalence problem was tractable.)

The equivalence problem for a *general* map ϕ is very hard. Often one can make substantial progress (particularly in the applied work) by restricting ϕ

in some way. For example, for *linear input-output systems* we will see that the material that the systems theorists know already reasonably well (Brunovsky canonical forms, Kronecker "indices") is a specialization. A recent paper by Jan Willems is another variant [1].

In [2] and [3], I have presented a new idea that was not present explicitly in Cartan's work, that of *extended vector fields*. (Another version of the same underlying geometric idea was developed more fully later on by various applied mathematicians, particularly Ibragimov and Anderson [4]. Unfortunately, they do not seem to understand how useful Cartan's and Ehresmann's ideas are for systematization of material like this.) Let $\mathcal{E}$ be an exterior differential system on a manifold X. Let

$$\mathcal{J}^r(Y,X)$$

be the space of r-jets of mappings of $Y \to X$ in the sense of Ehresmann. Let

$$\pi: \mathcal{J}^r(Y,X) \to X$$

be the "target" projection map. It is a fiber space map. An *extended vector field* is a *fiber preserving* map

$$V: \mathcal{J}^r(Y,X) \to T(X) \quad .$$

A one-parameter family

$$t \to \phi_t: Y \to X$$

of maps $Y \to X$ is then said to be an *orbit* of the *extended vector field* if the following condition is satisfied:

> For each $y \in Y$, the tangent vector to the curve $t \to \phi_t(y)$ is equal to $V(j^r(\phi_t(y)))$, where $j^r(\phi_t(y))$ is the r-jet of the map ϕ_t.

"Extended vector fields" thus may define curves (and even, in favorable cases, one-parameter transformation groups) in the space of maps $Y \to X$. V is said to be an *extended infinitesimal symmetry* of the exterior differential system $\mathcal{E}$ if the following condition is satisfied:

> ϕ_t is an integral map of $\mathcal{E}$ for all t if it is for one value of t.

Other key concepts in Cartan's theory are the *characteristic* and *singular solutions*. I do not want to go into this topic here. (The work of Robert Gardner [5 - 7] is the most complete in the recent literature. Much remains to be done.) The simplest sort are the *Cauchy characteristics*, i.e., the vector fields V on X such that

$$V \lrcorner \mathscr{E} \subset \mathscr{E} .$$

They turn out to be the key objects in analytical (Lagrangian and Hamiltonian mechanics), one-independent variable calculus of variations and the theory of first order partial differential equations in one dependent variable. (In fact, the theory of Cauchy characteristics is the geometric link between them!) Much of the material in Cartan's earlier book *Lecons sur les invariants integraux* is concerned with these ideas. Mathematicians and physicists often work with various special cases of this formalism without being aware of its full ramifications and setting, and thus "invent" many of the ideas for themselves. Lie also did much work on the relation between "symplectic-contact" manifold theory and group theory that is continually re-invented.

Yet another important concept is that of *conservation law*, i.e., a differential form θ such that

$$d\theta \in \mathscr{E} .$$

These objects have played an important role recently in the theory of nonlinear partial differential equations in two independent variables (Korteweg-de Vries, Sine-Gordon, etc.). They also appear in what physicists call "Noether's Theorem" in field theories. In my book *Lie Algebras and Quantum Mechanics* [8] I have described how they may be used in the quantum mechanics of particle and field theories. If θ is a zero-form, i.e., a function, it is a "constant of motion" in the classical sense. The work of Estabrook and Wahlquist [9] involves an important generalization--Lie-algebra valued one-forms θ on X such that

$$D\theta = d\theta + \frac{1}{2} [\theta,\theta] \in \mathscr{E} .$$

These are some sort of "generalized conservation laws". We suspect that there are many important still unknown properties of nonlinear partial differential equations which are related to these objects. They are also closely related to what the elementary particle physicists call "gauge fields", objects that have played a prominent role in recent speculations about the phenomonology of elementary particles.

EXTERIOR SYSTEMS

3. INPUT-OUTPUT SYSTEMS AS EXTERIOR DIFFERENTIAL SYSTEMS

Now we turn from these generalities to the material of most immediate interest in Systems Theory. An *input-output system* is a system of ordinary differential equations of the form:

$$\frac{dx}{dt} = f(x,u)$$
$$y = g(x,t) \tag{3.1}$$

The vectors $x = (x^1, \ldots, x^n)$ are coordinates on a manifold X, called *state space*. u are coordinates on a manifold U, called *input space*. t is the coordinate on a one-dimensional manifold. Set:

$$M = X \times U \times T \quad.$$

$\mathscr{E}$ = exterior differential system generated by the forms

$$dx - f(x,u)dt$$
$$y - g(x,t) \tag{3.2}$$

on M.

$\mathscr{E}$ is the basic object to study. For example, *controllability* [10] and *observability* [11] have already been related to certain differential geometric properties of $\mathscr{E}$. However, in this work I want to concentrate on other properties of systems which are less well-known and studied.

Consider another system

$$\frac{dx'}{dt'} = f'(x',u',t')$$
$$y' = g'(x',t') \quad. \tag{3.3}$$

It generates an exterior differential system $\mathscr{E}'$ on $M' = X' \times U' \times T'$. A *homomorphism* $M \to M'$ means that we give functions

$$\begin{pmatrix} x \\ u \\ t \end{pmatrix} \to \begin{pmatrix} t'(x,u,t) \\ x'(x,u,t) \\ y'(x,u,t) \end{pmatrix} \tag{3.4}$$

such that, for every solution

$$t \to x(t), \; u(t), \; y(t)$$

of (2.1), the following relations hold:

$$\frac{d}{dt}(x'(x(t),u(t),t)) = f'(x'(x(t),\ldots)\frac{dt'}{dt}$$

$$y'(x(t),\ldots) = g'(\cdots) \quad .$$

(3.5)

Among such transformations we find many special types which have been extensively studied (or at least considered) in the Systems Theory literature. For example,

a) State equivalence: x' is an invertible function $x'(x)$ of x alone. $y' = y$; $u' = u$; $t' = t$.

b) State feedback: $x' = x$, $u' = F(u,x)$, $t' = t$, $y' = y$.

and so on. Another popular condition is to suppose that both systems are linear and that the mapping (3.4) is also linear. This has recently been considered in a very interesting paper by Jan Willems [1]. In the time-independent case (which is really the only one which has been at all considered up to now) there are certain related algebraic invariants first considered in Systems Theory by Brunovsky [12] as a special case of the classical Kronecker theory of "pencils" of linear maps. Since this has been the case which has been most useful up to now (partly because it does appear so often, partly because it is really the only situation which is analytically tractable), I will now descend from the general Cartan point of view to consider the Brunovsky-Kronecker theory, which will lead into the work that Clyde Martin and I have done toward relating this to the algebro-geometric theory of *vector bundles*.

4. THE KRONECKER THEORY OF PENCILS OF LINEAR MAPS

Although the Kronecker theory is one of the highlights of 19th century algebra and invariant theory, it is very difficult to find complete expositions of it. The only one in recent times is in Gantmacher's magnificent *Theory of Matrices* [13], but even that becomes obscure at a key place. In [14] I translated Gantmacher's ideas into coordinate-free language, and (I believe) made it clearer, but did not fill in that expository gap. Here I want to present it in a different way, which is much more closely attuned to Systems Theory.

Let $V, W, \ldots$ be finite dimensional vector spaces over the real or complex numbers as field of scalars. (In fact, everything carries over to arbitrary scalar fields.) $L(V,W)$ denotes the vector space of linear maps

$$\alpha: V \to W$$

$L(V)$ denotes $L(V,V)$. $GL(V)$ denotes the group of $A \in L(V)$ such that A^{-1} exists. The product group $GL(W) \times GL(V)$ acts on $L(V,W)$:

EXTERIOR SYSTEMS

$$(g_1, g_2)(\alpha) = g_1 \alpha g_2^{-1} \tag{4.1}$$

for $g_1 \in GL(V)$, $g_2 \in GL(V)$.

The *orbits* of this group are readily described: They are the elements of $L(V,W)$ of a given rank and nullity.

dim (kernal α)

dim ($\alpha(V)$)

are the *invariants* of this action and label the orbits. Thus, we are dealing with a transformation group with only a finite number of orbits and with an algorithm to construct the invariants of the group action. It is a very difficult problem to do this for *general* transformation group actions--it is unknown for all but a few cases. (For example, it was a major labor of 19th century invariant theory to do this in *part* for algebraic actions of $SL(2,C)$, the group of 2×2 complex matrices of determinant one. Thus, invariant theory "died" before it could be extended to other groups, and to this day it has barely begun to revive.)

The Kronecker pencil theory fits in very well with this point of view. Consider $GL(W) \times GL(V)$ acting on $L(V,W) \times L(V,W)$ as follows:

$$(g_1, g_2)(\alpha_0, \alpha_1) = (g_1 \alpha_0 g_2^{-1}, g_1 \alpha_1 g_2^{-1})$$

for $g_1 \in GL(W)$, $g_2 \in GL(V)$, $\alpha_0, \alpha_1 \in L(V,W)$.

The Kronecker theory is an algorithm for computing orbits, invariants, and "canonical forms" for orbits for this particular transformation group action. This is one of the very rare "invariant theoretic" situations which can be described in this way. The theory of "quivers" due to Gabriel and Gelfand [15] offers one general insight into this!

Suppose $\alpha_0, \alpha_1 \in L(V,W)$. A pair $(V' \subset V, W' \subset W)$ of linear subspaces is said to reduce (α_0, α_1) if

$$\alpha_0, \alpha_1(V') \subset W' \quad .$$

Such a pair is said to be a *Kronecker reduction* if there is a map

$$\beta: V' \to W'$$

such that the following conditions are satisfied:

$$\alpha_1 = -\alpha_0 \beta \tag{4.2}$$

$$\beta \text{ is nilpotent}, \qquad (4.3)$$

i.e., $\beta^n = 0$ for n sufficiently large.

$$V' = \text{kernal}(\alpha_0) \oplus \beta(V')$$

Such a triple $(V',W'; \beta)$ is said to be an *elementary* Kronecker reduction if the following *additional* conditions are satisfied:

$$\dim(\text{kernel } \alpha_0) = 1 \qquad (4.4)$$

Here is the motivation for this definition in terms of the classical material. Consider the *pencil* of linear maps.

$$\alpha(s) = \alpha_0 + s\alpha_1 . \qquad (4.5)$$

(Thus, the term "pencil" refers to the fact that $\alpha(s)$ is a *linear* polynomial in s.) Consider vector-valued polynomials

$$s \to v_0 + v_1 s + \cdots + s^n v_n .$$

in

$$v_0, \ldots, v_n \in V$$

such that

$$\alpha(s)(v(s)) = 0 . \qquad (4.6)$$

<u>Theorem 4.1</u> (Kronecker). If $\alpha(s)$ is a singular pencil in the sense that

$$\text{kernel } \alpha(s) \neq 0$$

$$\text{for all } s ,$$

but

$$\text{kernel}(\alpha_0) \cap \text{kernel}(\alpha_1) = (0) ,$$

and if $v(s)$ is chosen to be the *minimal* degree polynomial such that (4.6) is satisfied, then the elements $v_0, \ldots, v_n$ are linearly independent. Further, if V' is a linear subspace of V' whose basis is $v_0, \ldots, v_n$, then

$$\alpha_0(V) = \alpha_1(V) = W' .$$

There are linear subspaces $V'' \subset V$, $W'' \subset W$ such that

EXTERIOR SYSTEMS 429

$$V = V' \oplus V''$$

$$W = W' \oplus W''$$

$$\alpha(s)(V'') \subset W'' \quad \text{for all} \quad s .$$

(4.7)

Thus, the process can be applied again to $\alpha(s)$ acting on V''.

In order to see where the nilpotent map β comes in, let us make (4.6) explicit:

$$\begin{aligned} 0 &= (\alpha_0 + s\alpha_1)(v_0 + sv_1 + \cdots + nv_n) \\ &= \alpha_0 v_0 + s\alpha_0 v_1 + \cdots + s^n \alpha_0 v_n + s\alpha_1 v_0 + s^2 \alpha_1 v_1 + \cdots \quad . \end{aligned}$$

Equating the coefficients of powers of s to zero gives the following relations:

$$\alpha_0(v_0) = 0 \tag{4.8}$$

$$\alpha_1 v_n = 0 \tag{4.9}$$

$$\alpha_0 v_1 + \alpha_1 v_0 = 0$$

$$\alpha_0 v_1 + \alpha_1 v_1 = 0$$

$$\vdots \tag{4.10}$$

$$\alpha_0 v_n + \alpha_1 v_{n-1} = 0$$

Let V' be the linear subspace of V spanned by the vectors $v_0, \ldots, v_n$. Define

$$\beta: V' \to V'$$

as follows:

$$\begin{aligned} \beta(v_0) &= v_1 \\ \beta(v_1) &= v_2 \\ &\vdots \\ \beta(v_n) &= 0 \end{aligned} \tag{4.11}$$

<u>Remark</u>. β is the map which operator-theorists call the *shift*. It plays an important underlying role in many systems-theoretic areas.

We can now rewrite relations (4.10) as:

$$\alpha_1 v_0 = -\alpha_0 \beta v_0$$

$$\alpha_1 v_1 = -\alpha_0 \beta v_1$$

$$\vdots \qquad (4.12)$$

$$\alpha_1 v_{n-1} = \alpha_0 \beta v_{n-1}$$

(4.9) means that

$$\alpha_1 v_n = \alpha_0 \beta v_n = 0$$

Thus, relations (4.9) - (4.10) can be summarized in the following relation:

$$\alpha_1 v' = \alpha_0 \beta v' \qquad (4.13)$$

for all $v' \in V'$.

Only relation (4.8), i.e., $v_0 \in \text{kernel } \alpha_0$ is not implied by this relation.

Notice that we can write the polynomial map $s \to v(s)$, which satisfies (4.6) in a more convenient basis-independent form:

$$v(s) = (1 - s\beta)^{-1}(v_0) \qquad . \qquad (4.14)$$

(This formula might be useful for generalization to infinite dimensional situations.)

We can now continue to split up V'' in this way, until there are no more polynomials with coefficients on V which annihilate $\alpha(s)$. Then, we work on the dual situation. Finally, we end up with a non-singular pencil. This process can be described in an overall way as follows.

<u>Theorem 4.2</u> (Kronecker). Let $\alpha_0, \alpha_1 : V \to W$ be a pair of linear maps. V and W can be split up into direct sums

$$V = V' \oplus V'' \oplus V'''$$

$$W = W' \oplus W'' \oplus W'''$$

such that:

$$\alpha_0, \alpha_1(V', V'', V''') \subset W', W'', W''' \quad ,$$

i.e., each of the pairs of linear spaces reduces the pairs (α_0, α_1) of linear maps. Further, there are *nilpotent* linear maps

$\beta': V' \to V'$

$\beta'': W'' \to W''$

such that the following conditions are satisfied:

$$\alpha_1 = \alpha_0 \beta' \quad \text{on} \quad V' \qquad (4.15)$$

$$\alpha_1 = \beta'' \alpha_0 \quad \text{on} \quad V'' \qquad (4.16)$$

$$V' = (\text{kernel } \alpha_0) \oplus \beta(V') \qquad (4.17)$$

$$\alpha_0(V') = W' \qquad (4.18)$$

$$W'' = \alpha_0(V'') + \text{kernel } \beta' \qquad (4.19)$$

For all but a finite number of s's, $\alpha_0 + s\alpha_1$ is an isomorphism $V''' \to W'''$, i.e., this component of the pencil is "regular". $\qquad (4.20)$

This marvelous theorem gives a complete description of the algebraic structure of pairs of linear maps. As I mentioned above, it is a very rare event in invariant theory when such a complete story is available. Luckily, many of the situations encountered in the theory of linear, time-independent one-dimensional input-output systems involve pairs of maps. We will now look at another, more geometric, way of analyzing the structure of pairs of linear maps.

5. THE VECTOR BUNDLE ASSOCIATED WITH PAIRS OF LINEAR MAPS

Continue with V, W as finite dimensional vector spaces, and with α_0, α_1 as a pair of linear maps $V \to W$. Suppose also that the complex numbers $\mathbb{C}$ are the field of scalars. $\mathbb{C}^2$ denotes the set of pairs (s_0, s_1) of complex numbers. $\mathbb{C}^{\#}$ denotes the multiplicative group of nonzero complex numbers. $\mathbb{C}^{\#}$ acts on $\mathbb{C}^{2,\#}$ as follows: ($\mathbb{C}^{2,\#}$ denotes the pairs $(s_0, s_1) \in \mathbb{C}^2$ with either s_0 or $s_1 \neq 0$.)

$$\lambda(s_0, s_1) = (\lambda s_0, \lambda s_1) \quad .$$

The orbit space $\mathbb{C}^{\#} \backslash \mathbb{C}^{2,\#}$ is $P_1(\mathbb{C})$, the one (complex) *dimensional projective space*. (Alternate names are the *projective line* and the *Riemann sphere*.)

For each $s = (s_0, s_1) \in \mathbb{C}^{2,\#}$, let

$$V(s) = \{v \in V : (s_0 \alpha_0 + s_1 \alpha_1)(v) = 0\}$$

Note that

$$V(\lambda s) = V(s)$$

for $\lambda \in \mathbb{C}^{\#}$.

Thus, the assignment

$$s \to V(s)$$

of a complex vector space to each $s \in \mathbb{C}^{2,\#}$ is constant on the orbits of $\mathbb{C}^{\#}$, hence, to each orbit $\pi(s) \in P_1(\mathbb{C})$ one can assign the vector space $V(s)$. This assignment defines a *complex vector bundle* E, [16,17,18] namely, the set of pairs $(\pi(s), V(s))$, $s \in \mathbb{C}^{2,\#}$. It is called the *kernel bundle* of the pair (α_0, α_1). It is *holomorphic* (even *algebraic*), i.e, defined by holomorphic functions. Of course, it may be *singular*, i.e., the dimension of the fibers may vary. If it is non-singular, a theorem of Grothendieck [19] gives the precise structure, a *direct sum of complex line bundles*. (A complex line bundle is a complex vector bundle with one-dimensional complex vector space fibers.)

For this special way of defining the bundle using the pair (α_0, α_1) of linear maps, another, purely algebraic, analysis of the bundle is available using the Kronecker theory outlined in previous sections. (Grothendieck's theorem uses "transcendental", non-algebraic tools.) For example, suppose the first part of the Kronecker reduction process succeeds in filling up V completely;

$$V'' = V''' = (0) \quad .$$

Then, the following conditions are satisfied:

$$\alpha_1 = \alpha_0 \beta$$

$$\text{kernel } \alpha_0 \oplus \beta(V) = V$$

with $\beta: V \to V$ a *nilpotent* map. For $s = (s_0, s_1) \in \mathbb{C}^{2,\#}$,

$$V(s) = \text{set of all } v \in V \text{ such that}$$

$$0 = (s_0 \alpha_0 + s_1 \alpha_0 \beta) v$$

$$= \alpha_0 (s_0 + s_1 \beta)(v) = 0 \quad ,$$

EXTERIOR SYSTEMS 433

or

$$(s_0 + s_1 \beta)(v) \in \text{kernel } \alpha_0 . \tag{5.1}$$

Now, if $s_0 \neq s$,

$$V(s) = \left\{ v \in V: \left(1 + \frac{s_1}{s_0} \beta\right)(v) \in \text{kernel } \alpha_0 \right\}$$

$$= \left(1 + \frac{s_1}{s_0} \beta\right)^{-1} (\text{kernel } \alpha_0)$$

For $s_0 = 0$,

$$V(s) = \{v \in V: \beta(v) \in \text{kernel } \alpha_0\}$$

$$= \text{, in view of (5.1),}$$

$$\{v \in V: \beta(v) = 0\}$$

$$= \text{kernel } \beta$$

Using (5.1) again to compute the dimension of kernel β:

$$\dim V = \dim \text{kernel}(\alpha_0) + \dim \beta(V)$$

$$= \dim \text{kernel}(\beta_0) + \dim V - \dim(\text{kernel } \beta) ,$$

or

$$\dim \text{kernel } \beta = \dim \text{kernel } \alpha_0 \tag{5.2}$$

This proves one of my main results.

<u>Theorem 5.1.</u> If $\alpha_0, \alpha_1: V \to W$ are linear maps, with only the "singular" Kronecker components present (i.e., the "cosingular" and "regular" components vanish), then the kernel bundle is non-singular.

We can use the nilpotent map β to exhibit the Grothendieck decomposition of the kernel bundle of (α_0, α_0) into line bundles. For $v_0 \in \text{kernel } \alpha_0$, $(s_0, s_1) \in \mathbb{C}^{2,\#}$, with $s_0 \neq 0$, set:

$$v_0(s_0, s_1) = \left(1 + \frac{s_1}{s_0} \beta\right)^{-1} (v_0) . \tag{5.3}$$

This defines a cross-section of E over the open subset

$P_1(\mathbb{C})$ - (the "point at infinity")

The one-dimensional linear singular space of $V(s_0, s_1)$ spanned by the value $v_0(s_0, s_1)$ of the cross-section can be continued to the point at infinity of $P_1(\mathbb{C})$, since

$$s_0^{n-1} v_0(s_0, s_1) = s_0^{n-1}\left(v_0 - \frac{s_1}{s_0} v_0 + \cdots + (-1)^{n-1}\left(\frac{s_1}{s_0}\right)^{n-1} \beta^{n-1} v_0\right)$$

$$\to (-1)^{n-1} s_1^{n-1} \beta^{n-1}(v_0)$$

$$\text{as } s_0 \to 0 \quad .$$

(n = least integer such that $\beta^n v_0 = 0$.) We obtain in this way a one dimensional sub-bundle of E. (It follows readily from this formula that n is the *Chern class* of this line bundle.) As v_0 varies over a basis of kernel α_0, we obtain in this way a "concrete" version of the Grothendieck decomposition.

6. THE KRONECKER-GROTHENDIECK INVARIANTS AND THE LIE-CARTAN EQUIVALENCE PROBLEM

It is trivial, but very instructive, to translate these algebraic-analytic results over to Cartan's formalism. Continue with V,W as finite dimensional complex vector spaces $\alpha_0, \alpha_1: V \to W$ linear maps. Set

$$\theta = \alpha_0 dv + \alpha_1 v\, dt \tag{6.1}$$

$$X = V \times R \quad .$$

θ is a vector-valued one-form on X. Its components generate an exterior differential system $\mathcal{E}$. The one-dimensional integral submanifolds are then essentially just the solution of the differential equations

$$\alpha_0 \frac{dv}{dt} + \alpha_1 v = 0 \quad . \tag{6.2}$$

We can then associate with $\mathcal{E}$ in a "natural" (i.e., "functional") way the pair (α_0, α_1) of linear maps. The Kronecker or Grothendieck theory sketched in previous sections then provides a description of "invariants" attached to $\mathcal{E}$. In certain situations (which I will go into at a later point), these "invariants" are enough to settle the general Lie-Cartan *equivalence problem* for $\mathcal{E}$, i.e., to decide when, given another system $\mathcal{E}'$ of the *same type*, there exists an equivalence ϕ between $\mathcal{E}$ and $\mathcal{E}'$.

Another interesting feature is that, to a certain extent, the "vector bundle" approach sketched in the previous section, which Martin and I

EXTERIOR SYSTEMS
435

introduced [16], is implicit in Cartan's work on exterior differential systems. He defines certain submanifolds of the Grassman bundles of X (i.e., the bundle whose fiber is the space of linear subspaces of the tangent bundle to X) as the "integral element". (Geometrically, they are the tangent spaces to the "integral submanifolds".) The vector bundles constructed in Section 5 are obviously related to these "integral element" bundles. Thus, I would suggest the study of the "algebro-differential geometric invariants" of the "integral element" bundles as a *new* way of describing *global* invariants of systems of differential equations, linear and nonlinear, ordinary and partial. This observation confirms my belief that there are many goodies for both pure and applied mathematics yet to be unearthed using Cartan's ideas on exterior differential systems.

7. THE HOLOMORPHIC VECTOR BUNDLE INVARIANTS OF LINEAR, TIME-INVARIANT INPUT SYSTEMS

Next we can return to material more familiar to systems theorists. Consider an input system

$$\frac{dx}{dt} = Ax + Bu \ . \tag{7.1}$$

Here (changing notation from that of previous sections so that it conforms with the more-or-less standard systems-theory notation),

$x \in X \equiv$ state space, a vector space

$u \in U \equiv$ input space

X, U are finite dimensional vector spaces; $A: X \to X$, $B: U \to X$ are linear maps. Set:

$V = X \times U$

$W = X \ .$

Write an element v of V as a partitioned column vector:

$$v = \begin{pmatrix} x \\ u \end{pmatrix} \ .$$

Define linear maps

$$\alpha_0, \alpha_1 : V \to W$$

as follows:

$$\alpha_0 \begin{pmatrix} x \\ u \end{pmatrix} = x$$

$$\alpha_1 \begin{pmatrix} x \\ u \end{pmatrix} = -(Au + Bu)$$

Thus Equations (7.1) take the following form:

$$\alpha_0 \frac{d}{dt} \begin{pmatrix} x \\ u \end{pmatrix} + \alpha_1 \begin{pmatrix} x \\ u \end{pmatrix} = 0 \qquad (7.2)$$

The pair (α_0, α_1) of linear maps then define a vector bundle E over $P_1(\mathbb{C})$, as described in previous sections.

For $s \in \mathbb{C}$,

$$E(s) = \left\{ \begin{pmatrix} x \\ u \end{pmatrix} : \alpha_0 \begin{pmatrix} x \\ u \end{pmatrix} + s\alpha_1 \begin{pmatrix} x \\ u \end{pmatrix} \right\}$$

$$= \left\{ \begin{pmatrix} x \\ u \end{pmatrix} : x - s(Ax + Bu) = 0 \right\}$$

$$= \left\{ \begin{pmatrix} x \\ u \end{pmatrix} : x = (sA - 1)^{-1} Bu \right\}$$

Thus, if $1/s$ is not an eigenvalue of A,

$$\dim E(s) = \dim U$$

$$E(\infty) = \left\{ \begin{pmatrix} x \\ u \end{pmatrix} : Ax + Bu = 0 \right\}$$

<u>Theorem 7.1.</u> If the input system (7.1) is controllable, then $s \to E(s)$ defines a non-singular vector bundle over $P_1(\mathbb{C})$, whose fibers have the same dimension as U.

This is proved in [16].

We can immediately use this to infer some important qualitative facts about controllable systems (7.1). The complex vector bundle is, following Grothendieck [19], a direct sum of line bundles. The degrees of these line bundles (i.e., their Chern classes evaluated on the generator of two-dimensional integral homology of $P_1(\mathbb{C})$) are then significant numbers for systems theory. (For example, they would remain invariant under both *complex analytic* deformations and *feedback transformations* of the systems (7.1).) Martin and I show that they are, in fact, the numbers in the Brunovsky canonical form. Their sum is then the first Chern class of the vector bundle

E, evaluated on the generator of $H_2(P_2(\mathbb{C}),Z)$. It is the *Macmillan degree*, i.e., the dimension of X. In particular, this shows that the Macmillan degree attached to the transfer function

$$T = C(s-A)^{-1}B$$

of a controllable, observable system is a *topological* invariant.

What purpose does this serve? Notice that we have taken the standard algebraic structure of linear systems and interpreted it *geometrically* in such a way that is much more amenable to dealing with more complicated systems. In [21] Martin and I have presented certain preliminary calculations which lead us to believe that the material does carry over to certain infinite-dimensional and time-varying systems. One obstacle to such applications is that the information we would need about the classification of holomorphic vector bundles on more complicated complex manifolds than $P_1(\mathbb{C})$ is not yet available from the work of the pure mathematicians, although it is a topic of intense development among algebraic geometers. There are also probably many more significant applied problems whose solution is dependent on information about the structure of holomorphic vector bundles. For example, the classification of holomorphic vector bundles on $P_3(\mathbb{C})$ has played a key role in the Atiyah-Singer-Ward theory of Yang-Mills "instantons", objects which appear naturally in *elementary particle physics*. C. Byrnes has constructed analogous bundles on $Z \times P_1(\mathbb{C})$, with Z a topological space, as useful gadgets for the study of the *delay systems*. In later sections I will show how certain N-dimensional systems (e.g., the Maxwell and Helmholtz equation) lead in a natural way to such vector bundles on higher dimensional complex manifolds.

There is also a possibility of an intermediate classification of systems by classifying the *structure groups* of their associated vector bundles. The structure group starts off as $GL(m,\mathbb{C})$, where m = dimension of input space. The Grothendieck theorem [19] says it can be reduced to the subgroup of diagonal matrices. However, it might be already given "by Nature" in such a way that the structure group is another subgroup of $GL(m,\mathbb{C})$. It appears that the systems occurring in circuit theory, analytical mechanics, etc., all have their typical and characteristic structure groups; in turn, this structure group can be related to properties of the Hanckel matrix of the system. For all of these reasons, I believe that people interested in applications should, in this case, relax their natural skepticism (which I share) about fancy mathematics and consider seriously the possibility that the theory of vector bundles gives a valuable way of describing unified properties of systems. Of course, again I should invoke Elie Cartan--he is certainly Godfather to the modern theory of fiber bundles. Future historians might well say he had the golden touch in *both* pure and applied mathematics, and that virtually every topic he treated was used in a significant way by succeeding generations. This is certainly a major reason for putting him in the same class with Gauss, Riemann and Poincaré. Another moral from this is that one of the talents required of an applied mathematician today is the ability to keep abreast of developments in pure mathematics, and choose those which best match the needs of the applied areas in which he is interested, *independent of his a priori prejudices about which part of mathematics he considers more applicable than another*. Of course, I do not claim that *every time* the more abstract and general will turn out to be the most applicable--indeed, I agree with the common sense feeling of most scientists and engineers that *usually* the more concrete the better. However, the exceptions to this common sense rule of thumb are extremely important, and indeed have the unique capability of completely changing the way we think about applied problems.

8. FEEDBACK EQUIVALENCE AND THE KRONECKER THEORY OF EQUIVALENCE OF PAIRS OF LINEAR MAPS

The work that Martin and I have done can be usefully interpreted in the language of category-functor theory. We attach to the "category" of (linear, time-invariant) input systems essentially *two* "functors". One is *geometric* with the category of holomorphic vector bundles, the other is *algebraic*, with the category of pairs of linear maps. The reader who is familiar with such things will recognize that this is a typical situation in such avante-garde branches of mathematics as algebraic topology and geometry. In these areas it has been found that it is precisely in *reconciling* the "geometric" and "algebraic" world view that some of the most significant and useful relations appear. Now, from the engineer's point of view, the natural "isomorphisms" of systems are the *feedback transformations*. The natural isomorphisms of pairs of linear maps

$$(\alpha_0, \alpha_1) \in L(V,W) \times L(V,W)$$

are the action of the group $GL(V) \times GL(W)$. The fact that they are the *same thing* is very worthwhile proving explicitly, because it plays such a fundamental role. (If it had been recognized earlier, say in 1970, it would have simplified the task of the authors of many papers.)

Now, it is obvious that feedback equivalence of systems implies equivalence under $GL(V) \times GL(W)$ of the corresponding pairs of linear maps. We must deal with the converse: Suppose then that

$$\frac{dx}{dt} = Ax + Bu \qquad (8.1)$$

$$\frac{dx'}{dt} = A'x + B'u \qquad (8.2)$$

EXTERIOR SYSTEMS 439

are two linear systems. Set

$$V = X \times U = \text{set of pairs } \begin{pmatrix} x \\ u \end{pmatrix}$$

$$W = X$$

$$\alpha_0 \begin{pmatrix} x \\ u \end{pmatrix} = x$$

$$\alpha_1 \begin{pmatrix} x \\ u \end{pmatrix} = -Ax - Bu$$

$$\alpha_0' \begin{pmatrix} x \\ u \end{pmatrix} = x$$

$$\alpha_1' \begin{pmatrix} x \\ u \end{pmatrix} = -A'x - B'u \quad .$$

(8.3)

Let us then suppose that (α_0, α_1) lies in the same $GL(V) \times GL(W)$ orbit as (α_0', α_1'), i.e.,

$$\alpha_0 = g_1 \alpha_0 g_2^{-1}$$

$$\alpha_1' = g_1 \alpha_1 g_2^{-1} \quad .$$

(8.4)

$$g_1 \in GL(X)$$

$$g_2 \in GL(X \oplus U) \quad .$$

Then,

$$\alpha_0 g_2 \begin{pmatrix} x \\ u \end{pmatrix} = g_1 \alpha_0 \begin{pmatrix} x \\ u \end{pmatrix} = g_1 x \quad . \tag{8.5}$$

Suppose g_1 is written in partitioned form:

$$g_2 \begin{pmatrix} x \\ u \end{pmatrix} = \begin{pmatrix} g_{2,11} & g_{2,12} \\ g_{2,21} & g_{2,22} \end{pmatrix} \begin{pmatrix} x \\ u \end{pmatrix}$$

$$= \begin{pmatrix} g_{2,11} x + g_{2,12} u \\ g_{2,21} x + g_{2,22} u \end{pmatrix}$$

(8.6)

$$g_{2,11} \in L(X,X); \quad g_{2,12} \in L(X,U);$$

$$g_{2,21} \in L(U,X); \quad g_{2,22} \in L(U,U) \quad .$$

Compare (8.5) and (8.6):

$$g_{2,11} x + g_{2,12} u = g_1 x \quad ,$$

hence

$$\boxed{g_1 = g_{2,11}} \tag{8.7}$$

$$\boxed{g_{2,12} = 0} \tag{8.8}$$

Plug (8.7) and (8.8) back into (8.4):

$$\alpha_1' g_2 \binom{x}{u} = \alpha_1 g_1 \alpha_1 \binom{x}{u}$$

or

$$\alpha_1' \begin{pmatrix} g_1 x \\ g_{2,21} x + g_{2,22} u \end{pmatrix} = -g_1 (Ax + Bu)$$

or

$$A' g_1 x + B' (g_{2,21} x + g_{2,22} u) = g_1 (Ax + Bu)$$

or

$$\boxed{g_1^{-1} A' g_1 = g_1^{-1} B g_{2,21} + A} \tag{8.9}$$

$$\boxed{B' g_{2,22} = g_1 B u} \tag{8.10}$$

These formulas can be summarized in the following way.

<u>Theorem 8.1</u>. The two input systems (8.1) lead to $GL(X) \times GL(X \oplus U)$ - equivalent pairs of maps if and only if the systems differ by *state feedback* and by *change of basis in input and state space*.

This result has two sorts of ramifications in Systems Theory. First, it enables the *state feedback classes* to be classified by the *Kronecker pencil theory*. In fact, it was Brunovsky [12] who did this in a way independent of the general Kronecker theory. Second, it indicates a major algebraic difference between *state feedback* on the one hand and output feedback and various sorts of "compensators" on the other hand. If the latter sort define a group (which is not always clear), usually the orbits of these groups on systems are *not* equivalent to the problem Kronecker solved, i.e., equivalence of pairs of linear maps under isomorphisms of domain and range spaces. As I have mentioned already, problems of ennumerating orbits are extremely difficult in our present state of knowledge of invariant theory if they do not reduce in one form or another to the Kronecker problem, or are closely related to it. It is then a lucky accident (?) that *state feedback* is governed by the Kronecker invariant theory.

9. DEFORMATION OF EXTERIOR DIFFERENTIAL SYSTEMS, SINGULAR PERTURBATION THEORY AND SIMPLIFICATION OF COMPLICATED SYSTEMS

In classical mathematics, we often find a study of systems (e.g., differential equations) which depend on "parameters". Of course, the physical or engineering situation from which these systems derive often provide such parameters (e.g., the velocity of light, Planck's constant, Reynold's number), and it is often of great interest to study the limiting values of these parameters. (As the velocity of light goes to infinity, Einsteinian mechanics goes over to Newtonian; as Planck's constant goes to zero, quantum mechanics goes over to classical; a variation of the Reynolds number is supposed to have something to do with the onset of turbulence, and so on.)

However, in the classical work, this process was not thought about systematically. One of the triumphs of post World-War II mathematics is that such a *systematic* theory has developed. I am familiar with it mainly through the work of Spencer and Kodaira (Spencer was my thesis advisor, and a very inspiring one) on the "deformation" of various sorts of *geometric* structures. Gerstenhaber, Nijenhuis and Richardson created a parallel (and equally beautiful) theory of deformation of *algebraic* structures. This theory has vast potential for application to a wide variety of applied disciplines; most of these possibilities are still untouched.

"Singular perturbation theory" is a significant topic among the bag of tools that are usually thought of as "applied mathematics". Nayfeh's book [22] is an excellent description of this area. I pointed out in [23] that there are extremely close links between "deformation theory" of *exterior differential systems* and the applied mathematician's "singular perturbation theory". Now, systems engineers use "singular perturbation theory" for two purposes:

a) To "simplify" complicated systems; for example, an engineer designing a system might "throw away" parts which he thought would not significantly effect certain performance criteria.

b) To distinguish between parts of complicated systems which operate on different time-scales. Biologists and chemists also use this idea.

M. Hazewinckel [24] has introduced an interesting mathematical motivation into this study--make the space of "systems" somehow into a *compact* topological space, so that the process of "singular perturbation" involves going from one chart of this space to another, much as the "hyperplane at infinity" is added to "affine space" to make "projective space", which is *compact*. I do not want to go into any detail here, but I do want to briefly indicate how the point of view of "deformation of exterior differential systems" subsumes Hazewinckel's framework.

Consider an input system

$$\frac{dx}{dt} = Ax + Bu \quad . \tag{9.1}$$

Convert it into an exterior differential system $\mathscr{E}(A,B)$, generated by the vector-valued one-forms

$$\omega(A,B) = dx - (Ax + Bu)dt \quad . \tag{9.2}$$

As A and B vary (and they are, of course, themselves parameterized by real numbers), one obtains the "deformation theory". For example, A might depend on the parameter λ in the following way:

$$A = A_0 + A_1 \lambda^{-1} \quad ,$$

where A_0, A_1 are linear maps $X \to X$. Let $E: X \to X$ be a linear map such that

$$EA_1 = 0 \quad .$$

Then,

$$E\omega(A,B) = Edx - (EA_0 x + EBu)dt \quad .$$

Also,

$$\lambda\omega(A,B) = \lambda dx - (\lambda A_0 x + A_1 x + \lambda Bu)dt \quad .$$

The limiting exterior differential system $\mathcal{E}_0$ (as $\lambda \to 0$) consists of the forms in $\mathcal{E}(\lambda)$ which have non-singular limits. In this case, they include

$$Edx - (EA_0 x + EBu) dt$$

$$A_1 x dt .$$

The integral manifold of this limiting system then involves going outside of the class of input systems. Perhaps there might be some *constraints* involved among inputs or outputs, changes of time-scale, etc.

It is actually easier to keep track of what is happening in terms of general differential geometry. Let M be a manifold and let $\mathcal{E}$ be an exterior differential system. Suppose we are interested in the integral manifolds of dimension n. Let

$$G^n(M)$$

be the *Grassmann bundle*, i.e., a point of $G^n(M)$ is a pair

$$(p, \gamma)$$

consisting of a point $p \in M$ and an n-dimensional linear subspace γ of the tangent space Mp. The map

$$(p, \gamma) \to p$$

defines $G^n(M)$ as a *fiber bundle* over M. The fibers are *compact*, the Grassmann manifolds of n-dimensional linear subspaces of R^n.

$\mathcal{E}$ defines a subspace $G^n(\mathcal{E}) \subset G^n(M)$, the n-*dimensional integral elements*, consisting of the n-dimensional linear subspaces of M which are tangent to n-dimensional integral manifolds of M. Let us suppose for simplicity that $G^n(\mathcal{E})$ is a *submanifold* of $G^n(M)$.

Now, let $\mathcal{E}$ vary, depending, say, on a parameter λ. $\lambda \to G^n(\mathcal{E}_\lambda)$ then varies with λ. In particular, the fibers over a fixed point $p \in M$ vary with λ, and are compact subsets of a compact space. There is a "natural metric" for such subsets, the *Hausdorff metric*. This gives us the technical means to discuss meaningfully the "continuous dependence of exterior differential systems on parameters". Thus, we would say that $\mathcal{E}_\lambda \to \mathcal{E}$ as $\lambda \to 0$, if, for each $p \in M$ the subsets of integral elements at p, $G^n(\mathcal{E}_\lambda)(p)$ converges, in the Hausdorff metric, to the set $G^n(\mathcal{E})(p)$ of n-dimensional integral elements of $\mathcal{E}$ at p.

Having defined appropriately a "continuity" notion for exterior differential systems, the heart of the matter is to analyze how the *integral submanifolds* vary. Some local results can readily be obtained using Cartan's methods. For the case $n = 1$ (which is that of most interest of "lumped parameter" systems) quite a bit can be done using the techniques of the theory of ordinary differential equations. (This is obviously just a geometric version of the material studied in the literature, e.g., in Nayfeh's book [22] on singular perturbation theory.) In any case, I suspect that this is an important field of research, a typical contribution of a *geometric* insight to the theory of differential equations. It is very curious to me why the "experts" in differential equations have not assimilated Cartan's methods, which after all, were completely developed by 1900, and give beautiful insights into the theory of differential equations and its applications. Astonishingly, I have seen them many times put down research into this area as "soft" and a mere "reformulation", which is not worthy of their lofty standards.

10. FINAL REMARKS

I hope I have given some solid evidence for my belief that ideas of "modern" differential and algebraic geometry are very significant for systems-theorétic problems. Of course, it seems to be a fundamental principle of applied mathematics that it is not a matter of straightforwardly applying Theorem A,B,C,... to applied problems $\alpha,\beta,\gamma,...$ There are always complex and subtle interconnections and feedback between "pure" and "applied". Further, systems theory itself is a vast and amorphous area, which can absorb almost any area of mathematics. The trick is to do it in such a way that it facilitates the understanding of the general concepts involved in applied science (and systems theory is, or should be, a sort of meta-engineering) and suggests more *powerful* techniques for handling them. After all, one of the main features of modern mathematics (in contrast to classical) is the development of a more powerful technique. I believe that the geometric approach is one of the few which can provide both benefits.

Bibliography

1. J. Willems, Topological Classification and Structural Stability of Linear Systems, preprint, Gronesgen University, 1977.

2. R. Hermann, E. Cartan's Geometric Theory of Partial Differential Equations", *Advances in Math.* 1 (1965), 265-317.

3. R. Hermann, *Geometry, Physics and Systems*, Marcel Dekker, 1973.

4. N.H. Ibragimov and R.L. Anderson, *Dolk. Akad. Nauk. SSSR* 227 (1970).

5. R. Gardner, Differential Geometric Viewpoints on the Development of Shock Waves, *Proceedings of the 1976 Ames Research Center (NASA) Conference on the Geometric Theory of Nonlinear Waves*, R. Hermann (ed.) "Lie Groups: History, Frontiers and Applications", Vol. 6, Math Sci Press, 53 Jordan Rd, Brookline, MA 1977.

6. R. Gardner, A Differential Geometric Approach to Characteristics, in *Proc. Symp. on Pure Math.* 18, AMS, Providence, R.I., 1970.

7. R. Gardner, A Differential Geometric Generalization of Characteristics, *Comm. Pure. Appl. Math.* 22 (1969), 597-626.

8. R. Hermann, *Lie Groups in Quantum Mechanics*, Addison-Wesley, Reading, MA 1970.

9. H.D. Wahlquist and F.B. Estabrook, The Prolongation Structures of Nonlinear Evolution Equations, *J. Math. Phys.* 16 (1975), 1-7.

10. R. Hermann, On the Accessibility Problem of Control Theory, *Proc. Symp. on Differential Equations*, A.J. Lasalle and S. Lefschetz (eds.), Academic Press, 1967.

11. R. Hermann and A. Krener, Nonlinear Controllability and Observability, *IEEE Trans. Aut. Cont.* AC-22 (1977), 728-740.

12. P. Brunovsky, A Classification of Linear Controllable Systems, *Kybenetika* 3 (1970), 173-187.

13. F. Gantmacher, *Matrix Theory*, Chelsea, New York, 1964.

14. R. Hermann, "Interdisciplinary Mathematics", Vol. 9, Math Sci Press, Brookline, MA, 1975.

15. M. Hazewinkel, Representation of Quivers and Moduli of Linear Dynamical Systems, *1976 Ames Research Center (NASA) Conf. on Geometric Control Theory*, C. Martin and R. Hermann (eds), "Lie Groups: History, Frontiers and Applications", Vol. 7, Math Sci Press, Brookline, MA, 1977.

16. C. Martin and R. Hermann, Topological and Holomorphic Invariants of Linear Systems (to appear, SIAM J. Cont.); as preprints: "Applications of Algebraic Geometry to Systems Theory, Parts 3 and 4".

17. D. Husemoller, *Fiber Bundles*, Springer-Verlag

18. J. Milnor and J. Stasheff, *Characteristic Classes*, Princeton Univ. Press, Princeton, NJ,

19. A. Grothendieck, *Am. J. Math.* 79 (1957), 121-138

20. R. Hermann, Remarks on the Geometry of Systems, *Proc. Nato Institute on Geometric and Algebraic Methods for Nonlinear Systems*, R. Brockett and D. Mayne (eds), D. Reidel, 1973.

21. R. Hermann and C. Martin, Applications of Algebraic Geometry to Systems Theory, Parts 5 and 6, in *Proc. 1976 Ames Research Center Conf. on Geometric Control Theory*, Math Sci Press, Brookline, MA, 1977.

22. A. Nayfeh, *Perturbation Theory*, Wiley, 1973.

23. R. Hermann, *The Geometric Theory of Differential Equations, Bäcklund Transformations, and Solitons*, Parts A and B ("Interdisciplinary Mathematics", Vols. 12 and 14), Math Sci Press, Brookline, MA 02146, 1976.

24. M. Hazewinkel, Degenerating Families of Linear Dynamical Systems, I, *Proceedings of 1977 IEEE Conf. on Decision and Control.*

SOME GENERAL REMARKS ABOUT GROUPS,
DIFFERENTIAL FORMS AND
THEIR GENERALIZATIONS

1. INTRODUCTION

"Modern" differential geometry has had its most striking successes when combined with the theory of Lie groups and transformations groups. This material often also has important physical ramifications. Having generalized the basic geometric notions (which physicists call "superspace") it is then natural to extend the group-theoretic ideas also. In fact, the basic work in this direction has been done by Berezin and Kac [1]; in this paper I will mainly expound and develop it further. One interesting possibility might be to work directly in a "quantum" geometric framework, generalizing the notion of Lie group even beyond that provided by Berezin and Kac [1].

2. SOME GENERAL IDEAS

A "group" is a well-known and studied mathematical entity. A key idea in modern "categorical" mathematics is to look for the "dual" of each mathematical notion. Now, it seems that the dual notion of "groups" is "Hopf algebras". (They are really "co-algebras".) In Section 7 this chapter I will briefly develop more of this material in order to serve as motivation for later developments.

First, here is some useful "generalized nonsense". Let S be a *finite* set. Let $F(S)$ denote the space of real-valued functions on S. Since such functions can be added and multiplied point-wise, they form (commutative, associative) aglebras.

Let S_1, S_2 be two finite sets, with $F(S_1), F(S_2)$ the corresponding algebra of functions. Let $S_1 \times S_2$ be the Cartesian product of the sets, and let $F(S_1 \times S_2)$ be the algebra of functions on this Cartesian product.

Consider $F(S_1)$, $F(S_2)$ as finite dimensional (with dimension equal to the number of elements in the set) vector spaces. Let

$$F(S_1) \otimes F(S_2)$$

denote their tensor product, as vector spaces. Construct a linear map

$$\phi: F(S_1) \otimes F(S_2) \to F(S_1 \times S_2)$$

by means of the following formula:

$$\phi(f_1 \otimes f_2)(s_1, s_2) = f_1(s_1) f_2(s_2) \qquad (2.1)$$

It is readily verified that ϕ is an isomorphism in the vector space sense. Further, it preserves the natural algebra structure: $F(S_1) \otimes F(S_2)$ is the tensor product of commutative associative algebras, (i.e., $(f_1 \otimes f_2)(f_1' \times f_2') = f_1 f_1' \otimes f_2 f_2'$) $F(S_1 \times S_2)$ an algebra via point-wise multiplication. Thus, we can essentially *identify* $F(S_1 \times S_2)$ with $F(S_1) \otimes F(S_2)$. (In the case where S_1 and S_2 are infinite sets--which is actually the case of most interest and usefulness--a "weak" version of this holds, typically with S_1, S_2 endowed with further geometric structure, e.g., as manifolds, vector spaces, etc.)

Here is one further important concept:

<u>Definition</u>. The differential form algebra $(\mathcal{D}, d, \alpha, \wedge)$ is said to be α-*commutative* if the following conditions are satisfied:

$$\omega_1^{\pm} \wedge \omega_2^{\mp} = \omega_2 \wedge \omega_1 \qquad (2.2)$$

$$\omega_1^{-} \wedge \omega_2^{-} = -\omega_2^{-} \wedge \omega_1^{-} \qquad (2.3)$$

$$\text{for} \quad \omega_1^+, \omega_2^+ \in \mathcal{D}^+ \; ; \quad \omega_1^-, \omega_2^- \in \mathcal{D}^- \; .$$

$\mathcal{D}^+$ and $\mathcal{D}^-$ are the eigenvectors with eigenvalues $+1$ and -1 where $\alpha: \mathcal{D} \to \mathcal{D}$ is a linear map such that $\alpha^2 = $ identity.

<u>Examples</u>. a) $\mathcal{D}$ = <u>differential form on a manifold</u> M.

$\wedge, d$ are usual exterior multiplication and derivative . Set:

$$\alpha(\omega) = (-1)^r \omega$$

if ω is a differential form of degree r.

b) $\mathcal{D}$ = <u>differential forms based on the "Boson" algebra</u>.

Let $\mathcal{A}$ be the free commutative associative algebra in the generators (x^i), $1 \leq i, j \leq n$. ($\mathcal{A}$ is then the algebra of "polynomials" in the "indeterminates" $x^1, \ldots, x^n$.) Introduce the "differentials", satisfying the following commutation relations:

$$x^i dx^j = dx^j x^i$$

$$dx^i \wedge dx^j = -dx^j \wedge dx^i \; .$$

GROUPS, DIFFERENTIAL FORMS AND THEIR GENERALIZATIONS   449

$\mathcal{D}$ is the space of linear combinations of objects of the form:

$$\omega = a dx^{i_1} \wedge \cdots \wedge dx^{i_r} ,$$

with $a \in \mathcal{A}$. d and $\wedge$ are defined as usual. $\alpha(\omega) = (-1)^r \omega$. In other words, $\mathcal{D}$ is the algebra of differential forms on R^n with *polynomial* coefficients. This ties "bosons" down to a specific geometric context.

c) $\mathcal{D}$ = differential forms based on the "Fermion" algebra.

Let $\mathcal{A}$ be the free associative algebra generated by (y^i), $1 \leq i,j \leq n$, with the following commutation relation

$$y^i y^j = -y^j y^i .$$

Introduce their "differentials", subject to the following commutation relations:

$$y^u dy^j = dy^j y^i$$

$$dy^i \wedge dy^j = dy^j \wedge dy^i$$

Again, elements of $\mathcal{D}$ are linear combinations of elements of the form:

$$a dy^1 \wedge \cdots \wedge dy^r .$$

The map α is now determined by the following rules:

$$\alpha(a) = (-1)^s a$$

if $a \in \mathcal{A}$ is a polynomial of degree s in the $y^i 1$

$$\alpha(dy^i) = dy^i .$$

Thus, $\mathcal{D}$ is really "higrated", namely, by the degree of a and the integer r. The "geometry" based on $\mathcal{D}$ is the "fermion" differential geometry   that has played the key role in the attempt   to extend quantum field-elementary particle ideas to "superspace" situations.

3. GRADED LIE ALGEBRAS DEFINED BY EXTERIOR DERIVATIVE RELATIONS IN DIFFERENTIAL FORM ALGEBRAS

Now, suppose $(\mathcal{D},d,\alpha,\wedge)$ is a differential form algebra which is α-commutative. Choose indices $1 \leq i,j,k \leq n$, and the summation convention on these indices (with a slight modification indicated below).

Let (ω^i) be a set of elements of $\mathcal{D}$, with *constants* (c^i_{jk}) such that:

$$d\omega^i = c^i_{jk} \omega^j \wedge \omega^k \tag{3.1}$$

Suppose that:

$$\alpha(\omega^i) = (-1)^{\alpha(i)} \omega^i \tag{3.2}$$

with $\alpha(i) = 0$ or 1 (*no summations* in this context!). Suppose also that:

$$d(\omega^i \wedge \omega^j) = d\omega^i \wedge \omega^j + \alpha(\omega^i) \wedge d\omega^j \quad . \tag{3.3}$$

Remark. So far, we have imposed no particular identity of this type.

Now, we have:

$$\omega^i \wedge \omega^j = (-1)^{\alpha(i)\alpha(j)} \omega^j \wedge \omega^i \quad . \tag{3.4}$$

Hence, we can suppose without loss in generality that

$$c^i_{jk} = (-1)^{\alpha(j)\alpha(k)} c^i_{kj} \quad . \tag{3.5}$$

Now, apply d to both sides of (3.1):

$$0 = c^i_{jk} d\omega^j \wedge \omega^k + c^i_{jk} \alpha(\omega^j) \wedge d\omega^k$$

$$= c^i_{jk} d\omega^j \wedge \omega^k + (-1)^{\alpha(j)} c^i_{jk} \omega^j \wedge d\omega^k$$

$$= c^i_{jk} d\omega^j \wedge \omega^k + (-1)^{\alpha(j)}(-1)^{\alpha(j)(\alpha(k)+1)} c^i_{jk} d\omega^k \wedge \omega^j$$

$$= c^i_{jk} d\omega^j \wedge \omega^k + (-1)^{\alpha(j)\alpha(k)} c^i_{jk} d\omega^k \wedge \omega^j$$

$$= 2 c^i_{jk} d\omega^j \wedge \omega^k$$

$$= \text{, using (3.1),}$$

$$2 c^i_{jk} c^j_{\ell m} \omega^\ell \wedge \omega^m \wedge \omega^k \quad . \tag{3.6}$$

GROUPS, DIFFERENTIAL FORMS AND THEIR GENERALIZATIONS 451

Let us suppose the elements $\omega^\ell \wedge \omega^m \wedge \omega^k$ are (aside from the symmetries (3.4)) linearly independent. (3.6) then implies a set of quadratic relations between c^i_{jk} composed of three terms

$$c^i_{jk} c^j_{\ell m} + (-1)^{\alpha(k)\alpha(m)} c^i_{jm} c^j_{\ell k} + (-1)^{\alpha(\ell)\alpha(k)} c^i_{j\ell} c^j_{km} = 0 \quad (3.7)$$

This is a generalization of the Jacobi identity of ordinary Lie algebra theory. In fact, it is just the "Jacobi identity" for Z_2-graded Lie algebras.

<u>Theorem 3.1</u>. Let L be an n-dimensional vector space with a basis consisting of elements

$$A_i \, .$$

Impose the following algebra structure on L:

$$[A_i, A_j] = c^k_{ij} A_k \, . \quad (3.8)$$

Set:

L^+ = subspace spanned by A_i with $\alpha(i) = 1$

L^- = subspace spanned by A_i with $\alpha(i) = 0$

This then, together with (3.8), defines a Z_2 graded Lie algebra.

Let us now continue the study of the geometric context of graded Lie algebras.

4. DERIVATIONS AND TWO-DERIVATIONS OF DIFFERENTIAL ALGEBRAS

Let $(\mathcal{D}, d, \alpha, \wedge)$ be a differential form algebra as before. As for every vector space, let

$$L(\mathcal{D})$$

denote the vector space of *linear* maps $\mathcal{D} \to \mathcal{D}$. $L(\mathcal{D})$ has two algebraic operations—one *commutative*, the other *skew-commutative*—the anti-commutator and the commutator:

$$(A_1, A_2) \to A_1 A_2 \pm A_2 A_1 = [A_1, A_2]_\pm$$

Let α act on $L(\mathscr{D})$ as follows:

$$A \to \alpha A \alpha$$

$$L^+(\mathscr{D}) = \{A: \alpha A \alpha = A\}$$

$$L^-(\mathscr{D}) = \{A: \alpha A \alpha = -A\}$$

Then,

$$L^+(\mathscr{D})(\mathscr{D}^\pm) \subset \mathscr{D}^\pm$$

$$L^-(\mathscr{D})(\mathscr{D}^\pm) \subset \mathscr{D}^\mp$$

$$\alpha [A_1, A_2]_\pm \alpha = [\alpha A_1 \alpha, \alpha A_2 \alpha]_\pm$$

Hence,

$$[L^\pm(\mathscr{D}), L^+(\mathscr{D})]_\pm \subset L^-(\mathscr{D}) \supset [L^+(\mathscr{D}), L^\pm(\mathscr{D})]_\pm$$

$$[L^-(\mathscr{D}), L^-(\mathscr{D})]_\pm \subset L^+(\mathscr{D})$$

<u>Definition</u>. $A \in L(\mathscr{D})$ is a *derivation* of $\mathscr{D}$ if

$$A(\omega_1 \wedge \omega_2) = A(\omega_1) \wedge \omega_2 + \omega_1 \wedge A(\omega_2) \tag{4.1}$$

for $\omega_1, \omega_2 \in \mathscr{D}$

$A \in \mathscr{L}(\mathscr{D})$ is an α-*derivation* if:

$$A(\omega_1 \wedge \omega_2) = A(\omega_1) \wedge \omega_2 + \alpha(\omega_1) \wedge A(\omega_2) \tag{4.2}$$

for $\omega_1, \omega_2 \in \mathscr{D}$

$$\alpha A = -A\alpha, \quad \text{i.e.,} \quad A \in L^-(\mathscr{D}) \quad . \tag{4.3}$$

Denote the derivations by $D(\mathscr{D})$, the α-derivations by $D(\alpha, \mathscr{D})$. It is well-known that

$$[D(\mathscr{D}), D(\mathscr{D})]_- \subset D(\mathscr{D}) \quad . \tag{4.4}$$

i.e., the derivations form a Lie algebra.

Theorem 4.1.

$$[D(\alpha,\mathcal{D}),D(\alpha,\mathcal{D})]_+ \subset D(\mathcal{D}) ,\qquad (4.5)$$

i.e., the subspace of $L(\mathcal{D})$ consisting of the sum of the derivations and the α-derivations forms a graded Lie algebra.

Proof. Suppose $A_1, A_2 \in D(\alpha,\mathcal{D})$.

$$A_1 A_2(\omega_1 \wedge \omega_2) = A_1(A_2\omega_1 \wedge \omega_2 + \alpha(\omega_1) \wedge A_2\omega_2)$$

$$= A_1 A_2 \omega_1 \wedge \omega_2 + \alpha A_2\omega_1 \wedge A_1\omega_2 + A_1\alpha\omega_1 \wedge A_2\omega_2$$

$$+ \alpha^2 \omega_1 \wedge A_1 A_2$$

$$= , \text{ using (4.4)},$$

$$A_1 A_2 \omega_1 \wedge \omega_2 - A_2 \alpha\omega_1 \wedge A_1\omega_1 + A_1\alpha\omega_1 \wedge A_2\omega_2$$

$$+ \omega_1 \wedge A_1 \omega_2$$

Hence,

$$(A_1 A_2 + A_2 A_1)(\omega_1 \wedge \omega_2) = [A_1,A_2]_+(\omega_1) \wedge \omega_2 + \omega_1 \wedge [A_1,A_2](\omega_2) ,$$

which proves (4.5).

5. DIFFERENTIAL FORM ALGEBRAS WITH GENERATING SUBALGEBRAS. "GEOMETRIC" ACTION OF GRADED LIE ALGEBRAS

Let $(\mathcal{D},d,\wedge,\alpha)$ continue as a differential form algebra. Suppose also that the algebra structure $\wedge$ has unit element 1. A subalgebra $\mathcal{A}$ is a *generating subalgebra* if it contains the unit element and if every element of $\mathcal{D}$ can be written as a linear combination of elements of the form

$$a \wedge da_1 \wedge \cdots \wedge da_r ,$$

with $a, a_1, \ldots, a_r \in \mathcal{A}$, and if

$$d(a_1 a_2) = (da_1)a_2 + a_1 da_2$$

for $a_1, a_2 \in \mathcal{A}$

For consistency with the notation of differential geometry, denote the algebra operation with an $\mathcal{A}$ by simple juxtaposition:

$$(a_1, a_2) \to a_1 a_2 \quad,$$

and denote the product between a $\in \mathcal{A}$ and $\omega \in \mathcal{D}$ also as juxtaposition, i.e., $a\omega, \omega a$ in place of $a \wedge \omega$, $\omega \wedge a$.

We now denote such a structure as

$$(\mathcal{D}, \mathcal{A}, d, \wedge, \alpha) \quad.$$

<u>Definition</u>. Let $GL = GL^+ + GL^-$ be a Z_2-graded Lie algebra. A *geometric action* of GL on $(\mathcal{D}, \mathcal{A}, d, \wedge, \alpha)$ is defined as a linear map

$$\rho: GL \to L(\mathcal{D})$$

such that

$$\rho(GL^+) \subset \text{derivations of } \mathcal{D} \tag{5.1}$$

$$\rho(GL)^- \subset \text{-derivations of } \mathcal{D} \tag{5.2}$$

$$\rho(GL)(\mathcal{A}) \subset \mathcal{A} \tag{5.3}$$

<u>Remark</u>. The differential geometric prototype of all of this is where:

M = manifold

$\mathcal{D}$ = differential form on M

$\mathcal{A}$ = C^∞ functions on M, i.e., 0-forms

GL = a "true" Lie algebra, i.e., $GL^- = 0$.

$\rho(GL)$ then consists of a "vector field" on M in the geometric sense. This may, of course, be thought of as an "infinitesimal" version of a Lie group action on M. Since these actions really play the distinguished role in all parts of mathematics and physics where "geometry" seems to be the key unifying principle, we see that we have available powerful heuristic principles for extending this work to "superspace" situations. Physically, this should, of course, be useful in understanding the geometric foundation of the theory of "fermions".

6. MAURER-CARTAN FORMS

As a Lie group, the "Maurer-Cartan forms" are any basis of the left-invariant one-forms. They satisfy

$$d\omega^i = c^i_{jk} \omega^j \wedge \omega^k \quad , \tag{6.1}$$

where (c^i_{jk}) are the structure constants of the Lie algebra. Let us extend this terminology a bit, and say that any set of one-forms on a manifold M which is a basis for one-forms and satisfies (6.1) for *some* set of constants c^i_{jk} is a *set of Maurer-Cartan forms*.

Let us now turn to the analogous notions for "generalized manifolds". Let $(\mathcal{D}, \mathcal{A}, d, \alpha, \wedge)$ be a differential form algebra, with generating sub-algebra $\mathcal{A}$ as defined in Section 5. We shall suppose that it is α-*commutative*, as defined earlier.

Let us define the *one-forms* of $\mathcal{D}$, denoted by $\mathcal{D}^1$, as those elements that can be written as linear combinations of elements of the form

$$(da_1)a \quad , \tag{6.2}$$

with $a, a_1 \in \mathcal{A}_1$.

Remark. Because of the postulated α-commutativity, elements of the form $a(da_1)a_2$ can be written as sums of elements of the form (6.2). The form (6.2) has been chosen (instead of ada_1, which might seem more natural, for reasons which will be evident soon.

Definition. A set (ω^i), $1 \leq i, j \leq n$, of elements of $\mathcal{D}^1$ is said to form a *basis* if each element $\omega \in \mathcal{D}^1$ can be written as

$$\omega = \omega^i a_i \quad , \tag{6.3}$$

with components $(a_i) \in \mathcal{A}$ uniquely determined by ω. (In other words, (ω^i) is a basis for $\mathcal{D}^1$ as an $\mathcal{A}$-module.)

Let us suppose that such a basis is given. Let us also suppose that

$$\alpha(\omega^i) = (-1)^{\alpha(i)} \omega^i \quad , \tag{6.4}$$

where $\alpha(i) = 0$ or 1. Let A_i be the map $\mathcal{A} \to \mathcal{A}$ such that:

$$da = \omega^i A_i(a) \tag{6.5}$$

Theorem 6.1. Each A_i defines a derivation or α-derivation of $\mathscr{A}$ and $\mathscr{D}$, depending on whether $\alpha(i) = 0$ or 1.

Proof. For $a, a' \in \mathscr{A}$,

$$d(aa') = (da)a' + ada'$$
$$= \omega^i A_i(a) a' + a\omega^i A_i(a') \qquad (6.6)$$

Now,

$$a = a^+ + a^-,$$

where $\alpha(a^+) = a^+$, $\alpha(a^-) = -a^-$. The postulated α-commutativity of $\mathscr{D}$ implies that

$$a^+ \omega^i = \omega^i a^+$$
$$a^- \omega^i = (-1)^{\alpha(i)} \omega^i a^-.$$

Hence,

$$a\omega^i = \omega^i a^+ + (-1)^{\alpha(i)} \omega^i a^-$$

$$= \begin{cases} \omega^i a & \text{if } \alpha(i) = 0 \\ \omega^i \alpha(a) & \text{if } \alpha(i) = 1 \end{cases}$$

Hence, (6.6) takes the following form:

$$d(aa') = \omega^i A_i(a) a' + \omega^i \begin{pmatrix} a \\ \text{or} \\ \alpha(a) \end{pmatrix} A_i(a')$$

Hence:

$$\boxed{\begin{aligned} A_i(aa') &= A_i(a)a' + aA_i(a') \\ \text{if } \alpha(i) &= 0 \end{aligned}} \qquad (6.7)$$

$$\boxed{\begin{aligned} A_i(aa') &= A_i(a)a' + \alpha(a)A_i(a') \\ \text{if } \alpha(i) &= 1 \end{aligned}} \qquad (6.8)$$

This shows that A_i is either a derivation or α-derivation of $\mathscr{A}$. It can now be extended to a derivation or α-derivation of $\mathscr{D}$:

$$A_i(ada_1 \wedge \cdots \wedge da_r) = A_i(a) da_1 \wedge \cdots \wedge da_r$$
$$+ adA_i(a_1) \wedge \cdots \wedge da_r + \cdots \quad (6.9)$$

if $\alpha(i) = 0$.

If $\alpha(i) = 1$:

$$A_i(ada_1 \wedge \cdots \wedge da_r) = A_i(a) da_1 \wedge \cdots \wedge da_r$$
$$+ \alpha(a) A_i(da_1 \wedge \cdots \wedge da_r) \quad (6.10)$$

$$A_i(da_1 \wedge \cdots \wedge da_r) = dA_i(a_1) \wedge \cdots \wedge da_r$$
$$+ \alpha(da_1) \wedge A_i(da_2 \wedge \cdots \wedge da_r) \quad (6.11)$$

etc.

Remark. In case (ω^i) are the left invariant Maurer-Cartan forms on a Lie group G, the A_i are the *left invariant vector fields on the Lie group* G.

Let us suppose now that (ω^i) define Maurer-Cartan forms for the differential form algebra $\mathcal{D}$. We have defined $A_i \in L(\mathcal{D})$, and show that they are either derivations or α-derivations. Let us now compute their commutators and anti-commutators. Now,

$$da = \omega^i A_i(a) \quad,$$

hence,

$$0 = d^2 a$$
$$= d(\omega^i A_i(a))$$
$$= d\omega^i A_i(a) - \omega^i \wedge dA_i(a)$$
$$= c^i_{jk} \omega^j \wedge \omega^k A_i(a) - \omega^i \wedge \omega^j A_j A_i(a)$$
$$= c^i_{jk} \omega^j \wedge \omega^k A_i(a) - \tfrac{1}{2}(\omega^i \wedge \omega^j A_j A_i(a) + \omega^j \wedge \omega^i A_i A_j(a))$$
$$= c^i_{jk} \omega^j \wedge \omega^k A_i(a) - \tfrac{1}{2}(\omega^i \wedge \omega^j A_k A_i(a) + (-1)^{\alpha(i)\alpha(j)} \omega^i \wedge \omega^j A_i A_j(a))$$
$$= c^i_{jk} \omega^j \wedge \omega^k A_i(a) - \tfrac{1}{2} \omega^i \wedge \omega^j (A_j A_i(a) + (-1)^{\alpha(i)\alpha(j)} A_i A_j(a))$$

Hence,

$$2c^i_{jk} A_i(a) = A_i A_i(a) + (-1)^{\alpha(i)\alpha(j)} A_i A_j(a) \quad .$$

This is clearly the identity which says that:

> The derivation and α-derivations A_i form a graded Lie algebra whose structure constants are the c^i_{jk}.

7. TAYLOR'S FORMULA AND THE HOPF ALGEBRA STRUCTURE FOR A LIE GROUP

Now, we turn our attention to the material in the fundamental article by Berezin and Kac [1]. The aim is to show how the usual sort of Lie group can be described in terms of its algebra of C^∞ functions, then use this description to generalize the notion of Lie group.

Recall that a Lie group G is a space with: a) a manifold structure and b) a group structure, such that the group operations are differentiable (say, C^∞) with respect to the manifold structure. Then, we should be able to recapture the group structure in terms of the properties of the algebra of C^∞, real-valued functions in G. To do this, let

$$\gamma: G \times G \to G$$

be the group multiplication map, i.e.,

$$\gamma(g_1, g_2) = g_1 g_2$$

Its dual is a map

$$\gamma^*: F(G) \to F(G \times G) \quad .$$

The projection maps

$$\pi_1: G \times G \to G \,, \quad \pi_1(g_1, g_2) = g_1$$

$$\pi_2: G \times G \to G \,, \quad \pi_2(g_1, g_2) = g_2 \,,$$

define maps

$$\pi_1^*: F(G) \to F(G \times G)$$

$$\pi_2^*: F(G) \to F(G \times G) \quad .$$

GROUPS, DIFFERENTIAL FORMS AND THEIR GENERALIZATIONS   459

Consider also the tensor product of these maps:

$$\pi_1^* \otimes \pi_2^*: F(G) \otimes F(G) \to F(G \times G) \quad .$$

Suppose that $\pi_1^* \otimes \pi_2^*$ is an isomorphism. (To make it such, one in fact has to "complete" the tensor product.) Then, we would obtain a map

$$(\pi_1^* \otimes \pi_2^*)^{-1} \gamma^*: F(G) \to F(G) \otimes F(G)$$

This is the "Hopf algebra". (It is really a "co-algebra".) The "Hopf algebra" of a Lie group has a special structure. The basic idea of the Berezin-Kac paper [1] is to analyze this appropriately, then to reinterpret it, replacing the algebra $F(G)$ by a more general algebra, and obtaining a "generalized Lie group". This chapter is basically just a reinterpretation of their work--I want to emphasize more the role played by the Maurer-Cartan forms, and use this insight to generalize the idea of a "Lie group".

Begin with the classical Taylor's formula of calculus, for functions of one real variable, $x \to f(x)$. Write it as follows:

$$f(x+y) = f(x) + f'(x)y + \frac{1}{2} f''(x) y^2 + \cdots \quad . \tag{7.1}$$

Regard the real numbers as a Lie group. Then the left hand side of (7.1) really involves the group-theoretic structure, i.e., essentially the "Hopf algebra" mentioned above.

Now, take a Lie group G. Suppose it has a coordinate system (x^i), $1 \le i,j \le n$ such that

$$x^i(1) = 0 \quad .$$

Let $f \in F(G)$ be a real-valued, C^∞ function on G. Then, there is (locally) an expansion of the following form:

$$f(g_1 g_2) = f(g_1) + f_i(g_1) x^i(g_2) + f_{ij}(g_1) x^i(g_2) x^j(g_2) + \cdots \tag{7.2}$$

The coefficient functions $f_i, f_{ij}, \ldots$ depend linearly on an f. Write them as follows:

$$f_i = A_i(f), \qquad f_{ij} = A_{ij}(f), \ldots$$

(It will be demonstrated below that the A_i, $A_{ij}, \ldots$ are differential operators.) Thus, we have:

$$f(g_1 g_2) = f(g_1) + A_i(f)(g_1) x^i(g_2) + A_{ij}(f)(g_1) x^i(g_2) x^j(g_1) + \cdots$$
$$\tag{7.3}$$

Theorem 7.1. The operators A_i are left invariant vector fields on the Lie group G. Further, they form a basis for the left invariant vector fields on G.

Proof. Let us first prove that they are vector fields, i.e., that:

$$A_i(f_1 f_2) = A_i(f_1)f_2 + f_1 A_i(f_1) \tag{7.4}$$

for $f_1, f_2 \in F(G)$.

To prove (7.4), apply (7.3) with $f_1 f_2$ replacing f:

$$(f_1 f_2)(g_1 g_2) = f_1(g) f_2(g) + A_i(f_1 f_2)(g_1) x^i(g_2) + \cdots$$

$$= \text{, also,}$$

$$f_1(g_1 g_2) f_2(g_1 g_2)$$

$$= (f_1(g_1) + A_i(f_1)(g_1) x^i(g_2) + \cdots)(f_2(g_2) + A_i(f_2) x^i(g_2) + \cdots)$$

$$= f_1(g_1) f_2(g_2) + (A_i(f_1)(g_1) f_2(g_1) + f_1(g_1) A_i(f_2)) x^i(g_2)$$
$$+ \cdots$$

whence (7.4).

Let L_{g_0} be the operation of left translation by $g_0 \in G$. Then, L_{g_0} is a linear map $F(G) \to F(G)$, defined as follows:

$$L_{g_0}(f)(g) = f(g_0 g) \tag{7.5}$$

Thus,

$$L_{g_0}(f)(g_1 g_2) = f(g_0 g_1) + A_i(L_{g_0}(f))(g_1) x^i(g_2) + \cdots$$

$$= f(g_0 g_1 g_2)$$

$$= f(g_0 g_1) + A_i(f)(g_0 g_1) x^i(g_2) + \cdots$$

whence,

$$L_{g_0}(A_i(f))(g) = A_i(f)(g_0 g_1)$$

$$= A_i(L_{g_0}(f))(g_1) .$$

whence:

GROUPS, DIFFERENTIAL FORMS AND THEIR GENERALIZATIONS   461

$$L_{g_0}(A_i(f)) = A_i(L_{g_0}(f)) \tag{7.6}$$

which (since (7.6) holds for *all* $g_0 \in G$) *means* that A_i is left invariant.

To finish the proof, we must show that the A_i are linearly independent. Since they are left-invariant, it suffices to show that they are linearly independent at the identity element of G, which we denote by "1". Now, apply the defining identity for $x^i = f$, $g_1 = 1$; we have:

$$x^i(g_2) = A_j(x^i)(e) x^j(g_2) + \cdots$$

i.e.,

$$A_j(x^i)(j) = \delta^i_j , \tag{7.7}$$

which shows that the A_j are linearly independent.

Remark. This argument enables us to identify the A_i more precisely; they are the unique set of left invariant vector fields on the Lie group G such that

$$A_i(1) = \frac{\partial}{\partial x^i}(1) . \tag{7.8}$$

Theorem 7.2. The operators $A_{ij\ldots}$ are differential operators, determined as follows:

$$A_{ijk\ldots} = A_i A_j A_k \cdots + \text{(lower order differential equations)} \tag{7.9}$$

Proof. The repeat of the above argument shows that these operators are invariant under left translation by G, hence that they are determined by their value at the identity element. Now,

$$f(g) = f(1) + A_i(f)(1) x^i(g) + A_{ij}(f)(1) x^i(g) x^j(g) + \cdots \tag{7.10}$$

Let us substitute into this relation the polynomials in the x^i. The identities obtained in this way show that $A_{ij}, A_{jk}, \ldots$ agree at the identity element with linear combinations of the partial derivative operators

$$\frac{\partial}{\partial x^i} , \frac{\partial}{\partial x^i} \frac{\partial}{\partial x^j} , \ldots \quad .$$

This shows that the $A_{ij}, \ldots$ are differential operators on all of G. Since they are left-invariant, they are *over all of* G linear combinations of products of the A_i. That it takes the form (7.9) is obvious from its values on the polynomial in the $\partial/\partial x^i$.

We can also see that the $A_{ij}, A_{ijk}, \ldots$ are differential operators using the more abstract, coordinate-free definition given in [7]. This is done by calculating their commutators with multiplication by a function, and showing (inductively) that they are lower order differential operators. Start again with the defining relation:

$$f(g_1 g_2) = f(g_1) + A_i(f)(g_1) x^i(g_2) + \cdots \qquad (7.11)$$

Substitute $f_1 f_2$ for f:

$$(f_1 f_2)(g_1 g_2) = f_1(g_1) g_2(g_2) + A_i(f_1 f_2)(g_1) x^i(g_2)$$
$$+ A_{ij}(f_1 f_2)(g_1) x^i x^j(g_2) + \cdots$$

$$= (f_1(g_1) + A_i(f_1)(g_1) x^i(g_2) + A_{ij}(f_1)(g_1) x^i(g_2) x^2(g_2) + \cdots$$

$$(f_2(g_2) + A_i(f_2)(g_1) x^i(g_2) + \cdots)$$

i.e.,

$$A_i(f_1 f_2) = A_i(f_1) f_2 + f_1 A_i(f_2) \qquad (7.12)$$

$$A_{ij}(f_1 f_2) = A_{ij}(f_1) f_2 + A_i(f_1) A_j(f_2) + A_j(f_1) A_i(f_2) \qquad (7.13)$$
$$+ f_1 A_{ij}(f_2)$$

and so forth.

For $f \in F(G)$ let m_f be the operation $F(G) \to F(G)$ of multiplication by the function f. Then, (7.12) takes the following form:

$$A_i(m_{f_1}(f_2)) = m_{A_i(f_1)}(f_2) + m_{f_1}(A_i(f_2)) \quad,$$

or

$$[A_i, m_{f_1}] = m_{A_i(f_1)} \quad, \qquad (7.14)$$

i.e., the commutator of A_i with multiplication by a function is a zero-th order differential operator. Hence, A_i itself is a first order differential operator, which we know already, of course.

Let us now analyze the higher order operators in a similar way. Rewrite (7.13) as:

$$A_{ij}(m_{f_1}(f_2)) = m_{A_{ij}(f_1)}(f_2) + m_{A_i(f_1)}(A_j(f_2))$$

$$+ m_{A_j(f_1)}(A_i(f_2)) + m_{f_1}A_{ij}(f_2) ,$$

or

$$A_{ij}m_{f_1} = m_{A_{ij}(f_1)} + m_{A_i(f_1)}A_j + m_{A_j(f_1)}A_i + m_{f_1}A_{ij} ,$$

or

$$[A_{ij}, m_{f_1}] = m_{A_{ij}(f_1)} + m_{A_i(f_1)}A_j + m_{A_j(f_1)}A_i \qquad (7.15)$$

This shows that the commutator of A_{ij} with multiplication by a function is first order differential operator, hence that A_{ij} *is a second order differential operator*. The argument can be continued to show that $A_{i_1...i_r}$ is an r-th order differential operator. The beauty of this argument is that it is remarkably algebraic and coordinate free!

We can now generalize this argument to more general algebras than $F(G)$. This will enable us (following Berezin and Kac [1]) to generalize the notion of a "Lie group".

8. GENERALIZED LIE GROUPS BASED ON AN ASSOCIATE ALGEBRA, FOLLOWING BEREZIN AND KAC

Let $\mathcal{A}$ be an associative algebra with a unit element "1". Let

$$(x^i) , \quad 1 \leq i,j \leq n$$

be a set of elements that generate $\mathcal{A}$. (Notice that we are not assuming $\mathcal{A}$ to be commutative.) Define a mapping

$$\gamma: \mathcal{A} \to \mathcal{A} \otimes \mathcal{A}$$

via the following formula:

$$\gamma(a) = a \otimes 1 + A_i(a) \otimes x^i + A_{ij}(a) \otimes x^i x^j + \cdots \qquad (8.1)$$

for $a \in \mathcal{A}$

where $A_i, A_{ij}, \ldots$ are linear maps: $\mathscr{A} \to \mathscr{A}$. (Of course, (8.1) must be understood in the more obvious "formal power series" sense. Alternately, one can complete $\mathscr{A} \otimes \mathscr{A}$ in an appropriate way, topologize this completion, and regard (8.1) as a rigorous statement about infinite series. At this stage, we are only interested in heuristics.)

Let $\gamma \otimes 1$, $1 \otimes \gamma$ be the maps

$$\mathscr{A} \otimes \mathscr{A} \to \mathscr{A} \otimes \mathscr{A} \otimes \mathscr{A}$$

defined as follows:

$$(\gamma \otimes 1)(a_1 \otimes a_2) = \gamma(a_1) \otimes a_2 \qquad (8.2)$$

$$(1 \otimes \gamma)(a_1 \otimes a_2) = a_1 \otimes \gamma(a_2) \qquad (8.3)$$

Definition. The map γ is said to be *co-associative* if

$$(\gamma \otimes 1)\gamma = (1 \otimes \gamma)\gamma \quad . \qquad (8.4)$$

Remark. The motivation for this definition comes from the fact [1,6] that, in case $\mathscr{A} = F(G)$, where G is a Lie group, (8.4) is equivalent to the associative law for the group.

Let us also suppose that the following condition is satisfied:

$$\gamma \text{ is an algebraic homomorphism} \qquad (8.5)$$

($\mathscr{A} \otimes \mathscr{A}$ is made into an algebra following the usual rule for associative algebras.)

Let us now work out the consequences of (8.4) and (8.5) with a γ of the form (8.1). First, (8.5):

$$\begin{aligned}
\gamma(a_1 a_2) &= a_1 a_2 \otimes 1 + A_i(a_1 a_2) \otimes x^i + \cdots \\
&= (a_1 \otimes 1 + A_i(a_1) \otimes x^i + \cdots)(a_2 \otimes 1 + A_i(a_2) \otimes x^i + \cdots) \\
&= a_1 a_2 \otimes 1 + A_i(a_1) a_2 \otimes x^i + a_1 A_i(a_2) \otimes x^i \\
&\quad + A_{ij}(a_1) a_2 \otimes x^i x^j + a_1 A_{ij}(a_2) \otimes x^i x^j \qquad (8.6) \\
&\quad + A_i(a_1) A_j(a_2) \otimes x^i x^j + \cdots
\end{aligned}$$

GROUPS, DIFFERENTIAL FORMS AND THEIR GENERALIZATIONS   465

We have made no assumptions about the generators x^i of $\mathcal{A}$. Such assumptions together with (8.6) determine certain relations among the $A_i, A_{ij}, \ldots$. They are generalizations of *derivations*, and indicate that the $A_i, A_{ij}, \ldots$ are some sort of *generalized differential operators*. Thus, we have hit on an interesting new idea--a method of generalizing the notion of "differential operator" to cover associative, but non-commutative algebras. (This should be useful and important in quantum mechanics, for example.)

For example, comparing the first order terms on both sides of (8.6), it seems that A_i might satisfy:

$$A_i(a_1 a_2) = A_i(a_2) a_2 + a_1 A_i(a_1)$$

i.e., that A_i is a *derivation* of $\mathcal{A}$. Now, in "fermion" geometry, we want the A_i to be "anti-derivations". This can be accomplished (as we shall see below) by *choosing a different way of defining the algebra structure for $\mathcal{A} \otimes \mathcal{A}$*.

Let us look at the co-associative law, (8.4)

$$\begin{aligned}
(\gamma \otimes 1)(\gamma)(a) &= (\gamma \otimes 1)(a \otimes 1 + A_i(a) \otimes x^i + \cdots) \\
&= \gamma(a) \otimes 1 + \gamma(A_i(a)) \otimes x^i + \cdots \quad (1 \otimes \gamma)(\gamma)(a) \\
&= (1 \otimes \gamma)(a \otimes 1 + A_i(a) \otimes x^i + \cdots) \\
&= a \otimes 1 \otimes 1 + A_i(a) \otimes \gamma(x^i) + \cdots
\end{aligned}$$

Hence

$$(a \otimes 1 + A_j(a) \otimes x^j + \cdots) \otimes 1 + (A_i(a) \otimes 1 + A_j(A_i(a)) \otimes x^j + \cdots) \otimes x^i$$

$$= a \otimes 1 \otimes 1 + A_i(a) \otimes (x^i \otimes 1 + A_j(x^i) \otimes x^j + \cdots)$$

$$+ A_{ij}(a) \otimes (x^i x^j \otimes 1 + A_k(x^i x^j) \otimes x^k + \cdots) + \cdots$$

Notice that the zero-th order terms in x^i are *automatically* equal. (This is an important check on the internal consistency!) Let us *equate* the first order terms, as an Ansatz. (They do not *necessarily* have to be equal; this depends on the relations that the algebra satisfies.) To do this, suppose that:

$$A_j(x^i) = c_j^i c_{jk}^i x^k + c_{jk\ell}^i x^k x^\ell + \cdots \qquad (8.7)$$

This gives:

$$c^i_j = \delta^i_j. \tag{8.8}$$

Notice that with (8.8) satisfied, the first order terms on the x's on both sides of (8.5) *do* cancel.

Now, let us equate the terms in $x^j \otimes x^i$. To do this, we must make an assumption about $A_k(x^i x^j)$. Let us assume it is of the following form:

$$A_k(x^i x^j) = A_k(x^i) x^j \pm x^i A_k(x^j). \tag{8.9}$$

Combining this with (8.7), (8.8) gives

$$A_k(x^i x^j) = \delta^i_k x^j \pm \delta^j_k x^i + c^i_{k\ell} x^\ell x^j \pm x^i c^j_{k\ell} x^\ell + \cdots \tag{8.10}$$

Equating terms in $x^j \otimes x^i$ in (8.6) then gives:

$$A_j A_i(a) \otimes x^j \otimes x^i = A_i(a) \otimes c^i_{jk} x^k \otimes x^j + A_{ij}(a) \otimes (\delta^i_k x^j \pm \delta^j_k x^i) \otimes x^k$$

Hence,

$$A_k A_\ell = A_i c^i_{\ell k} + A_{ik} \delta^i_\ell \pm A_{kj} \delta^j_\ell$$

or

$$A_{\ell k} \pm A_{k\ell} = A_k A_\ell - A_i c^i_{\ell k}. \tag{8.11}$$

Notice now that the terms of the form $(\ ) \otimes x^i x^j \otimes 1$ *automatically* cancel, which again is a check on the internal consistency.

This takes care of terms up to second order in the x's. We can continue in this way with the terms of third order. They obviously depend on the commutation relations being satisfied by the generators (x^i) of the algebra $\mathcal{A}$. We of course expect (on the basis of the analogy with Lie group theory) that these relations for the c^i_{jk} will be *like the Jacobi identity* of Lie algebra theory. (Recall that, in fact, the Jacobi identity for a Lie algebra is, in fact, the "infinitesimal" version of the associative law of multiplication of a Lie group.)

Exercise. Let us say that (x^i) is a *canonical coordinate system* for the Lie group G if the one parameter subgroups $t \to g(t)$ of G are *straight lines* in the coordinates x^i, i.e., if

$$\frac{d^2}{dt^2}(x^i(g(t))) = 0$$

Calculate the first few terms $A_{ij}, A_{ijk}, \ldots$ in terms of the generators of the enveloping algebra $A_i A_j, A_i A_j A_k, \ldots$, in such a canonical coordinate system.

9. THE TAYLOR SERIES FOR TRANSFORMATION GROUP ACTIONS

We can now profitably generalize the formulas to the case of a transformation group action. Let G be a Lie group, M a manifold, and let

$$(g,p) \to gp$$

be a transformation group action of G on M.

Let (x^i) be a coordinate system on G such that $x^i(1) = 0$. For $f \in F(M)$, Taylor's formula gives:

$$f(gp) = f(p) + A_i(f)(p)x^i(g) + A_{ij}(f)(p)x^i(g)x^j(g) + \cdots \quad (9.1)$$

with $A_i, A_j, \ldots$ linear maps: $F(M) \to F(M)$.

Let $\underset{\sim}{G}$ denote the Lie algebra of G. For $\underset{\sim}{g} \in \underset{\sim}{G}$, let

$$t \to \exp(t\underset{\sim}{g})$$

be the one-parameter subgroup of G generated by $\underset{\sim}{g}$. Suppose that the (x^i) are a *canonical coordinate system* for G, i.e.

$$\frac{d^2}{dt^2} x^i(\exp(t\underset{\sim}{g})) = 0 \quad (9.2)$$

for all $\underset{\sim}{g} \in \underset{\sim}{G}$.

Set:

$$a^i(\underset{\sim}{g}) = \frac{d}{dt} x^i(\exp(t\underset{\sim}{g}))\Big|_{t=0} \quad (9.3)$$

Then, we see that:

$$a^i(\underset{\sim}{g}) A_i(f)(p) = \frac{\partial}{\partial t} f(\exp(t\underset{\sim}{g}))\Big|_{t=0} \quad (9.4)$$

This formula shows that the operators A_i are the *infinitesimal generators of the action of* G, i.e., *the vector fields* on the manifold M resulting from the action of the Lie algebra on M.

Similarly,

$$\frac{d^2}{dt^2} f(\exp(tg)p)\Big|_{t=0} = 2A_{ij}(f)(p)a^i(g)a^j(g) \quad . \tag{9.5}$$

This formula (and those generalizing it, replacing d^2/dt^2 by d^r/dt^r) shows how the operators $A_i, A_{ij}, \ldots$ are determined by the transformation group action of G on M.

10. THE CO-ALGEBRAZATION OF TRANSFORMATION SEMI-GROUP ACTIONS, FOLLOWING BEREZIN AND KAC

The primitive concept of Lie's theory is that of *transformation semi-group action*. It can be defined, abstractly, as the following object:

A pair (G,M) of spaces, plus mappings
$\alpha_1: G \times G \to G$, $\alpha_2: G \times M \to M$ such that

$$\alpha_2(g_1, \alpha_2(g_2, p)) = \alpha_2(\alpha_1(g_1, g_2), p) \tag{10.1}$$

for $g_1, g_2 \in G$; $p \in M$

Let us suppose, for simplicity, that G, M are finite sets. Let

$F(G), F(M)$

denote the algebra (under pointwise addition and multiplication) of real-valued functions on G and M. Then $F(G \times G)$ can be "naturally" (i.e., in an appropriate "functional" way) be identified with

$F(G) \otimes F(G) \quad .$

Similarly, $F(G \times M)$ is identified with

$F(G) \otimes F(M) \quad .$

This is done as follows:

For $f_1, f_2 \in F(G)$

$$(f_1 \otimes f_2)(g_1, g_2) = f_1(g_1) f_2(g_2) \quad . \tag{10.2}$$

Bibliography

1. F. Berezin and G.I. Kac, Mat. Sb. USSR $\underline{82}$, 124 (1970); English translation 1970 (11), 311.

2. L. Corwin, Y. Ne'eman and S. Sternberg, Reviews of Modern Physics $\underline{47}$ (1975).

3. R. Hermann, *Quantum and Fermion Differential Geometry, Part A*, Math Sci Press, Brookline, MA 02146, 1977.

SOME FURTHER COMMENTS ABOUT DIFFERENTIAL FORMS OVER ASSOCIATIVE ALGEBRA

1. INTRODUCTION

Implicit in this development of differential geometry over arbitrary Grassmann algebras is the possibility of generalization of virtually every idea of traditional differential geometry, Lie theory, the calculus of variations, etc. Carried over to physics, this has the promise of providing new insights into the nature of elementary particles.

Now, I believe that the simplest and most direct route into these new areas is the use of Elie Cartan's methods. The characteristic feature of Cartan's work (as opposed, say, to classical tensor analysis) is the systematic use of "differential forms" and "moving frames of differential forms". Thus, any new situation where one can define a generalized sort of differential forms should provide a happy hunting ground for the Cartan ideas.

In this chapter I am preparing the way for applications to *a generalized calculus of variations formalism for quantum fields*.

2. DIFFERENTIAL FORMS OVER ARBITRARY ASSOCIATIVE ALGEBRAS

In Volume 16 I have briefly described how differential forms over Grassmann algebras may be defined in direct generalization of the method used by Cartan. Now, I want to show how it can be generalized.

Let $\mathcal{A}$ be an arbitrary associative algebra, say over the real numbers, as field of scalars. Suppose it has a unit element "1". We want to define objects labelled by an integer r, that we denote by

$$\mathcal{D}^r(\mathcal{A}) \;,$$

and call the r-*th degree differential forms*. As usual we will begin with $r = 1$ and 2, derive their algebraic rules, and attempt to define the three basic operations of "exterior multiplication", "exterior derivative", and "Lie derivative". In turn, differential-geometric ideas can be built up on the foundation of these three operations.

First,

$$\mathcal{D}^0(\mathcal{A}) = \mathcal{A} \;.$$

$\mathcal{D}^1(\mathcal{A})$ consists of the linear considerations of objects of the form

$$A_1(dA_2)A_3 \qquad (2.1)$$

for $A_1, A_2, A_3 \in \mathscr{A}$, with certain relations and algebraic rules of operation prescribed for these objects. The first relation is provided by the *derivation rule:*

$$d(A_1 A_2) = (dA_1)A_2 + A_1 dA_2 \quad . \qquad (2.2)$$

The next relation is provided by derivative of the associative law

$$A_1(A_2 A_3) = (A_1 A_2)A_3 \quad ,$$

namely:

$$dA_1(A_2 A_3) + A_1(dA_2 A_3 + A_2 dA_3) = ((dA_1)A_2 + A_1(dA_2))A_3 + (A_1 A_2)(dA_3) \qquad (2.3)$$

Since we want dA_1, dA_2, dA_3 to be independent, we would equate terms on both sides of (2.3) that involve them separately, i.e.,

$$(dA_1)(A_2 A_3) = ((dA_1)A_2)A_3$$

$$A_1(dA_2 A_3) = (A_1 dA_2)A_3$$

$$A_1(A_2 dA_3) = (A_1 A_2)dA_3$$

Similarly, there would be relations obtained by differentiating the commutativity properties (if any) of the algebra. For example, if $\mathscr{A}$ is commutative, i.e.,

$$A_1 A_2 = A_2 A_1,$$

we would require that

$$dA_1 A_2 + A_1 dA_2 = dA_2 A_1 + A_2 dA_1 \quad . \qquad (2.4)$$

Since one would like the dA_1, dA_2 to be independent, one would *require* the following relations,

$$A_1 dA_2 = (dA_2)A_1 \quad , \qquad (2.5)$$

which, of course, imply (2.4).

Forget about commutativity for the moment and proceed to second degree differential forms. One would define them as linear combinations of objects of the form

$$A_1(dA_2)A_3 \wedge (dA_4)A_5 \quad . \tag{2.6}$$

Again, there would be certain relations implied by associativity that I will not write down explicitly. The most interesting relation is that implied by requiring that

$$d^2 = 0 \quad . \tag{2.7}$$

(This condition is absolutely critical for applying Cartan's methods!) Define d on one-forms as follows:

$$d(A_1 dA_2 A_3) = (dA_1 \wedge dA_2)A_3 - A_1(dA_2) \wedge dA_3 \quad . \tag{2.8}$$

Thus,

$$d(d(A_1 A_2)) = d(dA_1 A_2 + A_1 dA_2)$$
$$= -dA_1 \wedge dA_2 + dA_1 \wedge dA_2 \quad ,$$

which is *automatically* zero. (In fact, that is why it was necessary to choose the minus sign on the right hand side of (2.8).)

Similarly, one defines r-th degree differential forms as linear combinations of objects of the form

$$A_1 dA_2 A_3 \wedge dA_4 \wedge \cdots \quad ,$$

containing r-wedge-products in all. Exterior differentiation is defined as follows:

$$d(A_1 dA_2 A_3 \; dA_4 \; \cdots) = dA_1 \wedge dA_2 A_3 \wedge dA_4 - A_1 dA_2 \wedge dA_3 \cdots$$
$$- \cdots \tag{2.9}$$

In order to have a richer structure, it is necessary to introduce some sort of "commutativity". Here is one general type.

3. "TWISTED" COMMUTATIVE ALGEBRAS

Continue with $\mathcal{A}$ as in Section 2.

<u>Definition.</u> $\mathcal{A}$ is *twisted commutative* if there is an algebra automorphism

$$\alpha: \mathcal{A} \to \mathcal{A}$$

such that

$$A_1 A_2 = \alpha(A_2) A_1 + A_2 \alpha(A_1) - \alpha(A_1 A_2) \tag{3.1}$$

for $A_1, A_2 \in \mathscr{A}$.

For example, for

$$\alpha = \text{identity}$$

(3.1) reduces to:

$$A_1 A_2 = A_2 A_1 \ . \tag{3.2}$$

i.e., $\mathscr{A}$ is commutative. As we have seen, the corresponding identity for differential forms is:

$$A_1 dA_2 = (dA_2) A_1 \ . \tag{3.3}$$

Let us extend α to a linear map

$$\mathscr{D}^r(\mathscr{A}) \to \mathscr{D}^r(A) \ , \qquad r = 1, 2, \ldots \ ,$$

so that:

α is an automorphism of exterior product (3.4)

$$\alpha d = - d\alpha \tag{3.5}$$

Now we will impose the conditions obtained from differentiating relation (3.1) and applying (3.5):

$$dA_1 A_2 + A_1 dA_2 = -\alpha(dA_2) A_1 + \alpha(A_2) dA_1 + dA_2 \alpha(A_1) - A_2 \alpha(dA_1)$$
$$+ \alpha(dA_1 A_2 + A_1 dA_2) \ .$$

Again, assuming dA_1, dA_2 are independent gives the following relation:

$$(dA_1) A_2 = \alpha(A_2) dA_1 - A_2 \alpha(dA_1) + \alpha(dA_1) \alpha(A_2) \ . \tag{3.6}$$

For example, if $\alpha = $ identity, (3.6) reduces to (3.3), as it should.

Here is the reason for choosing (3.5). Suppose given $\alpha = $ identity on $\mathscr{A}$, i.e., $\mathscr{A}$ is commutative. Differentiating (3.3) gives

$$dA_1 \wedge dA_2 = - dA_2 \wedge dA_1 \ , \tag{3.7}$$

which is the typical skew-commutativity property of exterior product for the *situation of "ordinary" differential geometry*. Notice that (3.7) *can be*

DIFFERENTIAL FORMS

written in exactly the same form as (3.1), *i.e.,*

$$dA_1 \wedge dA_2 = \alpha(dA_2) \wedge dA_1 + dA_2 \wedge \alpha(dA_1) - \alpha(dA_1 \wedge dA_2) \quad . \quad (3.8)$$

Hence, (3.1) becomes (for this case) an *identity in the whole algebra of differential forms*. I have not attempted to check for which other α's this is true.

4. Z_2-GRADED GRASSMANN ALGEBRAS AS TWISTED COMMUTATIVE ALGEBRAS

Suppose $\mathcal{A}$ and $\alpha: \mathcal{A} \to \mathcal{A}$ are as in Section 3. In addition, suppose that:

$$\alpha^2 = \text{identity} \quad . \quad (4.1)$$

Set

$$\mathcal{A}^+ = \{A \in \mathcal{A}: \alpha(A) = A\}$$

$$\mathcal{A}^- = \{A \in \mathcal{A}: \alpha(A) = -A\}$$

Then, $\mathcal{A}$ is a direct sum (as vector space) of $\mathcal{A}^+$ and $\mathcal{A}^-$. (3.1) implies that:

$$A_1^+ A_2^+ = A_2^+ A_1^+ \quad (4.2)$$

$$A_1^- A_2^- = A_2^- A_1^- \quad (4.3)$$

$$A_1^+ A_2^- = A_2^- A_1^+ \quad (4.4)$$

for $A_1^+, A_2^+ \in \mathcal{A}^+$; $A_1^-, A_2^- \in \mathcal{A}^-$.

These relations characterize a Z_2-*graded Grassmann algebra*. In Vol. 16 we have seen that they lead to the following relations:

$$A_1^+ dA_2^+ = dA_2^+ A_1^+ \quad (4.5)$$

$$A_1^- dA_2^- = - dA_2^- A_1^- \quad (4.6)$$

$$A_1^+ dA_2^- = dA_2^- A_1^+ \quad (4.7)$$

$$A_1^- dA_2^+ = dA_2^+ A_1^- \quad . \quad (4.8)$$

Applying exterior derivative to these relations gives the following commutation relations for exterior products:

$$dA_1^+ \wedge dA_2^+ = - dA_2^+ \wedge dA_1^+ \qquad (4.9)$$

$$dA_1^- \wedge dA_2^- = dA_2^- \wedge dA_2^- \qquad (4.10)$$

$$dA_1^+ \wedge dA_2^- = - dA_2^- \wedge dA_1^+ \qquad (4.11)$$

These relations can be written in the following form. Let $\mathcal{D}(\mathcal{A})$ be the direct sum of the $\mathcal{D}^r(\mathcal{A})$. Make $\mathcal{D}(A)$ into an algebra using exterior product as the algebra product. (In particular, $\mathcal{A}$ is $\mathcal{D}^0(\mathcal{A})$. Exterior multiplication is defined on $\mathcal{D}^0(\mathcal{A})$ so as to agree with the algebra product on $\mathcal{A}$.) Extend α to an algebra automorphism of $\mathcal{D}(\mathcal{A})$, such that

$$\alpha d = - d\alpha \qquad (4.12)$$

Set:

$$\mathcal{D}^+(\mathcal{A}) = \{\omega \in \mathcal{D}(A): \alpha(\omega) = \omega\}$$

$$\mathcal{D}^-(\mathcal{A}) = \{\omega \in \mathcal{D}(A): \alpha(\omega) = - \omega\}$$

Then the general commutation relations corresponding to (4.5) - (4.11) are the following:

$$\omega_1^+ \wedge \omega_2^+ = \omega_2^+ \wedge \omega_1^+ \qquad (4.13)$$

$$\omega_1^- \wedge \omega_2^- = - \omega_2^- \wedge \omega_1^- \qquad (4.14)$$

$$\omega_1^+ \wedge \omega_2^- = \omega_2^- \wedge \omega_1^+ \qquad (4.15)$$

We have seen already in Vol. 16 how these relations may be used in simply physical situations. We now turn to another example.

5. THE HEISENBERG-RELATION ASSOCIATIVE ALGEBRA. "QUANTUM" HAMILTONIAN THEORY

Let $\mathcal{A}$ be an associative algebra (with unit element "1") with generators labelled:

$$P_a, \quad Q^a, \qquad 1 \leq a, b \leq n ,$$

such that

DIFFERENTIAL FORMS

$$P_a Q^b - Q^b P_a = \delta_a^b i\hbar$$

$$0 = P_a P_b - P_b P_a$$

$$= Q_a Q_b - Q_b Q_a$$

(5.1)

Equations (15.1) are basically just the *Heisenberg commutation relations*. Construct the differential form spaces $\mathcal{D}^r(\mathcal{A})$ as in previous sections. We must however introduce relations into $\mathcal{D}(\mathcal{A})$ so as to accommodate the relations (5.1). This is done by applying d to both sides

$$A_1 dA_2 = A_2 dA_1 \qquad (5.2)$$

$$dA_1 \wedge dA_2 = - dA_2 \wedge dA_1 \qquad (5.3)$$

for $A_1, A_2 \in \mathcal{A}$.

Thus, the differentials for the "quantum" observables obey exactly the same algebra as the "classical" ones.

We can now develop "quantum" Hamiltonian theory following Cartan's method. Set

$$\omega = dP_a \wedge dQ^a \qquad (5.4)$$

It is an element of $\mathcal{D}^2(\mathcal{A})$. Let $h \in \mathcal{A}$. Define $h_a, h^a \in \mathcal{A}$ such that

$$dh = h_a dQ^a + h^a dP_a \qquad (5.5)$$

Add "t" as another independent variable to the P's and Q's. Introduce the "augmented" two forms

$$\omega' = \omega - dh \wedge dt$$

$$= dP_a \wedge dQ^a - h_a dQ^a \wedge dt - h^a dP_a \wedge dt$$

$$= (dP_a - h_a dt) \wedge (dQ^a - h^a dt) \qquad (5.6)$$

GRADED LIE GROUPS FROM CARTAN'S POINT OF VIEW

1. INTRODUCTION

With the recent development of the theory of "graded Lie algebras" in the mathematics and physics literature, the question has naturally arisen of whether there is a geometric object which stands to a "graded Lie algebra" as a "Lie group" stands to a "Lie algebra". My aim here is to look at this from another point of view, namely, Cartan's.

For Cartan, a Lie group corresponding to a (finite dimensional) Lie algebra L can be described as follows:

A manifold M, together with a basis (ω^i), $1 \leq i,j \leq m$, of one-differential forms on M such that

$$d\omega^i = c^i_{jk} \omega^j \wedge \omega^k , \qquad (1.1)$$

where (c^i_{jk}) are the *structure constants* of the Lie algebra L.

In IM, Volume 16, I have described how "differential forms" may be generalized. Thus, it is a reasonable idea to take (1.1) as a clue to how a "Lie group" may be generalized. That is the goal of this chapter.

2. EXTERIOR DERIVATIVE AND GRADED LIE ALGEBRAS

Return to (1.1). The key to L as a "Lie algebra" is the quadratic identity on (c^i_{jk}) which represents the Jacobi identity. This can be obtained by exterior derivation of both sides of (1.1) using the algebraic properties of differential forms on manifolds of the *classical type*. One may expect different algebraic identities to result from more general sorts of differential form algebras. In this section we shall investigate what is required to generate a "graded Lie algebra".

Here is a general algebraic context within which we shall work. $\mathcal{D}$ is an associative algebra (over the reals or complexes as field of scalars). We denote the algebra product by $\wedge$. There are two linear maps

$$d: \mathcal{D} \to \mathcal{D} \quad , \quad \alpha: \mathcal{D} \to \mathcal{D}$$

given which satisfy the following conditions:

$$
\begin{aligned}
d^2 &= 0 \\
\alpha^2 &= 1 \\
d\alpha &= -\alpha d \\
\alpha(\omega_1 \wedge \omega_2) &= \alpha(\omega_1) \wedge \alpha(\omega_2) \\
\text{for } \omega_1, \omega_2 &\in \mathcal{D}
\end{aligned}
\tag{2.1}
$$

Set:

$$\mathcal{D}^+ = \{\omega \in \mathcal{D} : \alpha(\omega) = \omega\}$$

$$\mathcal{D}^- = \{\omega \in \mathcal{D} : \alpha(\omega) = -\omega\}$$

It follows from (2.1) that:

$$\mathcal{D}^\pm \wedge \mathcal{D}^+ \subset \mathcal{D}^\pm \supset \mathcal{D}^+ \wedge \mathcal{D}^\pm$$

$$\mathcal{D}^- \wedge \mathcal{D}^- \subset \mathcal{D}^+$$

Let us also postulate the following commutation relations:

$$\omega_1 \wedge \omega^+ = \omega^+ \wedge \omega_1 \tag{2.2}$$

for $\omega_1 \in \mathcal{D}$, $\omega^+ \in \mathcal{D}^+$

$$\omega_1^- \wedge \omega_2^- = -\omega_2^- \wedge \omega_1^- \tag{2.3}$$

for $\omega_1^-, \omega_2^- \in \mathcal{D}^-$.

Choose the following range of indices and the summation convention:

$$1 \leq i,j,k \leq n \quad .$$

Let us suppose that a set (ω^i) of elements of $\mathcal{D}$ is given such that:

$$d\omega^i = c^i_{jk} \omega^j \wedge \omega^k \quad , \tag{2.4}$$

with scalar constants (c^i_{jk}). Suppose also that each (ω^i) belongs to either $\mathcal{D}^+$ or $\mathcal{D}^-$, i.e., there are relations of the form:

$$\alpha(\omega^i) = (-1)^{\alpha(i)} \omega^i \quad , \tag{2.5}$$

where

$$\alpha(i) = \begin{cases} 0 & \text{if } \omega^i \in \mathcal{D}^+ \\ 1 & \text{if } \omega^i \in \mathcal{D}^- \end{cases}$$

Thus,

$$\omega_i \wedge \omega_j = (-1)^{\alpha(i)\alpha(j)} \omega_j \wedge \omega_i \tag{2.6}$$

Thus, we can suppose without loss in generality, that the following relations are satisfied:

$$c^i_{jk} = c^i_{kj} (-1)^{\alpha(j)\alpha(k)} \quad . \tag{2.7}$$

Now, let L be an n-dimensional vector space with a basis labelled

$$A_i \quad .$$

Define an algebra product

$$L \times L \to L \quad ,$$

denoted by "bracket" [,], as follows

$$[A_i, A_j] = c^k_{ij} A_k \quad . \tag{2.8}$$

Set

L^+ = linear subspace of L spanned by the A_i with $\alpha(i) = 0$

L^- = linear subspace of L spanned by the A_i with $\alpha(i) = 1$.

Thus, we have

[,] is symmetric on L^+

[,] is skew-symmetric on L^- .

Theorem 2.1. Suppose the following identity is satisfied:

$$d(\omega^i \wedge \omega^j) = d\omega^i \wedge \omega^j + (-1)^{\alpha(i)} \omega^i \wedge d\omega^j \tag{2.9}$$

Suppose also that the elements

$$\omega_j \wedge \omega_k \wedge \omega_\ell$$

are (after the appropriate symmetry is taken into account) linearly independent. Then L, with this decomposition $L^+ + L^-$, and this bracket, is a Z^2-graded Lie algebra.

Proof. As for the traditional Lie algebra case, this follows from applying exterior derivative to (2.4).

$$\begin{aligned}
0 &= c^i_{jk} d\omega^j \wedge \omega^k + (-1)^{\alpha(j)} c^i_{jk} \omega^j \wedge d\omega^k \\
&= c^i_{jk} d\omega^j \wedge \omega^k + (-1)^{\alpha(j)} (-1)^{\alpha(j)(\alpha(k)+1)} c^i_{jk} d\omega^k \wedge \omega^j \\
&= 2c^i_{jk} d\omega^j \wedge \omega^k + (-1)^{\alpha(j)(\alpha(k)+2)} c^i_{kj} d\omega^j \wedge \omega^k \\
&= 2c^i_{jk} d\omega^j \wedge \omega^k \\
&= 2c^i_{jk} (c^j_{\ell m} \omega^\ell \wedge \omega^m) \wedge \omega^k \tag{2.10}
\end{aligned}$$

After this is made explicit in terms of the c's *alone*, one sees that it is equivalent to the graded Lie algebra version of the Jacobi identity.

REMARKS ABOUT BOSONS AND FERMIONS

1. INTRODUCTION

The aim of this chapter is to elucidate several points in standard quantum mechanics that I have always found obscure, but that every physicist is convinced that he understands perfectly well! My aim goes beyond sheer perversity; I believe there are important and basic geometric problems yet to be solved in connection with the classical and quantum mechanics of fermions. (For example, What is the Feynman path integral for the Dirac equation?) Another motivation is to understand the role that "graded Lie algebras" play in physics analogous to the role of "ordinary" Lie algebras which is described in my earlier books

2. ELEMENTARY REMARKS ABOUT THE FERMI EXCLUSION PRINCIPLE

Consider a particle moving in a one-dimensional space, parameterized by q. Quantum mechanically, it is described by a "Schrödinger wave function", i.e., a complex-valued function

$$q \to \psi(q)$$

such that

$$\int_{-\infty}^{\infty} |\psi(q)|^2 \, dq = 1 \quad . \tag{2.1}$$

For the sake of simplicity, we shall suppose that ψ is of differentiability class C^∞, and rapidly decreasing, so that differential operators--say with coefficients which are polynomials in q --can be freely applied to it. Let H denote the set of all C^∞ rapidly decreasing maps

$$q \to \psi(q) \quad .$$

Set

$$\langle \psi | \psi' \rangle = \int_{-\infty}^{\infty} \psi(q)^* \psi'(q) \, dq \tag{2.2}$$

With this inner product, H is a *pre-Hilbert space*. Let $PS(H)$ be the projective space associated with H. $PS(H)$ is the *state space* of the quantum mechanical system.

The classical mechanics state space is defined as the space of all (p,q), where p is the "momentum" coordinate of the particle. (Differential-geometrically, it is the cotangent bundle $T^d(Q)$ to the configuration space Q.)

Now, consider two particles with coordinates q_1, q_2. Let Q_1, Q_2 denote the space of these variables. In classical mechanics, putting these particles together to form a composite system amounts to defining the new configuration space as

$$Q = Q_1 \times Q_2 \quad.$$

Let H denote the pre-Hilbert space of all C^∞, rapidly decreasing complex valued function

$$(q_1, q_2) \to \psi(q_1, q_2) \quad,$$

with:

$$\langle \psi | \psi' \rangle = \int \psi(q_1, q_2)^* \psi'(q_1, q_2) \, dq_1 dq_2 \quad (2.3)$$

Let H_1, H_2 denote the analogous objects constructed from Q_1 and Q_2. There is a linear map

$$\phi: H_1 \otimes H_2 \to H \quad (2.4)$$

defined as follows:

> Consider elements $\psi_1 \in H_1$, $\psi_2 \in H_2$, i.e., functions $q_1 \to \psi_1(q)$, $q_2 \to \psi_2(q)$. Then,

$$\phi(\psi_1 \otimes \psi_2)(q_1, q_2) = \psi_1(q_1) \psi_2(q_2) \quad (2.5)$$

This construction reflects the following heuristic rules:

> Putting two systems together *classically* amounts to constructing the Cartesian product of their configuration spaces, *quantum mechanically* to constructing the tensor product (suitably completed) of their Hilbert spaces.

So far, there has been no necessary relation between Q_1 and Q_2 --they could be chosen as arbitrary spaces. Now, let us suppose that they are identical. One can then interchange coordinates, i.e., construct the permutation

$$(q_1,q_2) \to (q_2,q_1) \quad .$$

This is then just the finite group S_2 ($\equiv$ "symmetric group of two letters", see IM, Volume 1). It acts naturally on H:

$$\psi \to \psi' \quad ,$$

with

$$\psi'(q_1,q_2) = \psi(q_2,q_1) \quad .$$

One can decompose H under its action in the obvious way: Set

$$H^+ = \{\psi \in H: \psi(q_2,q_1) = \psi(q_1,q_2)\} \tag{2.6}$$

$$H^- = \{\psi \in H: \psi(q_2,q_1) = -\psi(q_2,2_1)\} \tag{2.7}$$

There is a corresponding decomposition of $H_1 \otimes H_2$ into *symmetric* and *skew-symmetric* tensors. Set:

$$H_1 \circ H_2 = \{\tfrac{1}{2}(\psi_1 \otimes \psi_2 + \psi_2 \otimes \psi_1): \psi_1 \in H_2, \ \psi_2 \in H_1, \ H_1 = H_2\} \tag{2.8}$$

$$H_1 \wedge H_2 = \{\tfrac{1}{2}(\psi_1 \otimes \psi_2 - \psi_2 \otimes \psi_1)\} \tag{2.9}$$

Thus,

$$H_1 \otimes H_2 = (H_1 \circ H_2) \oplus (H_1 \wedge H_2) \tag{2.10}$$

$$\phi(H_1 \circ H_2) \subset H^+ \tag{2.11}$$

$$\phi(H_1 \wedge H_2) \subset H^- \quad . \tag{2.12}$$

Let us think about the physical significance of H^+ and H^-. Suppose that $\psi_1 \in H_1$, $\psi_2 \in H_2$ are of norm one, i.e., they are *Schrödinger wave functions* for particles 1 and 2. According to the Born interpretation, the probability measures

$$|\psi_1(q_1)|\,dq_1$$

$$|\psi_2(q_2)|\,dq_2$$

determine the probabilities of finding particles 1 and 2, respectively, in regions of Q_1 and Q_2.

$$|\psi(q_1,q_2)|\,dq_1 dq_2$$

determines the probabilities of finding the combined particle position (q_1,q_2) in a region of $Q_1 \times Q_2$. Thus, if $\psi \in H^-$,

$$\psi(q_1,q_2) = 0 ,$$

the probability of finding particles close to each other is rather small.

The Fermi exclusion principle is that two identical particles, e.g., electrons, do not occupy the "same state". Thus, the skew-symmetry of the joint Schrödinger wave function is at least weakly related to the Fermi principle. (I do not want to go into it in more detail at the moment.)

Similarly, if $\psi \in H^+$, the probability measure $|\psi(q_1,q_2)|\,dq_1 dq_2$ will have a greater tendency to peak around the diagonal

$$q_1 = q_2 ,$$

i.e., symmetry under interchange is the appropriate group-theoretic formulation of the idea that Bose particles prefer to occupy the same state.

Now, in the modern elementary particle literature the intuitive, "physical" reasons and explanations for things like this tend (for better of worse-- probably the latter!) to disappear into the background and the roughly equivalent group-theoretic principle tends to take over. Since my goal is to develop the mathematics, I will follow this route, although I must admit to having puritanical feelings about this sort of mathematization of physics!

3. CONDITIONS ON DYNAMICAL INTERACTION

Let H_1, H_2, H, $\phi: H_1 \otimes H_2 \to H$, H^+, H^- be as in Section 2. We have agreed to say that "$\psi \in H^-$" has something to do physically with saying that "identical particles do not like to be in the same state", i.e., Fermi, and that "$\psi \in H^+$" has something physically to do with the Bose property, i.e., identical particles *prefer* to occupy the same state.

So far, this is a "static" or "kinematic" property. Its implications for "dynamics" are, however, their most important feature. In elementary quantum mechanics, the dynamics is determined as follows. A Hermitian operator

$$A: H \to H$$

is given, called the *Hermitian*, and the time evolution of $\psi \in H$ is given by the *Schrödinger equation*:

$$i\frac{\partial \psi}{\partial t} = A(\psi) \qquad (3.1)$$

or

$$\psi(t) = e^{-itA} \psi(0) . \qquad (3.2)$$

Now, if the split-up

$$H = H^+ \oplus H^-$$

into subspaces obeying Bose and Fermi "statistics" is to have a *dynamic* significance, we must have:

$$A(H^+) \subset H^+$$
$$A(H^-) \subset H^- \qquad (3.3)$$

These conditions guarantee that $\psi(t)$ belongs to H^+ or H^- for all t if $\psi(0) \in H^+$ or H^-.

We have identified the subset $\phi(H_1 \otimes H_2)$ in H. It is natural to try to choose A adopted to this decomposition, i.e., to suppose that A also acts as a Hermitian, linear map

$$A: H_1 \otimes H_2 \to H_1 \otimes H_2 .$$

Now, one way to construct Hermitian maps $H_1 \otimes H_2 \to H_1 \otimes H_2$ is to choose Hermitian maps

$$A_1: H_1 \to H_1$$
$$A_2: H_2 \to H_1$$

and form their tensor product:

$$A_1 \otimes A_2: \psi_1 \otimes \psi_2 \to A_1\psi_1 \otimes A_2\psi_2 . \qquad (3.4)$$

A is a sum of operators of this form:

$$A = A_1 \otimes A_2 + B_1 \otimes B_2 + \cdots , \qquad (3.5)$$

with $A_1, B_1, \ldots$ operators $H_1 \to H_1$, $A_2, B_2, \ldots$ operators $H_2 \to H_2$.

Now, let

$$\sigma: H_1 \otimes H_2 \to H_1 \otimes H_2$$

be the map defined by permuting particle 1 and particle 2:

$$\sigma(\psi_1 \otimes \psi_2) = \psi_2 \otimes \psi_1 . \qquad (3.6)$$

Thus, $\sigma^2 = 1$ and H^+ and H^- are the eigenvector decompositions. Thus, the condition that (3.3) is satisfied is that

$$\sigma A = A\sigma \quad . \tag{3.7}$$

Further, if $A_1 \otimes A_2$ is defined by (3.4), then

$$(A_1 \otimes A_2)(\psi_1 \otimes \psi_2) = \sigma(A_1\psi_1 \otimes A_2\psi_2)$$

$$= A_2\psi_2 \otimes A_1\psi_1$$

$$= (A_2 \otimes A_1)(\psi_2 \otimes \psi_1)$$

$$= (A_2 \otimes A_1)\sigma(\psi_1 \otimes \psi_2)$$

or

$$\sigma(A_1 \otimes A_2)\sigma^{-1} = A_2 \otimes A_1 \quad . \tag{3.8}$$

In particular, with (3.5) and (3.7) satisfied, we have

$$A_1 \otimes A_2 + B_1 \otimes B_2 + \cdots = A_2 \otimes A_1 + B_2 \otimes B_1 + \cdots \tag{3.9}$$

In summary, (3.9) is the abstract, general condition that the dynamics generated by a Hermitian operator A respect the Bose or Fermi property.

The simplest way to satisfy Equation (3.9) is as follows.

$$A = A_1 \otimes 1 + B_1 \otimes B_1 + 1 \otimes A_1 \quad .$$

The corresponding Schrödinger equation is

$$i\frac{d}{dt}(\psi_1(t) \otimes \psi_2(t)) = A_1(\psi_1) \otimes \psi_2 + B_1\psi_1 \otimes B_2\psi_2 + \psi_1 \otimes A_1(\psi_2) \tag{3.10}$$

Let us revert back to the case where $H_1 = C^\infty$, rapidly decreasing functions of q.

$$A_1 = \frac{d^2}{dq_1^2} + V(q_1)$$

$$A_2 = \frac{d^2}{dq_2} + V(q_2)$$

$$\psi(q_1, q_2) = (\psi_1 \otimes \psi_2)(q_1, q_2)$$

$$= \psi_1(q_1)\psi_2(q_2)$$

Then,

$$i\partial_t \psi = \partial^2_{q_1}(\psi) + \partial^2_{q_2}(\psi) + (V(q_1) + V(q_2))(\psi) + B_1\psi_1 \otimes B_1(\psi_1) \quad (3.11)$$

For example, one possible choice of B_1 might be

$$B_1\psi_1(q_1) = W(q_1)\psi_1(q)$$

With this choice, the Schrödinger Equation (3.11) takes the form

$$i\partial_t \psi = \left(\partial^2_{q_1} + \partial^2_{q_2}\right)(\psi) + (V(q_1) + V(q_2) + W(q_1)W(q_2)) \quad (3.12)$$

This is, of course, a Schrödinger-type equation for a particle moving in R^2. One reasonable choice is that where V,W are polynomials of degree at most two, (3.12) is then a "harmonic oscillator" type of Schrödinger equation.

A RELATION BETWEEN THE LR AND QR ALGORITHMS OF MATRIX NUMERICAL ANALYSIS AND LIE THEORY

1. INTRODUCTION

The influence of computers has revived numerical analysis as an active area of scientific interest and has suggested new ideas. In particular, the computer loves matrices and iterative operations on matrices, whereas classical hand calculation-oriented numerical analysis emphasized operations on scalars.

Now, mathematically, matrices have everything to do with *Lie theory* and *algebraic geometry*. It is surprising (and sad) that so little has been done to systematize this "new" numerical analysis in terms of those towering mathematical disciplines. The "sociological" reason for this lack of contact is clear-- the disdain of the "pure" types for the *seemingly* pedestrian numerical problems and the lack of mathematical breadth in the education of "applied" mathematicians. It is extremely difficult to make any sense of Lie theory and algebraic geometry without a considerable investment of time and effort.

Apparently, one of the most useful of these "new" algorithms are those called "LR" and "QR", for putting a square matrix into triangular form. Chapter 8 of Wilkenson's *The Algebraic Eigenvalue Problem* (which seems to be the Bible of this particular religion) gives the definitive treatment of this topic. (I must say that a depressing feature of all books on numerical analysis is that they bear the awful weight of their ancestral drudgery. Some have been tarted up with fancy functional analysis, which does not necessarily help.) Although I know almost nothing in detail about this subject, I want to take a few tentative steps toward bringing to the foreground the Lie theory which is so obviously there.

2. THE LIE THEORY OF THE LR ALGORITHM

Let G be a connected Lie group and let H and S be subgroups of G such that:

$$\boxed{\text{The complement of } H/K \text{ in } G \text{ is a union of a finite number of connected submanifolds of lower dimension.}} \quad (2.1)$$

$$\boxed{H \cap S = (1)} \quad (2.2)$$

Remark. H and S are not *invariant* subgroups, so that (2.1) is *not* a direct sum decomposition. The situation to keep in mind is:

$G = GL(n,\mathbb{C}) \equiv n \times n$ complex matrices of nonzero determinant.

$S =$ subgroup of lower triangular matrices with ones on the diagonal

$$A = \begin{pmatrix} 1 & 0 & \cdots & 0 \\ \times & 1 & & \\ & & \ddots & \\ \times & \times & \times & 1 \end{pmatrix}$$

$S =$ upper triangular matrix

$$B = \begin{pmatrix} \times & \times & & \\ & \times & \times & \\ & & \times & \\ 0 & & & \ddots \end{pmatrix}$$

Given one element $g \in G$, our "problem" is to find a $g_0 \in G$ such that

$$g_0 g g_0^{-1} \in S \quad . \tag{2.3}$$

In words, g is to be *conjugate to an element of* S.

Geometrically, this means that:

> g acting on the coset space G/S has a fixed point.

For, if

$$g(g_0^{-1} K) = g_0^{-1} s \quad ,$$

then

$$g g_0^{-1} = g_0^{-1} k \quad ,$$

whence (2.3), and conversely. Now, in the case of the "LR algorithm",

> G/S is a compact manifold of positive Euler characteristic.

Here G is connected, the *Lefschetz fixed point theorem* proves that the fixed point *exists*. The problem, and this is, of course, typical of numerical analysis, is to find it in a way implementable on a computer. This might involve finding a sequence of points $p_1, p_2, \ldots$ on the manifold G/S, each of which is "computable" in terms of some suitably efficient iterative algorithm such that

$$\lim_{g \to \infty} p_j = \text{fixed point of } g \text{ acting in } G/S .$$

Here is how the "LR algorithm" works, as explained by Wilkenson. Suppose g is written as

$$g = h_1 s_1 \qquad (2.4)$$

with $h_1 \in H$, $s_1 \in S$.

(Our hypothesis (2.1) implies that "almost all" g's can be written in this way. If G is an algebraic group, it would be appropriate to assume further that the set of all such g's is a *Zariski open subset* of G.)

Begin with (2.4) and rewrite it as

$$h_1^{-1} g h_1 = s_1 h_1 . \qquad (2.5)$$

Set:

$$\boxed{g_1 = h_1^{-1} g h_1} \qquad (2.6)$$

Note that

$$\boxed{g_1 \text{ is conjugate to } g \text{ via the subgroup } H .} \qquad (2.7)$$

Now, replace g by g_1, and begin again. Write

$$\boxed{\begin{aligned} g_1 &= h_2 s_2 \\ g_2 &= h_2^{-1} g_1 h_2 \\ &= s_2 h_2 \end{aligned}} \qquad (2.8)$$

Continuing in this way, we obtain a sequence

$$g, g_1, g_2, \ldots$$

of elements of g, *all of which lie on the same orbit of* Ad H. Note that each is formed from the preceding one by relatively simple computations; namely, inverting the map

$$H \times S \to G, \quad (h,s) \to hs,$$

and writing the product in the reverse order sh.

Wilkenson gives an elaborate proof (which I do not understand) that, under certain conditions,

$$\boxed{\begin{array}{ll} g_j \to g_\infty \in S & \text{as} \quad j \to \infty \\ \\ h_j \to 1 & \text{as} \quad j \to \infty \end{array}} \qquad (2.10)$$

Thus, we obtain an "approximate triangularization" of g.

$$\begin{aligned} g &= h_1 g_1 h_1^{-1} \\ &= h_1 h_2 g_2 h_2^{-1} h_1^{-1} \\ &= h_1 h_2 h_3 g_3 h_3^{-1} h_2^{-1} h_1^{-1} \\ &\quad \cdots \end{aligned}$$

Thus, for N sufficiently large,

$$h_N^{-1} \cdots h_3^{-1} h_2^{-1} h_1^{-1} g h_1 h_2 h_3 \cdots h_N$$

is *approximately triangular*.

Here is another, more geometric, way of thinking of the algorithm. Let

$$X = \text{coset space } G/S.$$

Let x_0 be the unit element. Then, (2.3) and (2.4) mean that:

> H acts with identity isotropy subgroup at x_0.
>
> The orbit Hx_0 is an open subset of X; the complement is a union of submanifolds of lower dimension. (2.11)

Thus, to write

$$g = hs$$

means to write

$$gx_0 \quad \text{as} \quad hx_0 \quad .$$

Set:

$$x_j = h_1 \cdots h_j x_0 \; , \qquad j = 1, 2, \ldots \tag{2.12}$$

Now,

$$g = h_1 \cdots h_j g_j h_j^{-1} \cdots h_1^{-1} \quad . \tag{2.13}$$

Hence, combining (2.12) and (2.13) gives:

$$gx_j = h_1 \cdots h_j g_j x_0$$

$$= \text{, using } g_j = h_{j+1} s_{j+1} \; ,$$

$$h_1 \cdots h_j h_{j+1} x_0$$

$$= x_{j+1} \quad .$$

Thus,

$$x_1 = gx_0 \; , \qquad x_2 = gx_1, \ldots \tag{2.14}$$

The points x_j are the orbits of the cyclic group generated by g. If

$$\lim_{j \to \infty} x_j = x_\infty \quad,$$

then

$$g x_\infty = x_\infty \quad,$$

i.e., x_∞ is the desired fixed point of x. Thus, *if*

$$\lim_{j \to \infty} x_j$$

exists, it is *plausible* that:

$$\lim_{j \to \infty} (h_1 \cdots h_j)$$

exists. (2.13) then *implies* that

$$\lim_{j \to \infty} g_j$$

exists, which is the LR algorithm.

To prove

$$\lim_{j \to \infty} x_j$$

exists is, at least for certain specially chosen g's, a matter of *Lie theory*. For example, write:

$$g = \exp(A) \quad,$$

with $A \in \mathcal{G}$, $\equiv$ Lie algebra of G and examine the orbits of the one-parameter group $t \to \exp(tA)$ on $X = G/S$. We shall see later methods for finding limits of such orbits. Later in this volume we shall examine this general question in *the context of the limits of solutions of matric Riccati equations*. Thus, there is a link-up between these two important areas of a "modern" applied mathematics, which had never before been noticed because of the mathematical "tunnel" in which applied mathematics is carried on nowadays. Gauss surely would have noticed and utilized this connection!

3. THE QR ALGORITHM

This just involves

$$G = KS \quad,$$

with K *compact*.

THE REDHEFFER *-OPERATION OF SCATTERING THEORY AND THE GEOMETRY OF GRASSMANNIANS AND LIE SYSTEMS

1. INTRODUCTION

R. Redheffer has introduced [1,2] an algebraic operation that plays a key role in the mathematical description of transmission lines, reactors, stochastic processes, etc. There has been recent interest in this operation by systems theorists [3].

One of the main mathematical features of Redheffer's operation is that it leads to matrix Riccati equations in a natural way. Now, these equations may be handled with the Lie theory by interpreting them as flows on Grassmann manifolds [4,5]. The aim of this note is to show how Redheffer's operation itself is defined in terms of Grassmannians. We shall also relate the general idea of systems which can be *linearized* to that of *Lie systems*, as defined by Lie and Vessiot in the 19th century.

We would like to thank Professor T. Kailath for mentioning this question to us.

2. WORKING BACKWARDS FROM THE LINEARIZATION

Redheffer points out [1,2] that his operation is a "twisting" of partitioned matrix multiplication. Physically, this means that the operation may be "linearized". We shall work backwards from this observation.

It is convenient to use the notation of coordinate free, "geometric" linear algebra. Let V, W denote finite dimensional real vector spaces. $L(V,W)$ then denotes the vector space of linear maps: $V \to W$. Thus, if $V = R^n$, $W = R^m$, $L(V,W)$ can be identified with the $m \times n$ real matrices.

$V + W$ denotes the direct sum vector space. It is the set of pairs $\begin{pmatrix} v \\ w \end{pmatrix}$ with addition defined as usual. Let $G(V)$ denote the set of all linear subspaces of the vector space V. For each integer $m \geq 0$, $G^m(V)$ denotes the subset of $G(V)$ consisting of the elements of dimension m. $G^m(V)$ is a *Grassmann manifold*.

Now, suppose that V_1, V_2, W_1, W_2 are given vector spaces. A linear map

$$A: V_1 \oplus V_2 \to W_1 \oplus W_2$$

can be written as a partitioned matrix

$$A = \begin{pmatrix} A_{11} & A_{12} \\ A_{21} & A_{22} \end{pmatrix}$$

$A_{11} \in L(V_1, W_1)$; $\quad A_{12} \in L(V_2, W_1)$

$A_{21} \in L(V_1, W_2)$; $\quad A_{22} \in L(V_2, W_2)$.

Then

$$A \begin{pmatrix} v_1 \\ v_2 \end{pmatrix} = \begin{pmatrix} A_{11} & A_{12} \\ A_{21} & A_{22} \end{pmatrix} \begin{pmatrix} v_1 \\ v_2 \end{pmatrix} \qquad (2.1)$$

where the usual matrix multiplication applies on the right hand side of (2.1). Write:

$$\begin{pmatrix} w_1 \\ w_2 \end{pmatrix} = A \begin{pmatrix} v_1 \\ v_2 \end{pmatrix} \qquad (2.2)$$

Thus,

$$\begin{aligned} w_1 &= A_{11} v_1 + A_{12} v_2 \\ w_2 &= A_{21} v_1 + A_{22} v_2 \end{aligned} \qquad (2.3)$$

Regard (2.3) as a set of *equations* for (v_1, v_2, w_1, w_2). Think of them as (implicit) *input-output relations*, with (v_1, w_1) as input, (v_2, w_2) as output. We can, of course, make then explicit:

$$v_2 = A_{12}^{-1}(w_1 - A_{11} v_1) \qquad (2.4)$$

$$w_2 = A_{22} A_{12}^{-1}(w_1 - A_{11} v_1) + A_{21} v_1 \quad . \qquad (2.5)$$

This can, of course, be written in matrix form as:

$$\begin{pmatrix} v_2 \\ w_2 \end{pmatrix} = \begin{pmatrix} -A_{12}^{-1} A_{11} & A_{12}^{-1} \\ A_{21} - A_{22} A_{12}^{-1} A_{11} & A_{22} A_{12}^{-1} \end{pmatrix} \begin{pmatrix} v_1 \\ w_1 \end{pmatrix} \qquad (2.6)$$

Thus, we have a correspondence

$$\begin{pmatrix} A_{11} & A_{12} \\ A_{21} & A_{22} \end{pmatrix} \leftrightarrow \begin{pmatrix} -A_{12}^{-1}A_{11} & A_{12}^{-1} \\ A_{21} - A_{22}A_{12}^{-1}A_{11} & A_{22}A_{12}^{-1} \end{pmatrix} \qquad (2.7)$$

between

$$L(V_1 \oplus V_2, \ W_1 \oplus W_2)$$

and

$$L(V_1 \oplus W_1, \ V_2 \oplus W_2) \quad .$$

Up to a slight change in notation, this is the basic formula (12.8) of [1], what is commonly called Redheffer's *-operation. In the technical language of algebraic geometry [6], this is a *rational map*.

3. INTRODUCTION TO GRASSMANIANS

Keep the notation of Section 2. Let V be the vector space

$$V_1 \oplus V_2 + W_1 \oplus W_2 \quad .$$

Each $A \in L(V,W)$ then determines--via Equation (2.3)--a linear subspace of V, namely the set of all pairs (v_1, v_2, w_1, w_2) which satisfy Equations (2.3). This determines a subset S_1 of $G(V)$. Similarly, one has a subset S_2 of $G(V)$ determined by linear relations of the form:

$$\begin{pmatrix} w_2 \\ w_2 \end{pmatrix} = B \begin{pmatrix} v_1 \\ w_1 \end{pmatrix}$$

Each element $\gamma \in S_1 \cap S_2$ then has as "parameters" *both* the matrices A and B. The relation between them is given by formula (2.7). We recognize that, in terms of manifold theory, we are dealing with different coordinate "charts" for an open subset of a manifold.

The group $GL(V)$ of all linear automorphisms of V acts on $G(V)$, hence on elements of S_1 and S_2. In particular, we can consider curves $S_1 \cap S_2$ which are orbits of one-parameter subgroups of $GL(V)$. These are the curves which satisfy "matrix Riccati" types of differential equations [4,5] which, in turn, will be the differential equations suggested by the physics of the situation.

4. LIE SYSTEMS AND LINEARIZATIONS

Many of the differential equations which appear in physics and engineering can be "linearized". It was recognized long ago by Sophus Lie [7] that this often has a group theoretic "explanation". We shall now suggest a general setting in which to consider this phenomena.

Let X be a manifold. (For notation and ideas of manifold theory, see [4].) Let G be a Lie group which acts on M as a Lie transformation group. Let $\underset{\sim}{G}$ be the Lie algebra of G. It acts on X as a Lie algebra of vector fields.

Let $t \to A(t)$ be a curve in $\underset{\sim}{G}$. It defines a set of (non-autonomous) differential equations in X, namely

$$\frac{dx}{dt} = A(t)(x(t)) \quad . \tag{4.1}$$

This system is called a *Lie system* relative to G.

It follows from general Lie theory that the solutions of (4.1) can be written in the form:

$$t \to x(t) = g(t)(x(0)) \quad , \tag{4.2}$$

where $t \to g(t)$ is a curve in G whose tangent vector is--when translated back to the identity element--just A(t). Thus, if G is (or can be realized as) a *group of matrices*, the curve $t \to g(t)$ in G is determined as a system of *linear* differential equations

$$\frac{dg}{dt} = A(t)g \quad . \tag{4.3}$$

(4.2) and (4.3) thus exhibit the *linearization* of the Lie system (4.1). It is, in fact, readily seen that all of the linearizations usually associated with the matrix Riccati equations (where X is a Grassmanian and G a subgroup of the linear automorphisms of the underlying vector space) are of this form.

Bibliography

1. R. Redheffer, "Difference Equations and Functional Equations in Transmission-Line Theory", in *Modern Mathematics for the Engineer*, E.F. Beckenbark, ed., McGraw-Hill, 1961.

2. R. Redheffer, "On the Relation of Transmission-Line Theory to Scattering and Transfer", *J. Math. and Phys.* 16, 1962, pp. 1-41.

3. L. Ljeing, T. Kailath, and B. Friedlander, "Scattering Theory and Linear Least Squares Estimation--Part 1: Continuous Time Problems", *Proc. IEEE* 64, 1976, pp. 133-139.

4. R. Hermann, *Differential Geometry and the Calculus of Variations*, Academic Press, New York, 1969.

5. R. Hermann, *Algebraic Topics in Systems Theory* (Interdisciplinary Mathematics, Vol. 2), Math Sci Press, Brookline, MA 1973.

6. I. Shafarevich, *Basic Algebraic Geometry*. Springer-Verlag, 1974.

7. S. Lie, *Transformationsgruppen*, Chelsea Publishing Co., New York, 1970.

RAYMOND H. FOGLER LIBRARY
DATE DUE